The Filamentous Fungi

Volume 1 Industrial Mycology

The Filamentous Fungi
Volume 2 Biosynthesis and Metabolism
is now in preparation

The Filamentous Fungi

Volume 1 Industrial Mycology

Edited by:
JOHN E. SMITH, D.Sc., F.I. Biol.
DAVID R. BERRY, Ph.D.
Department of Applied Microbiology, University of Strathclyde, Glasgow

EDWARD ARNOLD

First published 1975
by Edward Arnold (Publishers) Limited,
25 Hill Street, London, W1X 8LL

ISBN 0 7131 2467 9

Printed in Great Britain by
William Clowes & Sons, Limited, London, Beccles and Colchester

Preface

Fungi are having an ever increasing impact on mankind not only in the fields of alcoholic beverages and organic product formation, including antibiotics, but also in such fields as pollution, biodeterioration, flavour biosynthesis and biomass formation. The recent rapid advances in our understanding of the genetics, physiology and biochemistry of fungi has led to an explosive increase in the use of fungi in industry. The accompanying dramatic advances in technology of fungal fermentations that have occurred since the 1940's can be expected to lead to further utilization of the immense biosynthetic potential of the fungi.

It is our intention to limit the contents of the book mainly to the truly filamentous and multicellular fungi since the industrial applications of the unicellular yeast fungi have been comprehensively covered in recent volumes. The book will be directed mainly towards senior undergraduate and postgraduate students and to a wide range of industrial and government personnel involved in the pharmaceutical, chemical and food industries.

J.E.S.
D.B.

1974

Contents

List of Contributors

D. Allsopp,
Biodeterioration Information Centre,
Department of Biological Sciences,
The University of Aston,
Birmingham, England.

D.R. Berry,
Department of Applied Microbiology,
Strathclyde University,
Glasgow, Scotland.

J.A. Blain,
Department of Biochemistry,
Strathclyde University,
Glasgow, Scotland.

J.D. Bu'lock,
Microbial Chemistry Laboratory,
Department of Chemistry,
University of Manchester,
England.

W.H. Butler,
M.R.C. Toxicology Unit,
Medical Research Council,
Carshalton, Surrey, England.

H.O.W. Eggins,
Biodeterioration Information Centre,
Department of Biological Sciences,
The University of Aston,
Birmingham, England.

W.A. Hayes,
Department of Biological Sciences,
The University of Aston,
Birmingham, England.

J.R. Johnston,
Department of Applied Microbiology,
University of Strathclyde,
Glasgow, Scotland.

L.B. Lockwood,
Biology Department,
Western Kentucky University,
Bowling Green,
Kentucky, U.S.A.

P.G. Mantle,
Department of Biochemistry,
Imperial College of Science and Technology,
London, England.

L.M. Miall,
Pfizer Limited,
Sandwich,
Kent, England.

N.G. Nair,
Department of Biological Sciences,
The University of Aston,
Birmingham, England.

R.C. Righelato,
Philip Lyle Memorial Research Laboratory,
University of Reading,
Reading, Berkshire.

K. Singh,
Department of Microbiology,
Ayerst Research Laboratories,
Montreal, Quebec, Canada.

J.E. Smith,
Department of Applied Microbiology,
Strathclyde University,
Glasgow, Scotland.

G.L. Solomons,
The Lord Rank Research Centre,
High Wycombe,
Bucks., England.

W.B. Turner,
Imperial Chemical Industries Ltd.,
Pharmaceuticals Division,
Alderley Park, Macclesfield,
Cheshire, England.

C. Vézina,
Department of Microbiology,
Ayerst Research Laboratories,
Montreal, Quebec,
Canada.

B.J.B. Wood,
Department of Applied
Microbiology,
University of Strathclyde,
Glasgow, Scotland.

Yong Fook Min,
Institute of Standards
and Industrial Research,
Republic of Singapore.

CHAPTER 1

The Structure and Development of Filamentous Fungi

J. E. SMITH

1.1 Introduction

The fungi comprise a heterogeneous collection of heterotrophic organisms which exist as saprophytes or parasites or, less frequently, as symbionts in commensal association with other organisms. Fungi display a wide variety of morphological form ranging from the microscopic unicellular yeasts and certain water moulds to the multicellular macroscopic mushrooms. The essential vegetative structure of the fungi is the hyphal filament or its equivalent in the unicellular forms the yeast cell or the chytrid thallus.

Hyphae present a large surface area through which substances can be interchanged with the environment. Essential materials for the biosynthetic processes of growth and development are absorbed from the environment and waste products excreted. Following a period of vegetative growth reproduction normally takes place with the formation of sexual or asexual spores which when released from the parent body will, in a suitable environment, develop into new individuals. Ecologically, fungi are a highly successful form of life and are able to grow in environments hostile to most other forms of life.

This chapter will be concerned primarily with an examination of the factors which control the structure and development of the filamentous fungi. For a more comprehensive understanding of the physiology of fungi the reader should consult Burnett (1968) and the three volumes of *The Fungi* (edited by G.C. Ainsworth and A.S. Sussman, Academic Press).

1.2 The Fungal Spore

The fungal spore is a part of the fungus highly specialized for reproduction, survival and dispersal, is normally delimited from the main part of the fungus, and is characterized by a minimal metabolic turnover, low water content and lack of cytoplasmic movement (Gregory, 1966). Asexual fungal spores are produced by mitotic division while sexual spores are derived by meiosis and as such can provide new genetic recombinations. Apart from their obvious ecological role in nature in propagating the species spores are used in the laboratory for maintenance of cultures, as inoculum to produce biomass for experimental or industrial processes, and more recently for directly facilitating specific chemical transformations (see Chapter 10).

All fungal spores are considered to show some degree of dormancy since they are no longer engaged in the synthesis of new cellular material and have a much reduced metabolic activity. Basically, there are two types of dormancy prevalent in fungi: constitutional and exogenous. In constitutional dormancy, spores will not germinate when placed under suitable environmental conditions but first require an activation process to overcome an innate property of the dormant state which may be a barrier to the penetration of nutrients, a metabolic block, or the production of a self-inhibitor. On the other hand, in exogenous dormancy, the spores are inactive because of unfavourable chemical and/or physical conditions and will quickly resume growth in a suitable environment. In this type of dormancy germination is controlled by the need for essential external growth factors, whereas in constitutional dormancy there is a barrier preventing the environmental growth factors initiating the process of germination (Sussman & Halvorson, 1966).

When all dormancy barriers have been removed spores normally require a favourable moisture level and temperature with certain substances from the environment to begin the germination process. Aerobic respiration appears to be a normal requirement for most spores to initiate germination while some have a requirement for carbon dioxide. Most spores require an external carbohydrate source for germination while others with adequate organic reserve appear to need only water.

Germination basically comprises the many processes and changes that occur during the resumption of development or growth of a resting structure and its subsequent transformation to a morphologically different structure. In filamentous fungi this will involve the change from a non-polar spore to a polar germ-tube which will continue growth by extension at the tip. Spores can be considered as the beginning and the end of the developmental cycle of a fungus.

A general feature of most dormant spores is a sparse endoplasmic reticulum which rapidly increases during germination; an increase in mitochondria numbers is also associated with germination. Most fungal spores increase in size prior to germination. This should not be considered as a simple swelling phenomenon but rather as spherical growth. New cell wall material is laid down uniformly over the entire inner surface of the spore. However, under normal conditions of growth polarity is established and wall formation ceases at all but one or two areas which grow out and establish new hyphal filaments. The cell wall of the outgrowth is normally continuous with the innermost layer of the spore wall (Florance, Denison & Allen, 1972). A vesicular hypothesis for germ-tube outgrowth has been proposed and is similar to that involved in hyphal tip growth (Bartnicki-Garcia, 1973; see also p. 8).

The onset of germination has also profound effects on macromolecular synthesis, in particular, the synthesis of protein, RNA and DNA. Studies with several fungi have clearly shown that ungerminated spores do contain many if not all of the components of the protein synthesising apparatus:

ribosomes, tRNA, aminoacyl-tRNA synthetases and transfer enzymes and these components are functional when assayed *in vitro* with synthetic polyuridylic acid. The protein which is involved in the initial germ-tube outgrowth appears to be synthesised on a template of stable mRNA already present in the dormant spores. Alteration of this RNA in such a way that it is able to act as a template for protein synthesis could well be the first step in the germination process (Van Etten, 1969; Brambl & Van Etten, 1970; Hollomon, 1969).

1.3 The Vegetative State

The vegetative state of filamentous fungi consists of microscopic strands called hyphae. As the hyphae grow branches arise behind the tip area and the protoplasm continually moves forward occupying the newly formed areas and ultimately leaving behind an empty outer casing. For this reason filamentous fungi have been described as plasmodia which creep about in tubes. The mass of hyphae constituting the thallus of the fungus is called the mycelium.

Depending on the type of fungus the protoplasm may be continuous throughout the hypha or it may be interrupted at intervals by cross-walls dividing the hypha into cells. Such cross-walls are called septa and are in most cases incomplete having a minute central pore through which protoplasmic continuity can be maintained. For this reason the hyphae are not divided into a series of independent cells but retain continuity and are coenocytic. The role of such septa may well be that of mechanical strengthening. In the Phycomycetes the hyphae are clearly coenocytic and cross-walls are of rare occurrence and are mainly found sealing off old parts of the mycelium. Whereas in the Ascomycetes and Fungi Imperfecti the septum is normally a simple poroid disc which allows free movement of all cellular components, in the Basidiomycetes there is an annular thickening of the septum around the edge of the pore to give the 'dolipore' and on either side of this there is a dome-shaped structure, the parenthosome. Although permitting cytoplasmic continuity it is probable that the parenthosome restricts nuclear movement.

Filamentous fungi possess organized nuclei containing a nuclear membrane, a nucleolus and chromosomes. In the truly coenocytic condition the haploid or less occasionally diploid nuclei are to be found randomly embedded in the cytoplasm while in septate mycelium the individual cells may be uninucleate, binucleate (such as the dikaryotic state of most Basidiomycetes) or more often multinucleate.

The cell wall

In fungi the cell wall primarily serves to protect and separate the cell from the environment. Furthermore, the presence of a rigid cell wall determines to a large extent the cellular form of the filamentous fungi and it has been considered that morphological development within the fungi may be reduced to a question of cell wall morphogenesis (Bartnicki-Garcia, 1968). The cell wall is a complex dynamic structure, the site of diverse enzyme activity. Studies on the chemical composition of cell walls has provided information on the nature of the macromolecular components of the wall fabric while electron microscopy of wall material has revealed the spatial arrangement of many of the macromolecular aggregates.

The cell walls of filamentous fungi contain the structural macromolecules of chitin and cellulose together with many other polysaccharides and specific amounts of proteins and lipids. Using a shadow-casting technique on preparations which had previously been sequentially treated with enzymes such as laminarinase, pronase, cellulase and chitinase to degrade the cell wall from the outside inwards it has been possible to infer the co-axial distribution of some of the wall polymers (Hunsley & Burnett, 1970).

The cell wall is built up of interwoven microfibrils (chitin, cellulose) which are embedded or cemented by an amorphous matrix composed primarily of polysaccharides (mannans, galactans, glucans and hetero-polysaccharides) together with proteins. The polysaccharides which make up about 80–90% of the dry weight of the fungal cell wall are composed of hexoses, amino sugars, hexuronic acids, methylpentoses and pentoses. Glucose and N-acetylglucosamine are by far the most common building units in microfibril synthesis. The distribution of polysaccharides in the cell walls of fungi does not appear to be in a random manner. Bartnicki-Garcia (1968) has shown that by selecting dual combinations of the poly-saccharides that appear to be the main components of the cell wall of individual genera a classification of fungi into a number of cell wall categories can be achieved (Table 1.1).

Table 1.1 Cell wall composition and taxonomy of fungi[a]

Cell wall category	Taxonomic group	Representative genera
I Cellulose-Glycogen	Acrasiales	*Polysphondylium, Dictyostelium*
II Cellulose-β-Glucan	Oömycetes	*Phytophthora, Pythium, Saprolegnia*
III Cellulose-Chitin	Hyphochytridiomycetes	*Rhizidiomyces*
IV Chitin-Chitosan	Zygomycetes	*Mucor, Phycomyces, Zygorhynchus*
V	Chytridiomycetes	*Allomyces, Blastocladiella*
	Euascomycetes	*Aspergillus, Neurospora, Histoplasma*
	Homobasidiomycetes	*Schizophyllum, Fomes, Polyporus*
VI Mannan-β-Glucan	Hemiascomycetes	*Saccharomyces, Candida*
VII Chitin-Mannan	Heterobasidiomycetes	*Sporobolomyces, Rhodotorula*
VIII Galactosamine-Galactose polymers	Trichomycetes	*Amoebidium*

[a] For specific examples and references see Bartnicki-Garcia (1968). Seemingly, the β-glucan is always 1,3- and 1,6-linked. Deuteromycetes are included in their corresponding groups of perfect fungi.

The formation of the fungal cell wall is clearly a complex and elaborate process which must involve delicate interrelationships between enzymes involved in precursor formation, transport of precursors and the final formulation into the 3-dimensional form. How this is achieved has been the subject of extensive studies and only a brief summary will be attempted at this point.

Hyphal extension

That hyphae grow by the deposition of cell wall material at the apical tip has been demonstrated by microscopy (Robertson, 1965), autoradiography (Bartnicki-Garcia & Lippman, 1969; Gooday, 1971) and by fluorescent antibodies specific for wall components (Marchant & Smith, 1968). Hyphal extension of the apical growth process results in the characteristic cylindrical shape of the hyphae and represents the normal mode of development in the mycelial fungi. Localized sites of wall synthesis form at subapical zones giving rise to lateral branches where the apical dominance of wall synthesis can again prevail. Subapical sites can also function for a limited period of time and form septa. Although wall extension is mainly restricted to the apex of growing hyphae wall thickening may occur for a short distance behind the apex and occasionally at distal parts as in chlamydospore formation. The ability to control wall synthesis in filamentous fungi will undoubtedly be of major importance in many industrial processes but in particular where the mycelial biomass is to be used for human or animal feeding purposes. Here it may be necessary to control or vary the composition of specific cell wall components (see also Chapter 13).

Electron microscope studies have shown that a young hypha consists of three relatively distinct zones viz: an apical zone; a subapical zone; and a zone of vacuolation (Grove, Bracker & Morre, 1970; Grove & Bracker, 1970). The apical zone is characterized by an accumulation of cytoplasmic vesicles often to the exclusion of other organelles and ribosomes (Fig. 1.1). These vesicles are now generally accepted to have an important role in apical wall synthesis and in some cases it has been possible to see them fusing with the plasmalemma. Precursors for wall synthesis are undoubtedly conveyed in the vesicles to the apical synthetic zone where the microfibrillar skeleton of the wall is synthesized *in situ* either on the outer surface of the plasmalemma or within the wall fabric. The amorphous matrix material is probably synthesized in the cytoplasm behind the apex, transported there in the vesicles and discharged and directly anchored to the fibrillar material. The subapical zone is normally free of vacuoles but is rich in a variety of protoplasmic components such as nuclei, mitochondria, ribosomes, endoplasmic reticulum, dictyosomes, vesicles, microbodies and microtubules (Fig. 1.2). It is in this zone that the important wall synthesizing vesicles are generated and transported forward to the apex. At some distance from the apex the subapical zone merges into the zone of vacuolation. The degree of vacuolation increases with distance from the apex and is usually paralleled by an increase in lipid content (Fig. 1.3). It is possible that in this zone the important secondary metabolites discussed in Chapters 3 and 7 are produced. Clearly, the fungal hypha can never be considered as a homogeneous system but rather as a complex gradient of cytochemical changes which will vary depending on the environmental pressures.

Although it has now been unequivocably demonstrated that apical wall synthesis is dependent on the continuous flow of vesicles to the tip zone (Fig. 1.4) a detailed understanding of the biochemical mechanisms is only beginning to emerge. The importance of understanding the biochemical nature of hyphal tip growth cannot be overstated since it is in this area

Fig. 1.1

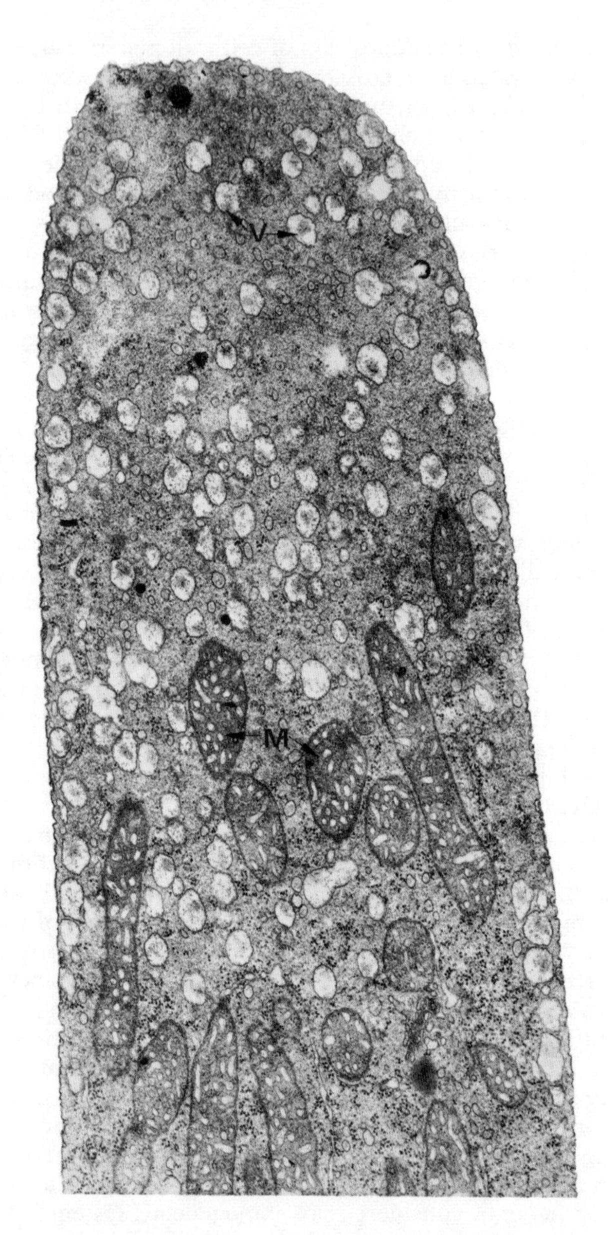

Figs. 1.1–1.3 Electron micrographs of longitudinal sections through a hypha of *Pythium ultimum*. Fig. 1.1. The apical zone is rich in vesicles (V), and the mitochondria (M) are clustered in the posterior portion of this zone. Note the scarcity of ribosomes (R) and ER compared with Fig. 1.2. Fig. 1.2. Subapical zone showing the complement of cell components and their distribution. A few lipid bodies (L) are found in this zone.
Fig. 1.3. Vacuolated zone (V) of a young hypha. All × 12 800 (Grove, Bracker & Morré, 1970).

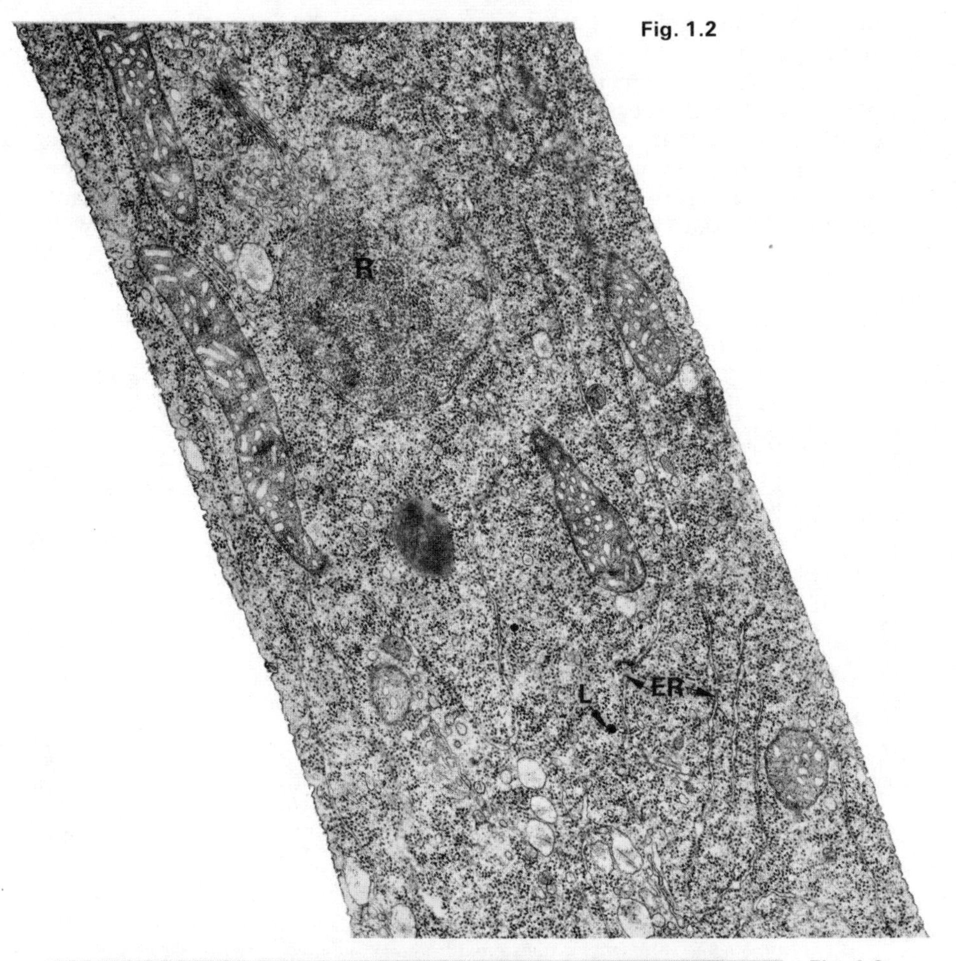

Fig. 1.2

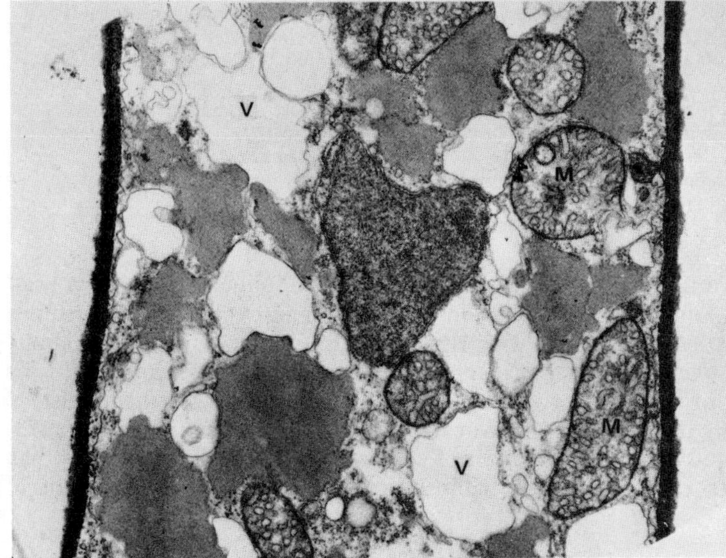

Fig. 1.3

that most morphological form is programmed by interaction with the environment.

It has long been considered that lytic enzymes may play an important part in the overall process of apical extension. An interesting and extremely relevant theory that would accommodate the presence of both synthetic and lytic enzymes at the apical growth zone has been proposed by Bartnicki-Garcia (1973). This theory does attempt to answer the important question of what controls the balance between wall synthesis and wall

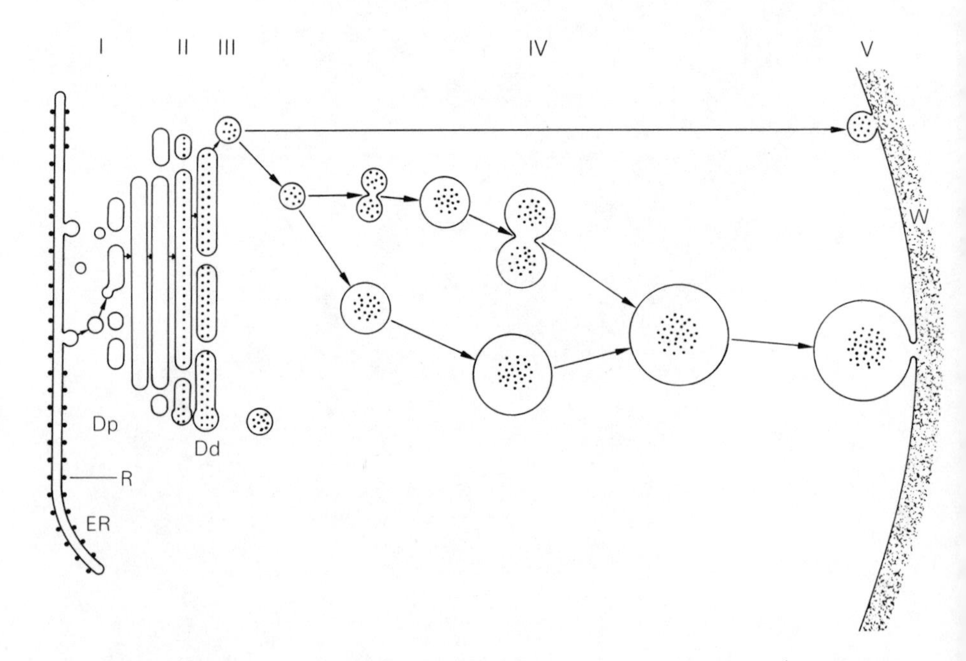

Fig. 1.4 Diagrammatic interpretation of the sequence leading to expansion of a hypha at the apex. I. Material is transferred from ER to dictyosome by blebbing of ER and refusion of vesicles to form a cisterna at the proximal pole of the dictyosome (Dp). II. Cisternal contents and membranes are transformed as the cisterna is displaced to the distal pole (Dd) by the continued formation of new cisternae. III. Cisternae vesiculates to form secretory vesicles as they approach and reach the distal pole. IV. Secretory vesicles migrate to the hyphal apex. Some may increase in size or fuse with other vesicles to form large secretory vesicles, while others are carried directly to the cell surface. V. Vesicles accumulate in the apex and fuse with the plasma membrane, liberating their contents into the wall region (Grove, Bracker & Morré, 1970).

lysis since wall synthesis in the absence of lysis could cause excessive wall thickening and possible arrestment of growth whereas lysis in the absence of synthesis would result in bursting of the hyphal tips. The harmonious balance between the factors responsible for wall synthesis and wall lysis must undoubtedly be involved not only in hyphal tip growth but also in determining the ultimate shape of the fungal cell. Bartnicki-Garcia (1973) has presented an intriguing mode of cell wall growth which considers that wall growth results from the cumulative action of minute hypothetical units of wall growth. The important feature of this hypothesis is that all types of wall growth can be more readily explained on the basis of the distribution of these units. The proposed mechanism involved in the increase in size of such a hypothetical minimum unit is shown in Fig. 1.5. For ease of interpretation the model considers that the fungal wall is composed of two major components: an amorphous substance and a microfibrillar skeleton.

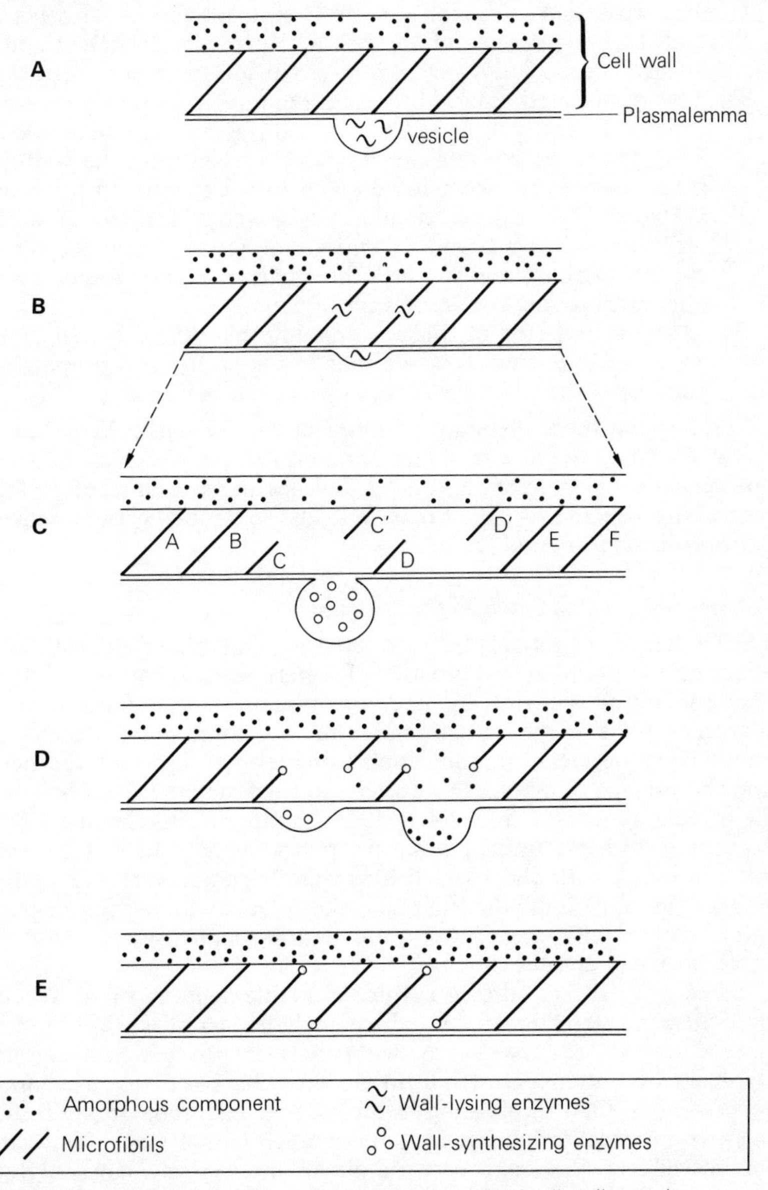

Fig. 1.5 Hypothetical representation of the events in a unit of cell wall growth. For detailed explanation see text (Bartnicki-Garcia, 1973).

In order to allow increase in wall growth five major events are considered to occur.

A. The process is initiated by the secretion of lytic enzymes from a vesicle, into the wall fabric.
B. The lytic enzymes attack the microfibrillar skeleton by splitting either inter or intra molecular bonds and so dissociating the microfibrils.
C. The weakened microfibrillar complex is not able to withstand the high turgor pressure of the cell and becomes stretched and less integrated, thus allowing an increase in surface area to occur.
D. The dissociated microfibrils are rebuilt by synthetase enzymes situated in the wall or in the outer surface of new plasmalemma. The synthetase enzymes are initially secreted into the wall fabric from vesicles and the soluble precursors are assumed to be transported across the plasmalemma. Vesicles containing the amorphous wall material in a largely or entirely preformed state deposit their contents against the wall and this material is then forced into the microfibrillar network by turgor pressure.
E. At the completion of this enzymic interplay the cell wall unit has expanded one unit area without losing its overall properties, in particular, the coaxial arrangement of wall polymers.

Perhaps the most intriguing feature of this vesicular hypothesis for apical growth is the lack of information on the factors which control the direction of movement of the vesicles. A fuller understanding of the factors controlling vesicle movement may well lead to a much greater control of morphological form and function.

Colony formation and multihyphal structures

The hypha should not only be considered as an individual element but also as one of many similar and functionally interrelated elements which comprise the mycelium. Most hyphae in a mycelium become interconnected at an early stage by the formation of numerous anastomoses which permit interconnection and communication along the entire mycelium, permitting the passage of cytoplasm, nuclei and food material from one part of the mycelium to another (Zalokar, 1965). An interesting feature of hyphal branching is that a branch from any given hypha does not necessarily become a cell with the same features as the parent cell. Growth and differentiation of individual hyphae takes place with varying degrees of co-ordination with adjacent hyphae and in some forms multihyphal aggregates may develop.

When growing on surface culture most filamentous fungi develop a heterogeneous growth pattern which includes aerial, surface and submerged hyphae each with its own particular physiological condition. Hyphae within such a growth form exhibit a characteristic differentiation depending on their localization in the culture. Autoradiographic studies, and enzyme localization *etc.* in individual hyphae of many fungi clearly demonstrate the heterogeneous nature of the hyphal filament. Such a pattern of differentiation of cellular activities along the hyphal filament is probably an inevitable consequence of the apical growth process.

When filamentous fungi grow in submerged liquid culture they encounter a more uniform environment and growth is normally 3-dimensional, either in the form of closely interwoven hyphal masses or pellets, or in the truly filamentous form. The importance of pH, inoculum size and the nature of the medium in determining the vegetative growth form of filamentous fungi has been extensively studied in academic and industrial laboratories. The growth kinetics of filamentous fungi in surface and submerged liquid culture is comprehensively treated in Chapter 5.

In many filamentous fungi multihyphal vegetative structures are formed that appear to involve a degree of co-ordination between the component hyphae that is greater than that found in a vegetative colony. The formation of such structures has been comprehensively examined by Butler (1966). Of particular interest from an industrial point of view are the multihyphal sclerotia in *Claviceps* since they are the source of important alkaloids (see Chapter 14).

In many fungi hyphae may become interwoven to form discrete hyphal aggregates of varying size which have the ability to withstand adverse environmental conditions such as drought and temperature variations for longer periods than the normal hyphae of the mycelium. These structures have been termed sclerotia and can be considered as a resting stage of the fungus and when the growing conditions return to normal can reproduce the fungus by the outgrowth of new mycelial growth and/or by the development of sexual or asexual spores (Willets, 1971).

Little is as yet known of the factors which regulate sclerotial development although a prerequisite for sclerotial initiation is the formation of a large number of hyphal branches. Anastomoses are a common feature of sclerotial development and help to form a communication bridge between hyphae. There are three stages of growth of the sclerotium: initiation or formation of small discrete initials; development and growth to full size; and maturation with surface delimitation, internal changes and pigmentation of the peripheral hyphae. The complexity of the sclerotium clearly indicates some degree of co-ordination during development. Different environmental conditions are also required for the initiation than for the further development of the sclerotium, and eventually the sclerotium becomes isolated both physiologically and nutritionally from the parent thallus (Willets, 1972). In *Claviceps* there is abundant evidence of the immense biochemical differences between sclerotia and normal mycelium. The commercial importance of many of the compounds present in *Claviceps* sclerotia has led to extensive studies on the possibility of fermenter cultivation of *Claviceps* mycelium to produce similar metabolites as are produced naturally in sclerotia. This has not been altogether an easy problem as shown in Chapter 14 and it is interesting to note that higher alkaloid production seems to parallel a growth form in liquid culture not unlike that occurring in early sclerotium formation. Clearly, cell to cell interrelationship may be of paramount importance in achieving the correct programme of metabolic production.

1.4 Reproduction

The ability of a filamentous fungus to change from a vegetative growth form to a reproductive form is the result of a combination of the genetic ability to produce the new growth and the effect of environmental factors. Thus a fungal mycelium genetically unable to reproduce will remain vegetative under all environmental conditions while a mycelium with a high potential for reproduction could be prevented from differentiating by sublethal environmental conditions.

In fungi reproduction commonly occurs after the phase of rapid vegetative growth. Although growth and cell multiplication on the one hand, and morphogenesis and differentiation on the other, are not necessarily completely incompatible, some antagonism can be expected if it is postulated that alternative metabolic pathways are involved in these processes.

There exists a wealth of information on the environmental factors which affect reproduction (Hawker, 1966). There is no apparent pattern underlying the complex effect of environmental factors on development and the enormous array of facts merely substantiate the principles set forth in 1898 by George Klebs and re-enunciated in modern terminology by Morton (1967): 'The conditions for vegetative growth and for reproduction are different. Some minimum period of vegetative development is required before the organism becomes competent to produce reproductive bodies during which it synthesizes specific metabolites, enzymes or food substances essential for reproduction. Reproduction is often induced when some external or internal factor, frequently a nutrient, becomes limiting for vegetative development. The external conditions inducing reproduction are usually narrower in range and more specific than those permitting vegetative growth'.

Thus it can be concluded that conditions permitting spore formation are almost invariably of a narrower range than those that will support vegetative growth; the requirements for sexual reproduction differ from those sufficient for vegetative growth far more than do the requirements for asexual reproduction in the same species (Hawker, 1957). If vegetative growth and reproductive processes do not occur simultaneously and are indeed separated by a definite metabolic shift, the point of change is probably associated with limitation of vegetative growth due to nutrient exhaustion. There are numerous examples among the fungi in which differentiation is associated with nutrient depletion (Smith & Galbraith, 1971). Unfortunately, little is known of the mechanism(s) by which perception of an external stimulus is translated into the visible initiation of the reproductive phase.

When vegetative growth becomes limited in many fungi under conditions which may or may not support reproduction, secondary metabolites many of which may be of industrial use frequently accumulate. These secondary metabolites are usually derived from primary metabolites which accumulate from primary synthetic processes and they are often highly specific and confined to one or a few fungal species (see Chapter 3).

Asexual reproduction

The types of mechanisms for spore production in the filamentous fungi are many and varied and some fungi possess more than one mechanism (Alexopoulos, 1962). In certain groups of the filamentous Phycomycetes asexual spore production takes place by the formation, within a sporangium, of naked cells, zoospores, which upon release swim away by means of flagella. Such spores eventually settle down, become encysted, and germinate forming a new filamentous thallus. However, in most filamentous fungi no motile spores are produced but rather the spores are covered with a rigid wall and are passively distributed by air and water. These spores are sometimes termed aplanospores in contradistinction to the motile zoospores or planospores.

In the coenocytic Mucorales such planospores are formed inside a sac or sporangium and are released upon the rupture or dissolution of the sporangial walls; such spores are called sporangiospores. In contrast, spores that arise as single separable cells of the mycelium are called conidia and may be formed by the fragmentation of the whole mycelium or of special hyphae into cylindrical, ovoid or spherical cells called oidia or arthrospores; or they may arise from the formation of terminal or lateral cells from special hyphae. The hyphae which produce conidia are called conidiophores and the conidiophores may be morphologically very similar to or very different from vegetative hyphae. The manner in which conidia are attached to their conidiogenous cells and the mechanisms by which they are released are not only of biological interest but also provide characters useful for classification (Alexopoulos, 1962; Smith, 1968).

Growth and reproduction under subaerial conditions represents the 'natural' mode of development of most filamentous fungi. Heterogeneity, which is an inherent characteristic of the filamentous form, is further magnified by overcrowding and compaction of hyphae within the macroscopic colonies which are formed under these conditions. While surface colonial growth is convenient for most genetical studies and for studies on the morphology of development it is unsuitable for biochemical studies (Smith & Galbraith, 1971). In recent years efforts have been made to improve the conventional surface method of spore production and now several techniques are available for the induction of conidiation under defined, relatively homogeneous conditions (Smith & Anderson, 1973; Chapter 9). Growth in submerged agitated batch culture clearly allows a much more precise examination of the effects of medium composition on spore formation than the static surface growth methods. However, the interpretation of development responses to particular cultural conditions is complicated in the batch system by the transient nature of the environmental conditions during growth. In the batch culture system it is not possible, for example, to determine whether spore formation results from nutrient limitation or from the limitation of growth rate imposed by this condition. Chemostat culture on the other hand permits a study of fungal populations at various growth rates and under various metabolic steady states.

It has been generally considered that growth and sporulation in fungi are mutually exclusive and that sporulation will only become manifest under conditions of nutrient exhaustion or limitation (Smith & Galbraith, 1971). In the batch system the substrate is consumed as the organism grows. Consequently, if there exists a growth limiting factor in the medium its concentration will also go on decreasing as will also the specific growth rate of the fungus until eventually both approach zero.

Studies with *Aspergillus niger* under chemostat conditions have shown that with citrate as the limiting carbon source conidiation can be induced by varying the dilution rate. When glucose was the limiting carbon source conidiation could not be induced to any extent by varying the dilution rate (Ng, Smith & McIntosh, 1973). Thus, although the experiments with varying dilution rates under citrate limitation demonstrate that conidiation can be controlled by growth rate the studies with glucose limitation show also that the composition of the medium must also be considered. Although conidiospore formation can occur in chemostat culture the intensity of conidiation is much less than in subaerial and batch liquid culture and may suggest that there is only a partial switching on of the sporulation mechanism.

A fuller understanding of the mechanisms of spore formation in filamentous fungi will undoubtedly be of considerable value to many industrial processes (see Chapter 9).

Sexual reproduction

Sexual reproduction in fungi involves the union of two compatible nuclei together with the recombination of genetic factors by meiosis. In the first instance plasmogamy brings the two nuclei together in one cell, karyogamy unites them into one diploid and meiosis then re-establishes the haploid condition. In the Basidiomycetes and in the higher Ascomycetes plasmogamy and karyogamy can be separated in time and space and nuclear fusion may not take place until much later in the life history of the fungus. The resultant binucleate, haploid phase in the life cycle is called the dikaryon and serves to increase the genetic recombination frequency per sexual fusion.

It is now considered that there are definite physiological mechanisms superimposed on the genetic control of sexuality. The secretion of sexual hormones which control and direct the initiation of sexually active organs to karyogamy is now well documented in several fungi (Gooday, 1973). However, with limited exceptions there is not much known about the biochemical and physiological changes that occur within an organism during sexual morphogenesis.

For a wider background on sexuality in the fungi reference should be made to *The Fungi* Volume 2 (1966; edited by G.C. Ainsworth and A.S. Sussman, Academic Press); Alexopoulos (1962); and Burnett (1968).

1.5 References

ALEXOPOULOS, C.J. (1962). *Introductory mycology*. New York: John Wiley and Sons.

BARTNICKI-GARCIA, S. (1968). Cell wall chemistry, morphogenesis and taxonomy in fungi. *Annual Review of Microbiology* 22, 87–108.

BARTNICKI-GARCIA, S. (1973). Fundamental aspects of hyphal morphogenesis. *Symposium Society General Microbiology* 23, 245–67.

BARTNICKI-GARCIA, S. & LIPPMAN, E. (1969). Fungal morphogenesis: cell wall construction in *Mucor rouxii*. *Science* **165**, 302–4.

BRAMBL, R.M. & VAN ETTEN, J. L. (1970). Protein synthesis during fungal spore germination. V. Evidence that the ungerminated conidiospore of *Botryodiplodia theobromae* contains messenger ribonucleic acid. *Archives of Biochemistry and Biophysics* 137, 442–52.

BURNETT, J.H. (1968). *Fundamentals of mycology*. London: Edward Arnold (Publishers) Ltd.

BUTLER, G.M. (1966). Vegetative structures. In *The Fungi* Vol. 2, pp. 83–112. Edited by G.C. Ainsworth and A.S. Sussman, New York and London: Academic Press.

FLORANCE, E.R., DENISON, W.C. & ALLEN, T.C. (1972). Ultrastructure of dormant and germinating conidia of *Aspergillus nidulans*. *Mycologia* **69**, 115–23.

GOODAY, G.W. (1971). An autoradiographic study of hyphal growth of some fungi. *Journal of General Microbiology* **67**, 125–33.

GOODAY, G.W. (1973). Differentiation in the Mucorales. *Symposium Society of General Microbiology* **23**, 269–94.

GREGORY, P.H. (1966). The fungus spore: what it is and what it does. In *The Fungus Spore*. Edited by M.F. Madelin, pp. 217–233. London: Butterworths.

GROVE, S.N. & BRACKER, C.E. (1970). Protoplasmic organization of hyphal tips among fungi: vesicles and spitzenkörper. *Journal of Bacteriology* **104**, 989–1009.

GROVE, S.N., BRACKER, C.E. & MORRÉ, D.J. (1970). An ultrastructural basis for hyphal tip growth in *Pythium ultimum*. *American Journal of Botany* **57**, 245–66.

HAWKER, L.E. (1957). *The physiology of reproduction in fungi*. London and New York: Cambridge University Press.

HAWKER, L.E. (1966). Environmental influences on reproduction. In *The Fungi*, Vol. 2, pp. 435–69. Edited by G.C. Ainsworth and A.S. Sussman. New York and London: Academic Press.

HOLLOMAN, D.W. (1969). Biochemistry of germination of *Peronospora tabacina* (Adam) conidia: evidence for the existence of stable messenger RNA. *Journal of General Microbiology* **55**, 267–74.

HUNSLEY, D. & BURNETT, J.H. (1970). The ultrastructural architecture of the walls of some hyphal fungi. *Journal of General Microbiology* **62**, 203–18.

MARCHANT, R. & SMITH, D.G. (1968). A serological investigation of hyphal growth in *Fusarium culmorum*. *Archiv für Mikrobiologie* **63**, 85–94.

MORTON, A.G. (1967). Morphogenesis in fungi. *Science Progress* **55**, 597–611.

NG. A.M.L., SMITH, J.E. & MCINTOSH, A.F. (1973). Conidiation of *Aspergillus niger* in continuous culture. *Archiv für Mikrobiologie* **88**, 119–26.

ROBERTSON, N.F. (1965). The mechanism of cellular extension and branching. In *The Fungi*, Vol. 2, pp. 613–23. Edited by G.C. Ainsworth and A.S. Sussman. New York and London: Academic Press.

SMITH, G. (1968). *An introduction to industrial mycology*. London: Edward Arnold (Publishers) Ltd.

SMITH, J.E. & ANDERSON, J. (1973). Differentiation in the Aspergilli. *Symposium Society General Microbiology* **23**, 295–337.

SMITH, J.E. & GALBRAITH, J.C. (1971). Biochemical and physiological aspects of differentiation in the fungi. *Advances in Microbial Physiology* **5**, 45–134.

SUSSMAN, A.S. & HALVORSON, H.O. (1966). *Spores: their dormancy and germination*. New York and London: Academic Press.

VAN ETTEN, J.L. (1969). Protein synthesis during fungal spore germination. *Phytopathology* **59**, 1060–4.

WILLETTS, H.J. (1971) The survival of fungal sclerotia under adverse environmental conditions. *Biological Reviews* **46**, 387–407.

WILLETTS, H.J. (1972). The morphogenesis and possible evolutionary origins of fungal sclerotia. *Biological Reviews* **47**, 515–36.

ZALOKAR, M. (1965). Integration of cellular metabolism. In *The Fungi* Vol. 1, pp. 377–426. Edited by G.C. Ainsworth and A.S. Sussman. New York and London: Academic Press.

CHAPTER 2

The Environmental Control of the Physiology of Filamentous Fungi

D. R. BERRY

2.1 Introduction

The filamentous fungi represent a physiologically diverse group of micro-organisms which are characterized by a saprophytic, or less frequently a parasitic, mode of nutrition. They may be grown on either solid or liquid media and in their natural environment are frequently found colonizing the surface of liquids and solids such that a large percentage of their hyphae are surrounded by air. Since such hyphae must acquire their nutrient from those hyphae which are in contact with the substrate, the transloca-tion mechanisms which exist in fungi must play an important role in controlling their development. In several fermentations the fungus is traditionally cultured on the surface of either a liquid medium, *e.g.* peni-cillin and citric acid production, or a solid medium, *e.g.* blue cheese or Koji production. In such fermentations the gaseous phase provides a ready supply of oxygen and the solid or liquid phase, the nutrients.

Growth in solid substrate may restrict gaseous exchange but usually provides an abundant supply of nutrients. These nutrients may be assimi-lated either directly or after breakdown by extracellular enzymes, *e.g.* amylases, proteases, which are secreted by many fungi. In a solid medium, the enzymes secreted remain concentrated in an area close to the fungus rather than being rapidly diluted, as occurs in liquid media. These reduced rates of diffusion in solid media do however mean that the nutrients in a given area can become exhausted. This presents no problem to the fila-mentous fungi which, unlike unicellular organisms, can easily move through solid media as a result of their mycelial growth form. Although in Western society solid state fermentations have been restricted to cheese and mushroom production they are much more widespread in Asia (Chapter 13).

In submerged liquid culture, nutrients when present are readily avail-able to the fungus by diffusion, and a relatively homogeneous environment can be maintained for the fungus by adequate agitation. However, not all substrates are readily soluble in aqueous solution. Oxygen in particular has a low solubility in water and high levels of aeration and agitation are required to maintain an adequate supply. The tendency of filamentous fungi to form pellets in liquid culture (Chapter 5) must be restricted if adequate aeration is to be maintained. The majority of fermentations at the present time are carried out in liquid culture using fermenters in which the culture conditions can be closely maintained and controlled.

Most industrial uses of fungi involve the manipulation of the physiology of the organism in order to stimulate production of the maximum amount of a primary or secondary metabolite, or an enzyme. Biomass production is perhaps the exception where the whole organism is the desired product. Fungal growth can be controlled by manipulating either the genotype or the environment (Fig. 2.1). The use of strain selection techniques to

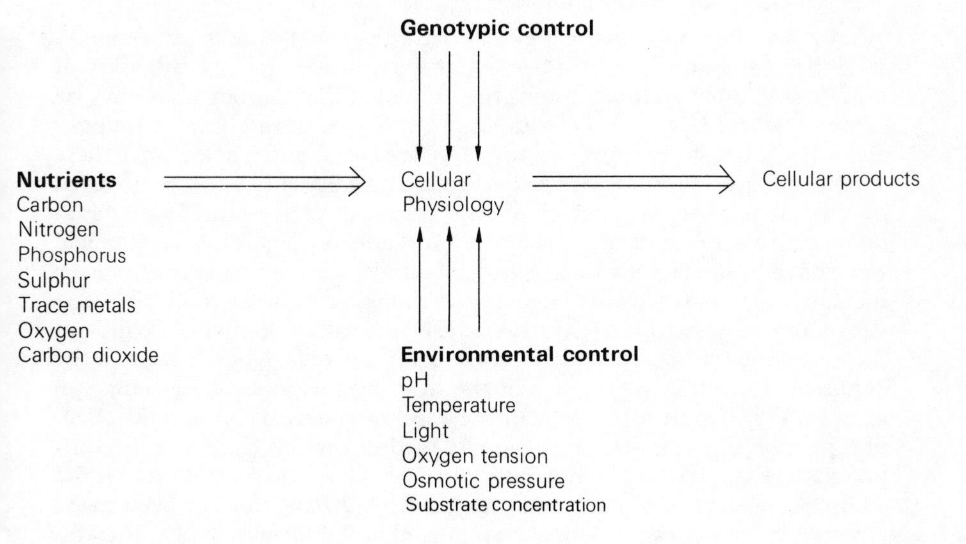

Fig. 2.1 Diagram illustrating the relationship between the genetic and environmental control of microbial metabolism.

manipulate the genotype is discussed in Chapter 4. The response of fungi to environmental factors is extremely varied since each species can be considered to exhibit a unique range of responses particularly with respect to differentiation and secondary metabolite formation. Many of the recent advances in fermentation technology have been concerned with improving the design of the culture vessel so giving a greater degree of homogeneity, improved oxygen supply, improved monitoring techniques and in general a greater degree of control over the fermentation. However, such studies should not be allowed to obscure the fact that in most instances the fermenter is only an instrument which permits the operator to control the state of the growth medium, and it is the state of the medium, the pH, temperature, nutrient status and degree of aeration, which controls the growth of the organism. It is with this aspect of fungal physiology, namely the control of the organism by manipulation of the growth medium, that this Chapter will be concerned.

Two exceptions to this are however particularly important in fungi. The shearing action of the impeller in a fermenter plays an important role in preventing pellet formation in fungal fermentations. In addition to reducing the inhomogeneity resulting from pellet formation this shearing action may also be an essential requirement for exponential growth in fungal fermentations (Chapter 5).

The use of continuous culture techniques to control the mean residence time of cells in a fermenter provides a mechanism by which the growth rate of a population of cells can be controlled. In *Penicillium chrysogenum*, it has been shown that RNA content increases with growth rate (Chapter 5).

Primary, secondary and intermediary metabolism

Maximal growth rates are achieved when the growth rate is restricted by the inherent characteristic of the organism rather than the availability of nutrients. Under such conditions the growth of filamentous fungi may be exponential (Chapter 5). A reduction in the growth rate can be brought about by limiting the supply of any of the essential nutrients. Under these conditions growth is no longer balanced and nutrients become diverted into metabolic pathways which are not essential for growth. Those metabolic pathways which occur when the organism is growing at maximal rates have been referred to as *primary* pathways, as opposed to *secondary* metabolic pathways which become functional or at least more active at sub-maximal growth rates (Bu'Lock, 1967; see also Chapter 3). Which of these secondary metabolic pathways become active is dependent upon the organism and upon the nature of the limiting nutrient. Most fungi can grow on very simple media containing a simple source of carbon, nitrogen, phosphorus, sulphur and trace metals. Within the cell these simple compounds are converted through a series of metabolic pathways into all the complex molecules which constitute a cell. A distinction has been made between those pathways which are involved in the formation of substrates for several biosynthetic pathways—intermediary metabolism—and those which convert intermediary metabolites into other, usually more complex, compounds (see p. 21). Many of the reactions of intermediary metabolism are also involved directly or indirectly in the generation of ATP and reduced pyridine nucleotides which are also essential for the biosynthetic and metabolic activities of the cell. The majority of the carbon, nitrogen and phosphorus which becomes incorporated into the structure of the cell, or into extracellular products is assimilated initially into a limited number of intermediary metabolites, *e.g.* glucose 6-phosphate, glutamic acid, ATP *etc.* The problem of medium development becomes more precisely a problem of manipulating the medium such that intermediary metabolites are diverted preferentially into the desired metabolic pathways.

The extensive literature which has accumulated on the effects of different environmental stimuli on fungal physiology and differentiation is relevant to this problem (Foster, 1949; Hawker, 1950; Cochrane, 1966; Ainsworth & Sussman, 1965, 1966).

2.2 Carbon Metabolism

Assimilation

Although carbohydrates are usually the major source of carbon for fungi, fatty acids and organic acids, particularly those produced by the deamination of amino acids are also readily metabolized. Most fungi can utilize a range of monosaccharides, oligosaccharides and polysaccharides although uptake is normally restricted to monosaccharides. Disaccharides and polysaccharides are usually broken down by hydrolytic enzymes, which are either secreted into the surrounding medium or are present at the cell surface, *e.g.* amylases, cellulases, invertase. These enzymes are frequently inducible, *e.g.* α-amylase. Glucose, fructose and mannose are readily utilized by most fungi. Galactose, however, is less readily metabolised and many fungi cannot use it as a sole carbon source.

Uptake of carbohydrates is an active process which may involve a phosphorylation reaction. Free monosaccharides do not accumulate in the cell during uptake but most hexoses are converted to glucose-6-phosphate or fructose-6-phosphate before being metabolized in glycolysis. The activity of hexokinase has been shown to increase with increasing dilution rate in continuous cultures of *Aspergillus nidulans* (Dean, 1972). Several pentoses support growth, *e.g.* xylose and arabinose, and are initially incorporated into the hexose monophosphate pathway.

The rate of assimilation of carbohydrates frequently limits fungal growth rates. When lactose is provided as the sole carbon source, it is metabolized slowly and growth is restricted in *Penicillium chrysogenum*. Starch has also been used for this purpose. The rate of assimilation of a given carbohydrate is frequently affected by other medium components. Glucose suppresses the uptake of less readily metabolized sugars such as galactose, and in *Aspergillus nidulans* acetate has been reported to inhibit glucose uptake although it has no effect on sucrose assimilation (Romano & Kornberg, 1968). A reduction in the carbohydrate feed rate has been shown to stimulate autolysis in *Aspergillus niger* (Lahoz & Miralles, 1970).

Catabolism

Carbohydrate catabolism is traditionally divided into three phases, glycolysis, pyruvate metabolism and the tricarboxylic acid cycle, and oxidative phosphorylation. The details of the enzymology of these processes are well documented for a range of organisms and the available evidence indicates that these pathways in the fungi are essentially the same as in other organisms (Blumenthal, 1965; Niederpruem, 1965; Lindmeyer, 1965).

Glycolysis has been defined as the physiological conversion of glucose to pyruvate or lactate without regard to mechanism (Blumenthal, 1965). In fungi, three pathways have been described, the Embden-Meyerhof-Parnas Pathway (EMP), the hexose monophosphate pathway (HMP) and the Entner Doudoroff pathway (ED) (Fig. 2.2). The relative importance of the different pathways varies with the species (Table 2.1).

The EMP is the major pathway in most species with the HMP being the alternative minor pathway. The ED pathway has only been described in a few species of fungi and in these it is the dominant pathway (Blumenthal, 1965).

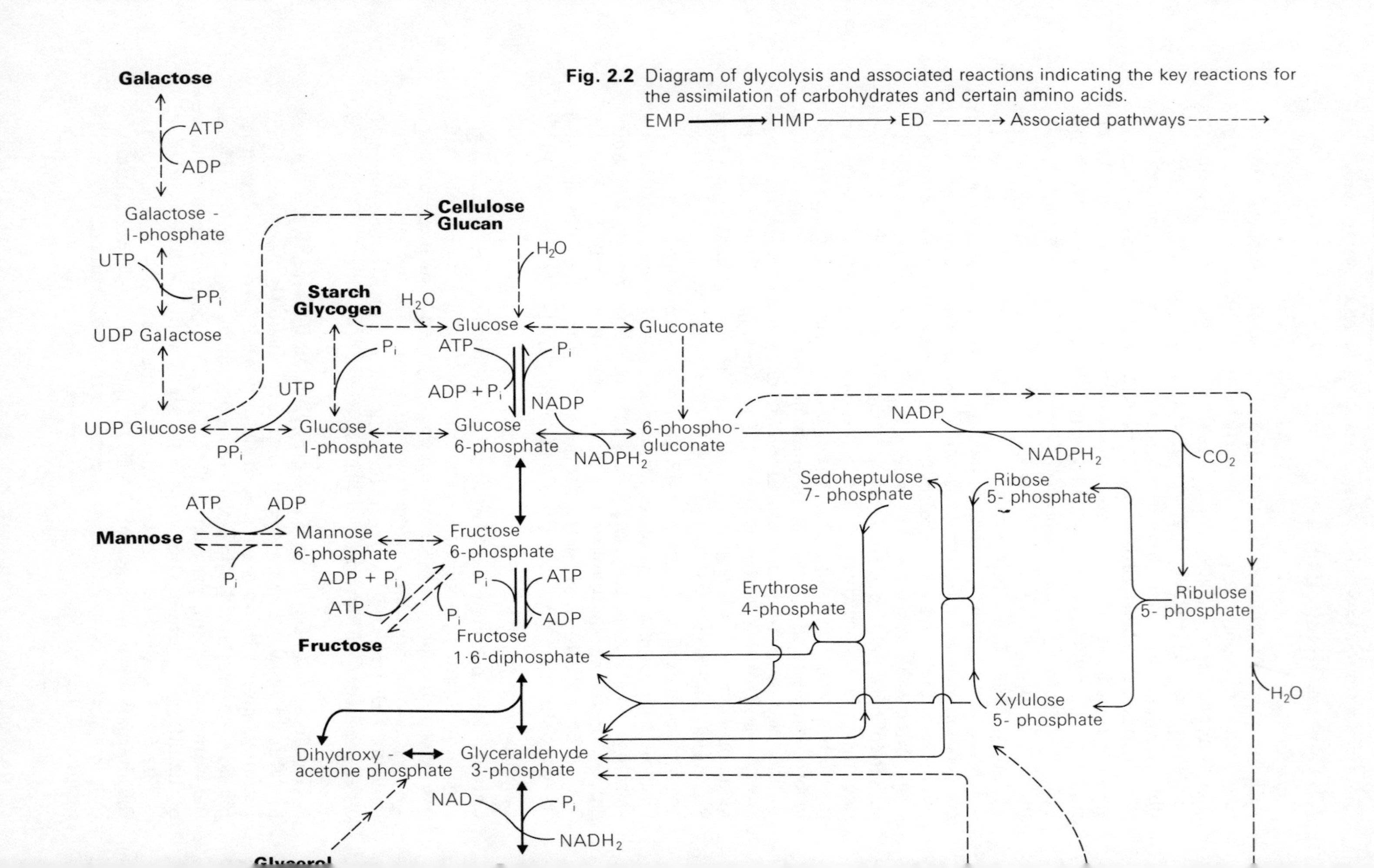

Fig. 2.2 Diagram of glycolysis and associated reactions indicating the key reactions for the assimilation of carbohydrates and certain amino acids.
EMP ⟶ HMP ⟶ ED - - - → Associated pathways - - - →

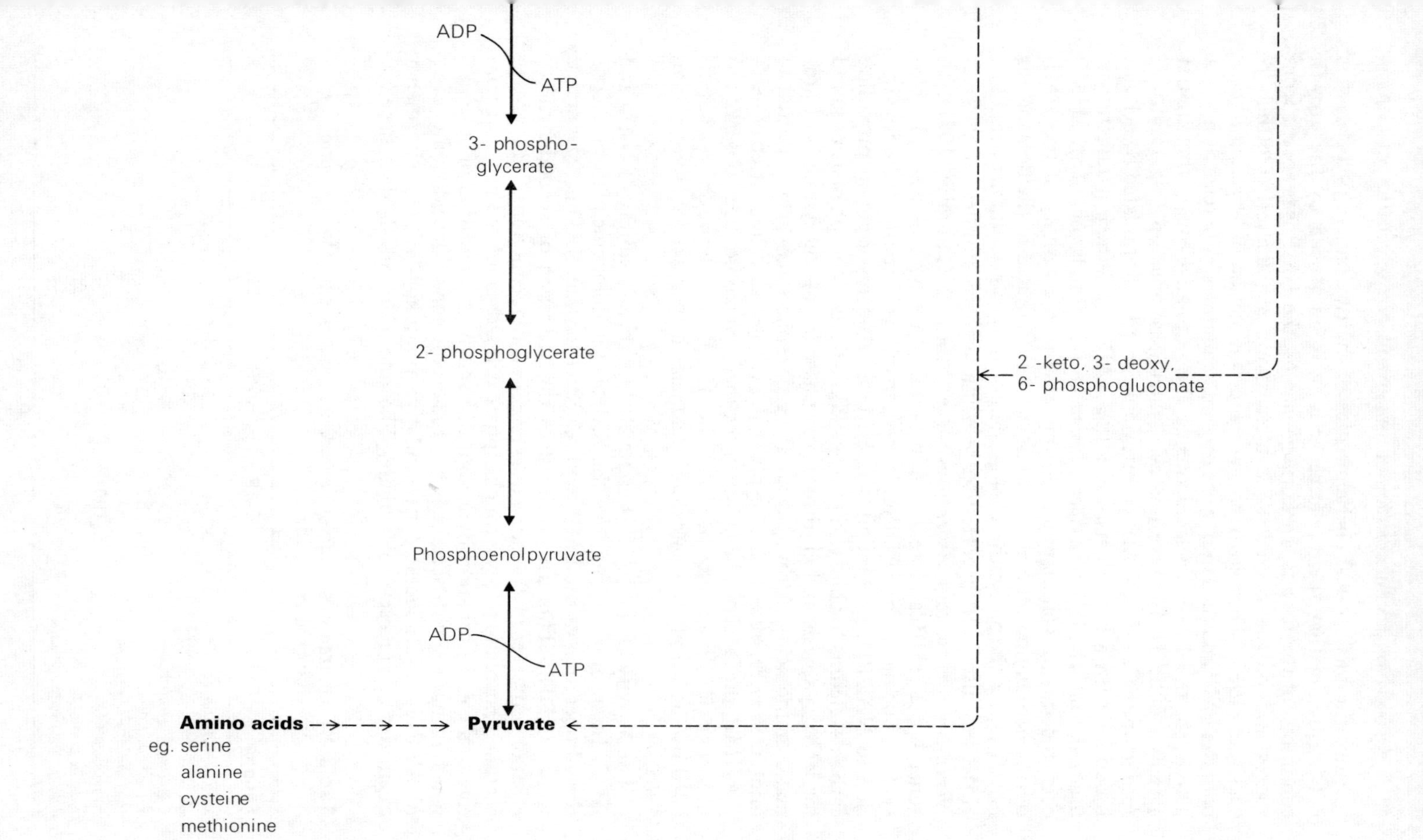
ADP
ATP
3- phospho-
glycerate
2- phosphoglycerate
Phosphoenolpyruvate
ADP
ATP
Pyruvate
Amino acids
eg. serine
alanine
cysteine
methionine
2 -keto, 3- deoxy,
6- phosphogluconate

Since the biochemical characteristics of the two pathways differ, it is not surprising that their relative activity is influenced by the growth conditions of the fungus. In *Ustilago maydis*, the HMP was absent from spores and only appeared 12 h after spore germination (Gottlieb & Caltrider, 1963). During the growth and conidiation of *Aspergillus niger* both the EMP and the HMP were present at all times. Whereas the highest activity of the EMP occurred during the vegetative stage, the highest activity of the HMP were observed during conidiation (Ng. Smith & Anderson, 1972). The HMP pathway appears to be stimulated by cellular functions with a high demand for the $NADPH_2$. When glutamate was used instead of NH_4 in *Aspergillus niger* no stimulation of the HMP occurred during conidiation. The increase in the activity of the HMP during conidiation on NH_4 medium can be attributed to the demand for $NADPH_2$ for the biosynthesis of glutamate from α-ketoglutaric acid. The HMP has also been reported to be stimulated by nitrate reduction in *Neurospora crassa* and *Aspergillus niger* (Smith & Galbraith, 1971). In general, the HMP is more active in rapidly growing cells when the biosynthetic demand for $NADPH_2$ is high, than it is in resting cells. In addition to $NADPH_2$ generation, the HMP is also important in providing carbon skeletons for biosynthetic reactions. Ribose phosphate is required for RNA synthesis and erythrose phosphate for the biosynthesis of aromatic amino acids. Although inorganic phosphate has been reported to inhibit, *in vitro*, glucose-6-phosphate dehydrogenase, the first enzyme in the HMP, the level of phosphate in the medium has not been observed to influence HMP activity *in vivo*. The HMP may however be stimulated by the supply of pentoses as the major carbon source (Blumenthal, 1965).

The activity of the EMP is controlled by the availability of NAD and ADP. In anaerobic conditions the EMP is the dominant pathway. Continued operation of the EMP requires however that NAD is regenerated from $NADH_2$. This usually involves the reduction of acetaldehyde to ethanol. However in *Saccharomyces cerevisiae* the reduction of dihydroxyacetone phosphate to glycerol phosphate can be stimulated as an alternative to alcohol formation by a high pH and by the addition of sulphite, which combines with acetaldehyde. In aerobic conditions adequate levels of ATP can be generated by oxidative phosphorylation from smaller quantities of glucose-6-phosphate. The rate of glycolysis then becomes

Table 2.1 Utilization of different glycolytic pathways by selected fungi (Blumenthal, 1965)

Fungus	EMP	HMP	ED
Aspergillus niger	78	—	—
Caldariomyces fumago	—	35	65
Claviceps purpurea	90–96	4–10	—
Fusarium lini	83	17	—
Penicillium chrysogenum	56–70	46–30	—
Rhizopus MX	100	—	—
Tilletia caries spores	—	—	100
Tilletia caries mycelium	66	34	—

restricted to a level which provides a sufficient supply of pyruvate to the TCA cycle. This reduction in the rate of glucose consumption, and the concomitant reduction in the rate of glycolysis in aerobic conditions, known as the Pasteur effect, has been explained on the basis of substrate competition for ADP and/or Pi between the ATP generating steps in glycolysis and oxidative phosphorylation. More recent studies suggest that such a control mechanism is supported by feedback inhibition of glycolysis by glucose-6-phosphate, ATP and citrate (Sols, Gancedo & Delafuente, 1970).

The overall efficiency of utilization of carbohydrates, expressed as dry weight of fungus produced per unit weight of carbohydrate consumed, has been referred to as the Economic Coefficient. It is influenced by the strain of fungus used, the level of carbohydrate in the medium, the level of trace metals, especially zinc, and the pH (Perlman, 1965). An extreme example of inefficient utilization of carbohydrates would be the citric acid fermentation in which the conditions are designed to give maximum conversion of glucose to citric acid. During the industrial production of secondary metabolites economy demands that the utilization of nutrients for biomass production is kept to a minimum.

The second stage in the oxidation of carbohydrates in aerobic cells involves the oxidation of pyruvate to carbon dioxide via the tricarboxylic acid cycle (TCA).

Under normal conditions acetate, which is produced by the oxidative decarboxylation of pyruvate, is the main substrate of the TCA cycle. However, some exogenously supplied TCA intermediates can be metabolized via the TCA cycle and the carbon skeletons of amino acids such as glutamate and aspartate, are also readily metabolized. In animal cells it is frequently possible to demonstrate the presence of TCA cycle activity by feeding intermediates and observing the subsequent increase in respiration. That this response has rarely been observed in fungi can probably be attributed to the fact that fungal cell membranes are only permeable to TCA acids at low pH. Amino acids are more readily assimilated and may provide a more satisfactory method of feeding the corresponding carbon skeleton. TCA intermediates are readily assimilated as esters. The metabolism of fatty acids and of lipids which are broken down to acetate, is dealt with below. The efficient operation of the TCA cycle requires an adequate supply of oxalacetate as a substrate for citrate synthetase, and NAD as a hydrogen acceptor. TCA cycle activity can be limited by a shortage of either of these. NAD is generated from $NADH_2$ by the process of oxidative phosphorylation which is ultimately controlled by the demand for ATP and potentially the level of available oxygen. It is possible that complex media may contain compounds which uncouple oxidative phosphorylation and so release the dependence on ATP generation. The TCA cycle is inactive in anaerobic conditions.

The level of oxaloacetate can be reduced by the drain of intermediates such as α-ketoglutarate and oxaloacetate into amino acid biosynthesis. This loss is made good by one of two mechanisms. The series of reactions known as the glyoxylate cycle permit the synthesis of two molecules of oxaloacetate from one of isocitrate and one of acetate. Alternatively oxaloacetate can also be synthesized by the carboxylation of either phosphoenol pyruvate or pyruvate (Fig. 2.3). Recent studies on *Neurospora crassa* (Flavell & Woodward, 1971) suggest that these two mechanisms operate under different conditions. In the presence of an adequate supply of glucose, the level of TCA intermediates is maintained by carboxylation reactions from phosphoenol pyruvate, and the glyoxylate cycle is non-functional. In contrast, during growth on acetate, the level of intermediates

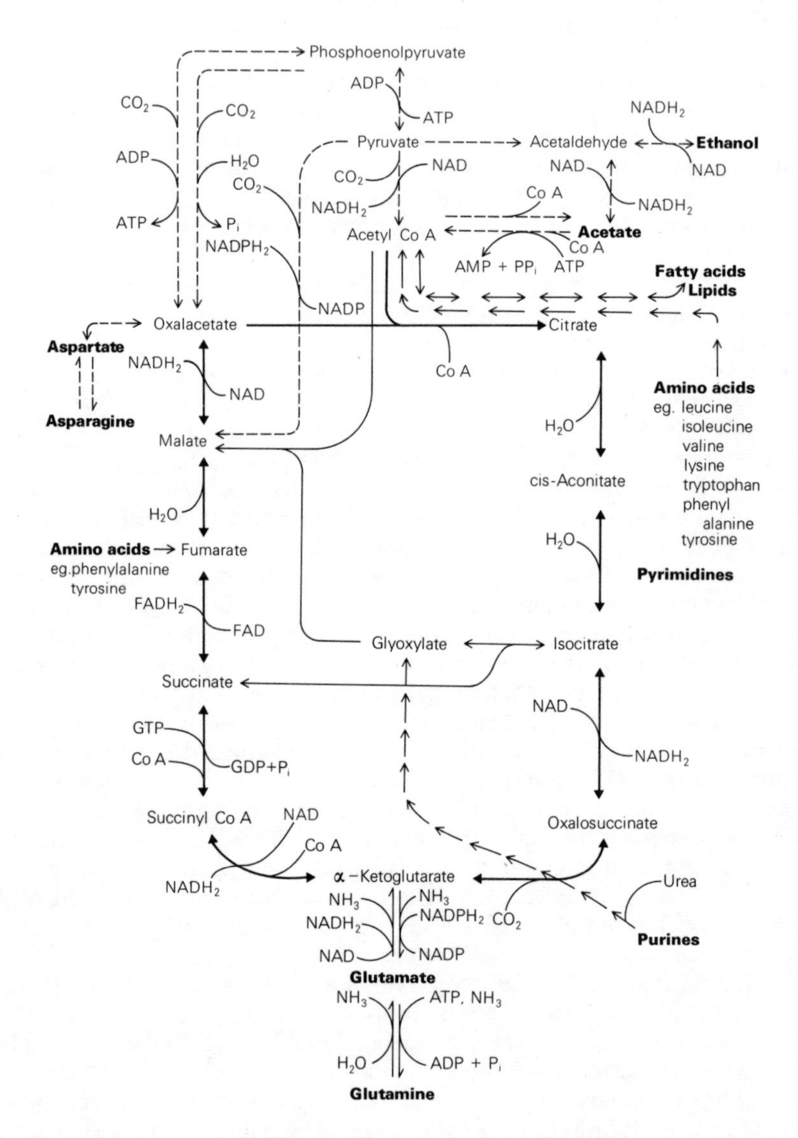

Fig. 2.3 Diagram of the tricarboxylic acid cycle, the glyoxylate cycle and associated pathways indicating the key reactions for the utilization of amino acids, bases, lipids and organic acids as carbon substrates.
Tricarboxylic acid cycle ——→ Glyoxylate cycle ——→
Associated pathways - - - -→ Sequence of reactions →→→

is maintained wholly by the activity of the glyoxylate cycle. In *Coprinus lagopus*, the enzymes involved in the glyoxylate cycle have been found to occur in glyoxysomes and to be repressed in the presence of glucose (Sullivan & Casselton, 1973). When acetate is the sole carbon source the hexose and pentose molecules required for structural carbohydrates are derived from acetate by the process of gluconeogenesis. This process involves the decarboxylation of oxaloacetate to phosphoenol pyruvate and a reversal of glycolysis. Thus the switch from glucose metabolism to acetate metabolism causes a dramatic change in the activity of the glyoxylate cycle, and in the role of the EMP. It has been shown that pyruvate kinase has a very low activity during gluconeogenesis and a high activity during glycolysis (Stewart & Moore, 1971).

2.3 Inorganic Nutrients

Nitrogen metabolism

In view of their saprophytic mode of nutrition, it is not surprising that fungi can utilize a wide range of nitrogenous compounds. All fungi can utilize ammonia as the sole nitrogen source and most can also utilize nitrate or nitrite. Certain species of the Blastocladiales, Saprolegniales and Basidiomycetes are exceptional in not utilizing nitrate. Proteins and amino acids provide an excellent nitrogen source and many fungi can grow using a single amino acid (Cochrane, 1966). Asparagine in particular has been found to be a good source of nitrogen for fungi. The initial stages of assimilation of both nitrate and nitrite involve their reduction to ammonia (Nicholas, 1965). The enzymes involved in this pathway have been shown to be sensitive to the presence of Fe, Cu, Mn, Na, Zn in the medium (Table 2.2). The effect of the demand for $NADPH_2$ in nitrate reduction on the HMP pathway has been referred to above. The capacity of fungi to utilize nitrite is dependent upon their relatively high resistance to nitrite toxicity, for example growth of *Neurospora crassa* is unaffected by 10^{-3}M nitrite. When both NH_4^+ and NO_3^- are present in the medium, then not only is NH_4^+ taken up preferentially but in addition the uptake of NO_3^- is also repressed. Nitrite uptake does not appear to be subject to repression by ammonium ions. The preferential uptake of ammonium ions from ammonium nitrate results in a rapid drop in pH.

Amino acids may be assimilated and incorporated directly into proteins; however, it is apparent that in some fungi deamination of amino acids

Table 2.2 Effect of metal deficiencies on nitrite hyponitrite and hydroxylamine reductases in *Neurospora crassa* (Media, A. & Nicholas, D.J.D. 1957. *Biochim. Biophys. Acta*, **25**, 138–41)

Metal omitted	Fe	Cu	Mn	Mo	Zn
Reductase Nitrite	22	36	53	100	68
Hyponitrite	51	53	60	100	100
Hydroxylamine	100	100	57	100	95
Weight of Mycelium	32	43	41	40	59

occurs, resulting in an accumulation both of ammonia and of organic acids in the medium. In the batch penicillin fermentation, the subsequent rapid metabolism of organic acids results in a marked increase in the pH of the medium (Hockenhull, 1963). Proteins are broken down to amino acids by extracellular proteases before being utilized as a nitrogen source. The slow breakdown of proteins and the subsequent deamination of the amino acids released can provide a steady supply of ammonia in a fermentation. In *Aspergillus nidulans*, an extra-cellular protease has been isolated which is repressed by ammonia (Cohen, 1972). If this is typical, then it seems likely that proteins are only used as a nitrogen source when any ammonia present in the medium has been exhausted.

Glutamate synthesis appears to be the major path of incorporation of ammonia into amino acids. Both $NADH_2$ and $NADPH_2$ dependent glutamate dehydrogenases are present in fungi, but only the $NADPH_2$ dependent enzyme is active in biosynthesis. Ammonia can be incorporated directly into the amide groups of glutamine and asparagine and it is possible that other amino acids such as aspartate, glycine, alanine and serine may be synthesized by direct amination in some fungi (Burnett, 1968).

Both the level of nitrogen in the medium and the nature of the nitrogen source are important in controlling fungal development. Exhaustion of nitrogen from the medium provides a stimulus for the induction of conidiation in *Penicillium griseofulvin*, *Neurospora crassa*, *Aspergillus niger* and many related species (Smith & Galbraith, 1971). In *A. niger*, conidiation can also be induced by carbon limitation. Under these conditions, the presence of nitrate stimulates, whereas the addition of ammonium ions inhibits conidiation. In many fungal species, nitrogen limitation may also lead to autolysis. The importance of nitrogen in the control of the gibberellic acid fermentation is discussed in Chapter 3.

Phosphate metabolism

Phosphate uptake is an energy dependent process which only occurs in the presence of either aerobic or anaerobic respiration. Since only the monobasic phosphate ion is assimilated, the rate of uptake declines rapidly above pH 6.0. Maximal rates of phosphate uptake also require that the acid-base balance of the cell is maintained. Electrical neutrality is maintained during phosphate uptake by the excretion of OH^- ions causing an increase in the acidity of the cell and a corresponding decrease in the medium. In balanced media, such pH changes are nullified by the concomitant uptake of cations such as K^+ and excretion of H^+. Much higher levels of both K^+ and $H_2PO_4^-$ are assimilated when both are present in the medium (Burnett, 1968).

Inorganic phosphate is involved in many enzymatic reactions in the fungal cell. Detailed studies on the importance of phosphate levels in regulating cellular processes *in vivo* are not available, but phosphate limitation has been reported to favour gluconic acid synthesis in *Aspergillus niger* and the biosynthesis of ergot alkaloids by *Claviceps sp.* (see Chapter 14).

Phosphate may be stored in the cell as polyphosphates when taken up in excess.

Sulphur metabolism

Most fungi can utilize inorganic sulphur as sulphate, persulphate, thiosulphate, sulphide and disulphide as well as organic forms such as cysteine, methionine and thiourea. The Saprolegniales and the Blastocladiales however require organic forms of sulphur for growth. The sulphur requirements of fungi are low and a concentration of 0.0001–0.0006 mol dm^{-3} is normally adequate. Oxidized forms of sulphur are reduced to sulphide by the cell and are incorporated initially into cysteine (DeRobichon-Szulmajster & Surdin-Kerjan, 1971).

The observed toxicity of barium ions to *Aspergillus niger* can be attributed to the removal of sulphate from the medium as insoluble barium sulphate. Barium is not toxic if an alternative sulphur supply is provided (Cochrane, 1966).

Other inorganic nutrients

An extensive literature has been assembled on the requirements of fungi for inorganic nutrients both for growth and differentiation (Cochrane, 1966; Hawker, 1950; Foster, 1949). It is apparent that the trace elements, K, Fe, Cu, Mn, Mg, Zn, Mo and Ca are essential for growth. A requirement for a range of other elements which are present in most media, Na, V, Ni, Sn, Cr, Cl, I, Br. Se, B has not yet been demonstrated in the fungi although several have been shown to be essential for other groups of organisms (Hutner, 1972). The difficulties of providing an adequate inorganic nutrient feed have been emphasized in the recent review by Hutner (1972). Storage of trace element components as a dry mixture, a practice widely used in animal nutrition, has been advocated in preference to the storage of one or more solutions of salts. An example of such a mixture is shown in Table 2.3. Precipitate formation can be minimized by reducing the level of free ammonium, phosphate and iron. Unfortunately, this cannot be readily achieved on an industrial scale in view of the

Table 2.3 Trace element dry mixture (Hutner, 1972) Amounts of salt calculated for 1000 litres final medium

Element	Final medium (Mg dm^{-3})	Salt	Gravimetric weight	
			Factor	Salt (g)
Fe	6.0	$Fe(NH_4)_2(SO_4)_2.6H_2O$	7.0	42.0
Mn	5.0	$MnSO_4.H_2O$	3.0	15.5
Zn	5.0	$ZnSO_4.7H_2O$	4.4	22.0
Mo	2.0	$(NH_4)_6MoO.O_{24}.4H_2O$	1.8	3.6
Cu	0.4	$CuSO_4$ (anhydrous)	2.5	1.0
V	0.2	NH_4VO_3	2.3	0.46
Co	0.1	$CoSO_4.7H_2O$	4.8	0.48
B	0.1	H_3BO_3	5.7	0.57
Ni	0.1	$NiSO_4.6H_2O$	4.5	0.45
Cr	0.1	$CrK(SO_4)_2.12H_2O$	9.6	0.96
			Total	87.02

cheapness of ammonia and inorganic phosphate relative to organic substitutes, and the extensive use of iron vessels. In alkaline media, Ca, Mg and Mn readily precipitate as carbonates if the CO_2 level in the fermentation rises.

The uptake of trace metals from the medium may be influenced by other medium components. Cobalt facilitates the uptake of iron by stimulating the production of an iron binding agent in *Neurospora*. In the same organism however, cobalt uptake is inhibited by magnesium ions (Hutner, 1972). The uptake of alkali metals in general appears to be affected by both pH and oxygen tension (Burnett, 1968).

Most of the essential metals are involved as prosthetic groups of enzymes, *e.g.* Fe, Cu, Zn, or as enzyme activators, *e.g.* Mn, Zn, Co, Fe, Ca. The effect of trace metals on the biosynthesis of secondary metabolites in bacteria and fungi has been reviewed recently (Weinberg, 1970). The presence or absence of certain trace metals has been found to have dramatic effects on certain fungal fermentations. Citric acid production is inhibited by iron in the medium. A requirement for zinc has been established for fumarate synthesis (Chapter 8) and zinc appears to play a major role in carbohydrate metabolism, since the level of zinc has a marked effect on the efficiency of utilization of carbohydrates (Perlman, 1965). Zinc also affects the griseofulvin fermentation. In the presence of excess zinc, fulvic acid accumulates in place of griseofulvin (Weinberg, 1970).

2.4 Gaseous Requirements

Oxygen

All filamentous fungi appear to be strictly aerobic. Although precise requirements for oxygen vary with the species and the growth conditions, most fungi grow as well at an oxygen pressure of 20–44 mm Hg as at atmospheric pressure (160 mm Hg). However, *Blastocladiella pringshiemia* and *Achlya prolifera* grow at very low levels of oxygen whereas *Aspergillus oryzae* appear to require high oxygen tensions for maximal growth (Cochrane, 1966). The importance of aeration and agitation in maintaining a high input of oxygen into aqueous solution is common to all fermentations.

Accurate techniques for studying oxygen uptake have only recently been developed and few examples of fungal systems have been studied intensively (Harrison, 1973). In most micro-organisms studied in continuous culture, it has been found that the actual *in situ rate* of respiration was lower than the maximum *potential rate* possible under conditions of excess nutrient and oxygen. The oxygen uptake was usually independent of oxygen tension from 20–150 mm Hg but dropped off rapidly below this range, the lower limit of which is referred to as the critical dissolved oxygen tension.

In studies on *Aspergillus nidulans* (Carter & Bull, 1971) the specific rate of O_2 uptake has been studied at different dissolved oxygen levels using a continuous system at a dilution rate of 0.05 h^{-1}. The critical dissolved oxygen tension was found to be as low as 1.75 mm Hg. This value is apparently influenced by the growth conditions of the mycelium. This was studied by transferring cells grown at 156 mm (A) or 2 mm Hg of oxygen (B) in a fermenter into an oxygen electrode cell. The critical dissolved oxygen tension was found to be 7 mm Hg in A, and 3 mm in B. The potential respiration rate was also affected since the value in B was twice that found in A. A small decrease in the specific rate of oxygen uptake occurred at oxygen levels above the critical dissolved oxygen tension. This may not be significant in *Aspergillus nidulans*, but inhibition of growth at high oxygen tensions has been observed in other species, *e.g. Ophiobolus graminis* (Cochrane, 1966).

The morphology of *Aspergillus nidulans* was not affected by O_2 tensions above 18 mm Hg; however between 3.5 and 18 mm Hg the frequency of unicells increased with decreasing oxygen tension. Unfortunately, information is lacking on the effects of oxygen tension on specific enzymes in filamentous fungi although in *Saccharomyces cerevisiae* it has been shown that alcohol dehydrogenase and pyruvate decarboxylase are both partially repressed above an oxygen level of 75 mm Hg. In *Aspergillus niger* the rate of oxygen consumption has been shown to increase up to 650 mm Hg pressure of oxygen. This exceptional response may be related to oxygen consumption in citric acid production since the optimum O_2 level for this process has been shown to be higher than for growth (Rose, 1968).

The major role of oxygen in the cell is as a hydrogen acceptor in aerobic respiration. In *Saccharomyces cerevisiae*, cytochromes aa_3, b_1, c and c_1 are absent from anaerobically grown cells and presumably synthesized after the onset of aerobic conditions. In *S. carlsbergenesis* on the other hand the cytochromes are present both in aerobic and anaerobic conditions (Sols *et al.*, 1971). Although few other reactions which require molecular oxygen have been described in fungi (Dixon & Webb, 1964) many have been described in higher animals, particularly in the field of steroid metabolism. Oxygen tension may be critical in determining the activity of such reactions where they occur in fungi.

Oxygen demand varies with different physiological states in the same organism. Only one example will be given here. The QO_2 for conidiation in *Neurospora crassa* is 18.9 compared with 12.8 for growth (Turian, 1969).

Carbon dioxide

The effects of carbon dioxide on growth are varied and not readily amenable to critical study. Carbon dioxide dissolves in aqueous solutions forming the weak acid, carbonic acid which dissociates in dilute solution to form bicarbonate and carbonate ions.

$$CO_2 + H_2O \rightleftharpoons H_2CO_3$$
$$H_2CO_3 \rightleftharpoons H^+ + HCO_3^-$$
$$HCO_3^- \rightleftharpoons H^+ + CO_3^{2-}$$

When carbon dioxide is bubbled through an aqueous solution, carbonic acid, bicarbonate and carbonate ions are generated (Ponnamperuma, 1967). Recent studies indicate that fungi are influenced by the concentration of the bicarbonate ion (MacCauley & Griffin, 1969 *a*, *b*). This is extremely sensitive to pH since the concentration of bicarbonate ions increases by tenfold for every decrease in 1 pH unit. In view of this extreme pH sensitivity it is not easy to interpret earlier work in which the pH was not controlled. Carbon dioxide at less than 2.1 atm was found to have no effect on linear growth rates of most fungi in aerobic conditions (MacCauley & Griffin, 1969*a*). However, inhibition of growth by CO_2 has been observed in several species, *e.g. Sclerotium rolfsii, Penicillium nigricans* (Griffin, 1972).

Burgess & Fenton (1953) divided fungi into those which were carbon dioxide sensitive and those which were tolerant of high carbon dioxide concentrations. The effect of carbon dioxide was found to be more conspicuous if growth was measured in terms of dry weight increase in liquid culture rather than as linear colony growth. *Fusarium oxysporium* and *F. solani* have an exceptionally high tolerance to carbon dioxide and high partial pressures of carbon dioxide have been used to isolate these species from the soil.

Carbon dioxide concentration plays an important role in controlling differentiation in many fungal species. In *Chaetomium globosum*, perithecium formation and ascospore germination are both dependent upon carbon dioxide and in *Fusarium solani* chlamydospore germination and hyphal growth are stimulated by carbon dioxide but chlamydospore formation is repressed (Griffin, 1972). The requirements for carbon dioxide for sporangium formation in *Blastocladiella emersonii* and for yeast-like growth in *Mucor rouxii* is well known (Smith & Galbraith, 1971).

In addition to being a major excretion product of fungi, carbon dioxide is also a key substrate. The importance of carboxylation reactions in the formation of TCA acids has already been referred to so it would not be unexpected if the level of carbon dioxide influenced the activity of the TCA cycle and its associated pathways. Glutamate and aspartate are rapidly labelled in yeast fed with $^{14}CO_2$, and in *Neurospora crassa* carbon dioxide has been reported to stimulate the activity of the glyoxylate cycle during conidiation (Smith & Galbraith, 1971). Carbon dioxide fixation also occurs in the synthesis of malonyl CoA during fatty acid formation and the synthesis of carbamyl phosphate required for arginine synthesis (De Robican-Szulmajster & Surdin-Kierjan, 1971). Although carbon dioxide fixation does not occur in the major pathways of biosynthesis of isoprenoid compounds carbon dioxide has been shown to stimulate gibberellic acid biosynthesis (Chapter 7). A mechanism for the biosynthesis of β-carotene from leucine, and which involves a carboxylation reaction, has been described (Burnett, 1968).

2.5 pH

The influence of pH on fungal growth and metabolism is complex. Fungi are characteristically tolerant of low pH and most have an optimum pH of between 5.0 and 7.0 for growth. This acid tolerance has frequently been used to isolate fungi from mixed cultures of fungi and bacteria. The high pH of certain acid fermentations reduces the risk of bacterial contamination. Unfortunately, it is not easy to establish whether changes in pH affect the organism directly or only indirectly as a result of changes in the medium. The internal pH of the cell is controlled independently of the medium at around pH 5.0–6.0 so the effects of changes in the external pH are probably restricted to permeability and other surface phenomena.

The importance of the effect of pH on the solubility of salts and the ionic state of substrates such as phosphate and organic acids have been referred to elsewhere. The increased uptake of ammonia with increased pH has been attributed to the entry by passive diffusion, of the undissociated molecule. In contrast the effect of pH on the process of nitrate uptake (which has a pH optimum of 6.0) can be attributed to its effect on the enzymes involved in nitrate assimilation.

Besides being affected by the pH of the medium, the fungus also has a marked effect on the pH. Differential uptake of anions and cations, the excretion of organic acids or the release of ammonia can all produce rapid changes in the pH of the medium. In the majority of industrial fermentations the pH is controlled either by acid or alkali additions or by the controlled addition of nutrients such as ammonia, ammonium sulphate and sugars.

The manipulation of the nature and quantity of the common nutrients of a growth medium plays an important role in controlling the physiology of the cell. The emphasis placed on this aspect is not however intended to minimize the importance of specific substrates, inducers or inhibitors in the control of any one fermentation but rather reflects a limit on the space available. The improvements in the design of fermenters over the past decade has given mycologists the opportunity to control the growth conditions of his organisms more precisely. However, the problem remains as to which set of conditions are most suited to making a particular organism enter the correct physiological state to produce maximal amounts of a desired product. The unique nature of each organism and the complicated interactions which occur between different environmental factors make this a task of considerable complexity and subtlety.

2.6 References

AINSWORTH, G.C. & SUSSMAN, A.S. (1965, 1966). *The fungi*, Vol. 1, 2. New York and London: Academic Press.

BLUMENTHAL, H.J. (1965). Glycolysis. In *The fungi*, Vol. 1, pp. 229–68. Edited by G.C. Ainsworth and A.S. Sussman, New York and London: Academic Press.

BU'LOCK, J.D. (1967). *Essays in biosynthesis and microbial development*. London and New York: John Wiley and Sons.

BURGESS, A. & FENTON, E. (1953). The effect of carbon dioxide on the growth of certain soil fungi. *Transactions of the British Mycological Society* **36**, 104–8.

BURNETT, J.H. (1968). *Fundamentals of mycology*. London: Edward Arnold (Publishers) Ltd.

CARTER, B.L.A. & BULL, A.T. (1971). The effect of oxygen tension in the medium on the morphology and growth kinetics of *Aspergillus nidulans*. *Journal of General Microbiology* **65**, 265–73.

COCHRANE, V.W. (1966). *Physiology of fungi*. London and New York: John Wiley and Sons.

COHEN, B.C. (1972). Ammonia repression of extracellular protease in *Aspergillus nidulans*. *Journal of General Microbiology* **71**, 293–9.

DEAN, A.C.R. (1972). Influence of environment on the control of enzyme synthesis. *Journal of Applied Chemistry and Biotechnology* **22**, 245–59.

DIXON, M. & WEBB, E.C. (1964). *The enzymes*. London and New York: Academic Press.

FLAVELL, R.B. & WOODWARD, D.O. (1971). Metabolic role, regulation of synthesis, cellular localization and genetic control of the glyoxylate cycle enzymes in *Neurospora* crassa. *Journal of Bacteriology* **105**, 200–10.

FOSTER, J.W. (1949). *Chemical activities of fungi*. London and New York: Academic Press.

GOTTLIEB, D. & CALTRIDER, P.G. (1963). Synthesis of enzymes during the germination of fungus spores. *Nature, London* **197**, 916–17.

GRIFFIN, D.M. (1972). *Ecology of soil fungi*, London: Chapman Hall.

HARRISON, D.E.F. (1973). Growth, oxygen and respiration. *CRC. Critical Reviews in Microbiology* **2**, 185–228.

HAWKER, L.E. (1950). *Physiology of fungi*. London: University of London Press.

HOCKENHULL, D.J.D. (1963). Antibiotics. In *Biochemistry of micro-organisms*, pp. 227–299. Edited by C. Rainbow and A. H. Rose. London: Academic Press.

HUTNER, S.H. (1972). Inorganic nutrition. *Annual Review of Microbiology* **26**, 313–46.

LAHOZ, R. & MIRALLES, M. (1970). Influence of the level of the carbon source on the autolysis of *Aspergillus niger*. *Journal of General Microbiology* **62**, 271–6.

LINDENMAYER, A. (1965). Terminal oxidation and electron transport. In *The Fungi*. Vol. 1 pp. 301–48. Edited by G.C. Ainsworth and A.S. Sussman, New York and London: Academic Press.

MACCAULEY, B.J. & GRIFFIN, D.M. (1969a). Effect of carbon dioxide and oxygen on the activity of some soil fungi. *Transactions of the British Mycölogical Society* **53**, 53–62.

MACCAULEY, B.J. & GRIFFIN, D.M. (1969b). Effect of carbon dioxide and the bicarbonate ion on the growth of some soil fungi. *Transactions of the British Mycological Society* **53**, 223–38.

NG, W.S., SMITH, J.E. & ANDERSON, J.G. (1972). Changes in carbon catabolic pathways during synchronous development of *Aspergillus niger*. *Journal of General Microbiology* **71**, 495–504.

NICHOLAS, D.J.D. (1965). Utilisation of inorganic nitrogen compounds and amino acids by fungi. In *The fungi*, Vol. I, pp. 369–76. Edited by G.C. Ainsworth and A.S. Sussman. London and New York: Academic Press.

NIEDERPRUEM, D.L. (1965). Tricarboxylic acid cycle. In *The fungi*, Vol. 1, pp. 269–300. Edited by G.C. Ainsworth and A.S. Sussman. New York and London: Academic Press.

PERLAMN, D. (1965). The chemical environment for fungal growth. 2, Carbon sources. In *The fungi*, Vol. 1, pp. 479–89. Edited by G.C. Ainsworth and A.S. Sussman. London and New York: Academic Press.

PONNAMPERUMA, F.N. (1967). A theoretical study of aqueous carbonate equilibria. *Soil Science* **103**, 90–100.

ROMANO, A.H. & KORNBERG, H.L. (1968). Regulation of sugar utilization by *Aspergillus nidulans*. *Biochimica et Biophysica Acta* **158**, 491–3.

DE ROBICHON-SZULMAJSTER, H. & SURDIN-KERJAN, Y. (1971). Nucleic acid and protein synthesis in yeasts. In *The yeasts*, Vol. II, pp. 335–418. Edited by A.H. Rose and J.S. Harrison. London. Academic Press.

ROSE, A.H. (1968). *Chemical microbiology*. London: Butterworth.

SMITH, J.E. & GALBRAITH, J.C. (1971). Biochemical and physiological aspects of differentiation in the fungi. *Advances in Microbial Physiology* **5**, 45–134.

SOLS, A., GANCEDO, C. & DELAFUENTE, G. (1971). Energy yielding metabolism in yeast. In *The yeasts*, Vol. II, pp. 271–307, Edited by A.H. Rose and J.S. Harrison. London: Academic Press.

STEWART, G.R. & MOORE, D. (1971). Factors affecting the level and activity of pyruvate kinase from *Coprinus lagopus* (Sensu Buller) *Journal of General Microbiology* **66**, 361–70.

SULLIVAN, J.O. & CASSELTON, P.J. (1973). Subcellular localization of glyoxylate cycle enzymes in *Coprinus lagopus* (Sensu Buller) *Journal of General Microbiology* **75**, 333–7.

TURIAN, G. (1969). *Differentiation fongique*. Paris: Masson et Cie.

WEINBERG, E.D. (1970). Biosynthesis of secondary metabolites; role of trace metals. *Advances in Microbial Physiology* **4**, 1–44.

CHAPTER 3

Secondary Metabolism in Fungi and its Relationships to Growth and Development

J. D. BU'LOCK

3.1 Introduction

The ability of fungi to grow quickly on a variety of nutrients has been one of the prerequisites for their technical and industrial exploitation. With a few notable exceptions (of which various biomass processes are the most typical) this facility of rapid growth is mainly used to build up the mass of cells which are needed to carry out the desired process in a subsequent production phase. This is particularly true of fungal processes for the production of the so-called 'special' or 'secondary' metabolites. For some of these processes there may still be a substantial amount of growth during the production phase, while in others this later growth is negligible, but in either case, from the standpoint of production control, the important feature is that conditions which are optimal for rapid growth are seldom optimal for the production phase, and vice versa. In the simplest of batch processes, the initial conditions are so arranged that the actual activity of rapid growth also modifies the conditions towards those which are optimal for a subsequent production phase. This is obviously a compromise solution and even in batch processes there is usually scope for a considerable degree of controlling manipulation of the culture conditions as the fermentation develops. It is this relatively commonplace aspect of fermentation science which is the least obviously intelligible in terms of classical biochemistry, and will consequently receive most attention in this chapter. However, it is necessary to begin with some general account of the phenomenon of secondary metabolism in fungi.

Since many individually important systems are fully dealt with in subsequent sections of this book, illustrative material in the present chapter will be drawn from a rather restricted range of examples, often selected because they have been preferred subjects for exploratory research rather than for their practical importance. Moreover, since a degree of chemical variety seems to be fundamentally intrinsic to the biological phenomena being considered, we have not tried to restrict our analysis simply to those examples which man currently believes to be of profitable use. The coverage given to major topics elsewhere in this book does not, however, wholly excuse my extensive use of examples from our own laboratory experience, which is at least equally due to bibliographic idleness.

3.2 Secondary Metabolites of Fungi

Secondary metabolites are chemically very diverse substances which are formed by pathways of far greater variety than the normal categories of comparative biochemistry will accommodate. Comparative biochemistry deals with differences in primary metabolic processes (*i.e.* in the 'unity of biochemistry') between major taxonomic groups, whereas secondary metabolites characterize sub-species, species, or groups of species, *i.e.* they are genotypically specific. As a group, the fungi have proved to be particularly prolific in their secondary metabolites, and since only a minority of species have so far been investigated we may be confident that the real range of products is even wider than is yet known. Of the thousand or so fungal metabolites listed in Turner's recent compilation (which is admittedly incomplete for certain rather numerous metabolite groupings) relatively few are known to be produced by more than three or four species (Turner, 1971). A less immediately obvious characteristic of secondary metabolites is one which soon becomes apparent when practical problems of obtaining them are approached, *viz.* they are also phenotypically specific, in that their production is extremely sensitive to culture conditions and previous history.

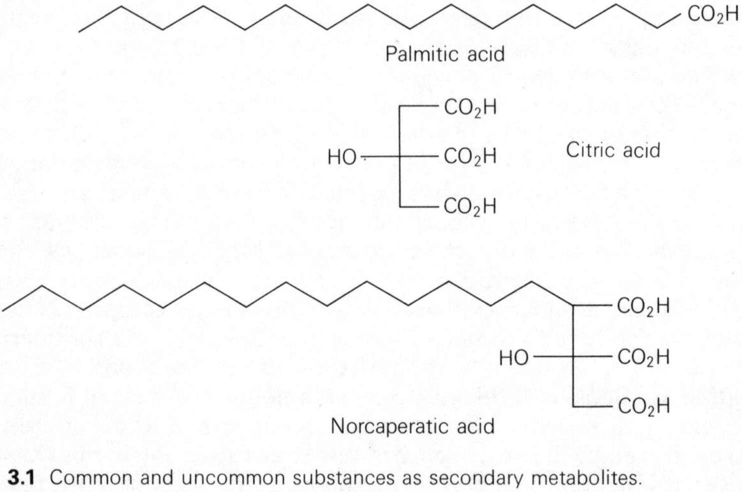

CO_2H

Palmitic acid

HO— CO_2H / CO_2H / CO_2H

Citric acid

CO_2H

HO— CO_2H

CO_2H

Norcaperatic acid

Fig. 3.1 Common and uncommon substances as secondary metabolites.

These considerations apply most clearly to metabolites of admittedly 'unusual' chemical structure, which to that extent are obviously of a specific nature, but in practice they apply with almost equal force to instances where 'secondary' metabolism takes the form of an extraordinary accumulation of some quite ordinary cell component. For example, consider the three substances of Fig. 3.1. Probably all fungi produce citric acid in the course of their normal intermediary metabolism and palmitic acid as a component of their structural lipids. Moreover, many species will accumulate such tricarboxylic acid cycle intermediates and/or lipids in response to certain environmental histories, and in some cases these will take the particular form of accumulations of citric acid or of a palmitate-rich fat. However, very high conversions of substrate into either product are only characteristic of a much smaller number of species, and the culture conditions may now be quite critical; thus these accumulations of 'ordinary' compounds are also both genotypically and phenotypically specific. They then become quite comparable with, for example, those few species of Basidiomycetes which by one special reaction link fatty acid synthesis with citrate formation and, under the right environmental conditions, produce the 'extraordinary' metabolite, norcaperatic acid.

The genotypic diversity of secondary metabolites

As already noted, the secondary metabolites of fungi are very diverse. Their variety can only be reduced to significant order by analysis of the biosynthetic pathways by which they are produced, and which therefore link them to the comparatively uniform network of the primary biochemical processes. From the considerable volume of structural and biosynthetic studies which has been carried out with fungal products during the last twenty-five years it emerges that the great majority of them are formed by pathways which branch off from primary metabolism at a relatively small number of points, and which comprise a relatively small number of reaction types (Bentley & Campbell, 1968). It is also increasingly apparent that the key precursors for these secondary pathways are also key intermediates in the primary processes and that they are therefore the natural foci of very general and important regulatory mechanisms (Bu'Lock, 1961, 1965*a*, *b*). For example, in the fungi the series of secondary metabolites formed from acetyl-CoA is particularly varied and numerous (Turner, 1971), and this reflects the key position of this substance in the primary network of metabolism in these organisms. It links sugar metabolism and the tricarboxylic acid cycle, each of which is to some extent independently regulated but both of which contribute to co-ordinated anabolic processes; to a large extent it occupies a balancing position between fermentative and aerobic catabolism and between cytoplasmic and mitochondrial processes.

The general situation is usually summed up in terms of the main flow of carbon metabolism, in schemes such as that of Fig. 3.2. However, it is always important to bear in mind that the carbon flow represents only one of the several intersecting networks in metabolism, and in particular not to overlook: (a) the flow of nitrogen from the nutrient source through amination and transamination reactions into amino acids, nucleotides, and macromolecules; (b) the cycling of phosphate through selectively 'reactive' and 'unreactive' anhydrides and esters ('energy charge'); and (c) the redox or electron transfer processes in which a few key steps generate reduced coenzymes which are re-oxidized in a much larger variety of reactions. Each or all of these may be as important as the carbon flow when the overall regulation of secondary metabolism is being considered, and schemes such as in Fig. 3.2 are not really adequate for such purposes.

From key points in the primary metabolic network, the genotypic diversity of secondary metabolites is attained by branching series of

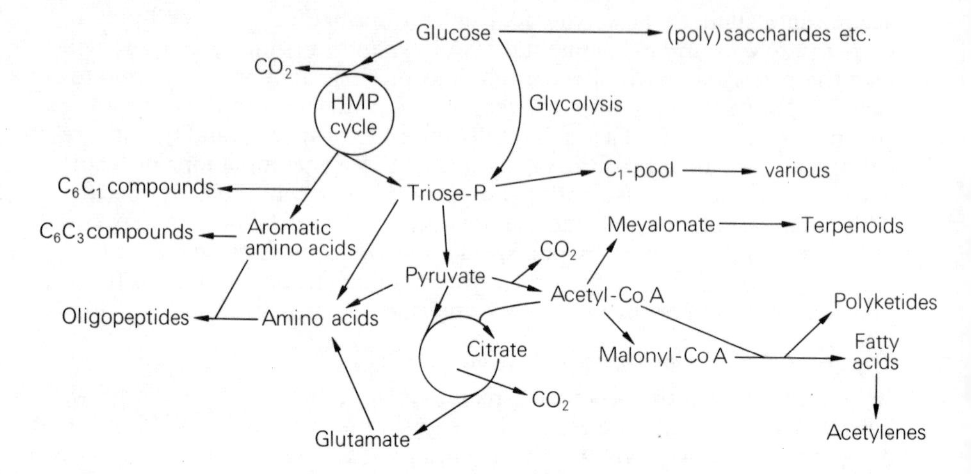

Fig. 3.2 The principal pathways of sugar catabolism in fungi and the major categories of fungal secondary metabolites.

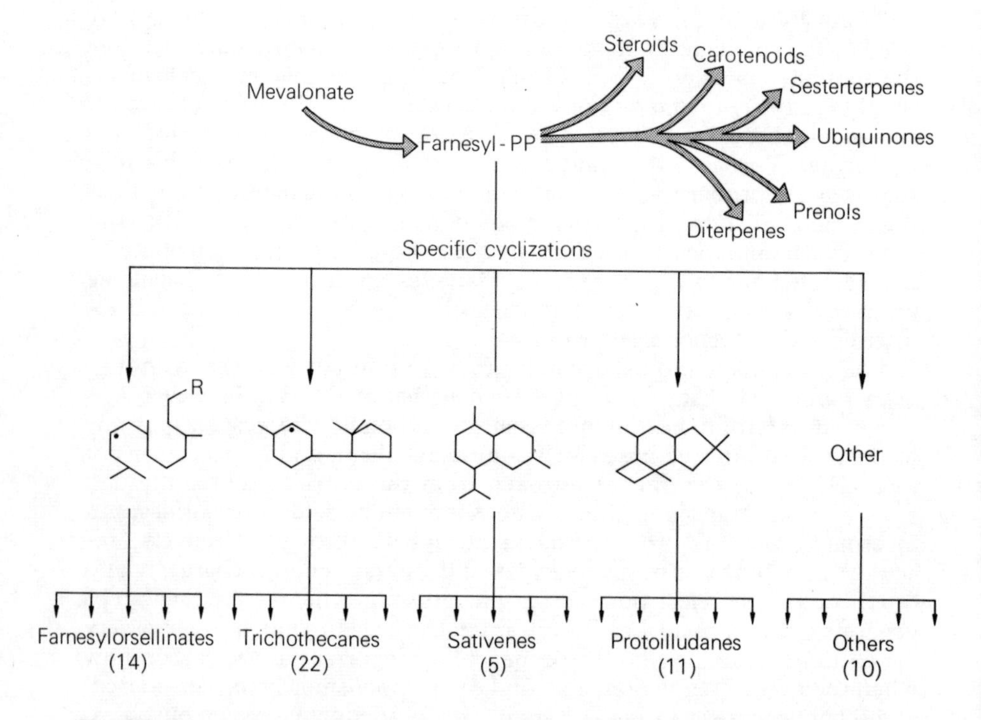

Fig. 3.3 Branching pathways of secondary biosynthesis, exemplified by the fungal sesquiterpenes. Numbers in parentheses are the numbers of products in each class as listed by Turner (1971).

reactions (Bu'Lock, 1965 *a*, *b*). The earlier steps of these series, which are thus common to several ultimately distinct pathways, define broad biosynthetic categories and are frequently associated in their occurrence with taxonomically recognizable species groupings, while later steps are increasingly characteristic of individual products and also of particular organisms. This general type of branching pattern can be exemplified by the fungal terpenes, and in particular by a single category such as the 60 or so sesquiterpenes which are listed by Turner (1971) as products from some 35 fungal species. The pathways marked in Fig. 3.3 by heavy lines are those of widespread occurrence, for example those leading to steroids and to prenols (this does not preclude their involvement in secondary processes). From farnesyl pyrophosphate, the intermediate in this main sequence whose involvement defines the sesquiterpenes as a group, there are several group-specific cyclization reactions, for example that to 'protoilludane', which occurs in a range of Basidiomycete species. The first cyclization products are then transformed by steps of progressively greater genotype specificity to give the final variety of products. An understanding of this type of pattern is not merely an aid to classification, but is clearly essential if we are to interpret regulatory mechanisms; the mechanisms of most general occurrence will be those controlling steps early in the branching sequences, *e.g.* in the present case the overall supply of mevalonic acid and, more selectively, the activity of the cyclization enzymes which tap off intermediates in the prenylation sequence. For this reason, there are also instructive parallels between this pattern of genotypic diversity and the pattern of metabolites which can be obtained by manipulation of the conditions for a given species.

Mechanisms of secondary biosynthesis

Like any other biochemical process, secondary metabolites are formed by consecutive series of enzyme-catalysed reactions, and they are subject to all the kinetic and regulatory effects which this implies. For example, in some *Aspergillus* species the main pathway of secondary metabolism comprises the addition of one special enzyme to the normal tricarboxylic acid cycle, effecting the decarboxylation of cis-aconitate to give itaconic acid.

$$\begin{array}{l} CO_2H \\ CO_2H \\ CO_2H \end{array} \longrightarrow \begin{array}{l} \\ CO_2H + CO_2 \\ CO_2H \end{array}$$

The enzyme can be isolated and its detailed mechanism studied in the conventional manner (Bentley & Thiessen, 1957); the only differences from enzymes of the 'normal' cycle are that (a) it is only produced by a very few species, (b) the production of this enzyme during rapid growth of *Aspergillus terreus* on a rich nutrient medium is minimal, and (c) the organism has no obvious 'use' for the product.

Similarly it has been possible to isolate the enzyme system which in *Penicillium urticae* is responsible for the synthesis of 6-methylsalicylic acid from acetyl-CoA, malonyl-CoA, and NADPH. In this case the biosynthesis is carried out by a multienzyme complex with many features which are similar to the more generally-known type of fatty acid synthetase complex, and in these terms its properties appear to be quite normal even though the detailed mechanism is not yet worked out (Dimroth, Walter & Lynen, 1970; Dimroth, Greull, Seyffert & Lynen, 1972). However, we

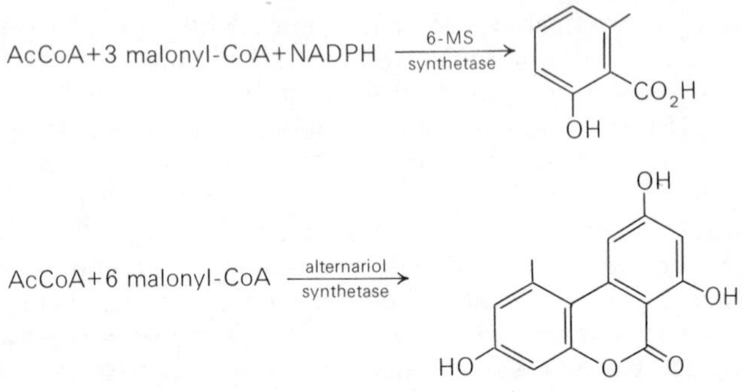

AcCoA+3 malonyl-CoA+NADPH $\xrightarrow[\text{synthetase}]{\text{6-MS}}$ CO_2H OH

AcCoA+6 malonyl-CoA $\xrightarrow[\text{synthetase}]{\text{alternariol}}$ OH OH HO O O

should also observe that other fungi produce polyketide synthetase complexes in which almost identical overall mechanisms are employed to produce quite different products, for example alternariol (Sjoland & Gatenbeck, 1966).

In primary metabolism, processes of biosynthesis are normally carried out by enzymes of high specificity; only one substrate is accepted and only one product is formed. Clearly these enzymes have evolved because the precise nature of the product is a matter of real consequence for the organism, and it is significant that in the realm of secondary biosynthesis the equivalent degree of specificity is not always encountered. This fact is also of technical importance since it may allow us to use a system to produce quite new structures simply by providing unnatural substrates.

Perhaps the best-known example of this situation is the enzyme system which in *Penicillium chrysogenum* catalyses the exchange of the L-α-aminoadipyl sidechain of isopenicillin-N for other acyl groups; this has a high specificity for the 6-aminopenicillanic acid moiety but it will readily accept a very wide range of substituted acetic acids, both endogenous and added. The industrial exploitation of this system (see Chapter 7) is well-known.

This lack of absolute specificity in some secondary biosynthetic sequences can have other consequences. Both the substrate and the product for one enzyme in a sequence may serve equally well as substrate for another enzyme, so that the precise order of steps in a sequence may be of no great consequence and a multiplicity of pathways and intermediates may result. This can be seen, for example, in the 'pathway' to the very common fungal steroid, ergosterol. Starting from lanosterol (Fig. 3.4) this involves changes in four regions of the molecule, namely (a) in the side-chain, methylation and hydrogen transfers, (b) at C-14, the removal of the methyl group, (c) at C-4, the stepwise removal of two methyl groups and (d) double bond migrations and dehydrogenation in ring B. The balance of present evidence (Mulheirn & Ramm, 1972) is that these four regions of the molecule are transformed more or less independently. Though within each region the relevant enzymes catalyse highly stereo-selective transformations, there is no absolute requirement by the enzymes effecting changes in one region for changes in the other regions to have reached any particular stage. The effect is that there is no unique pathway to ergosterol, but some pathways are doubtless more favoured than others, and indeed it may be that the balance of sterol composition in a particular fungus partly reflects such *kinetic* (rather than absolute) enzyme specificities. One obvious consequence of systems of this kind is that an organism can be used to carry out regionally specific one- or two-step transformations of unnatural sterol substrates, and this has of course acquired considerable practical importance (Chapter 9).

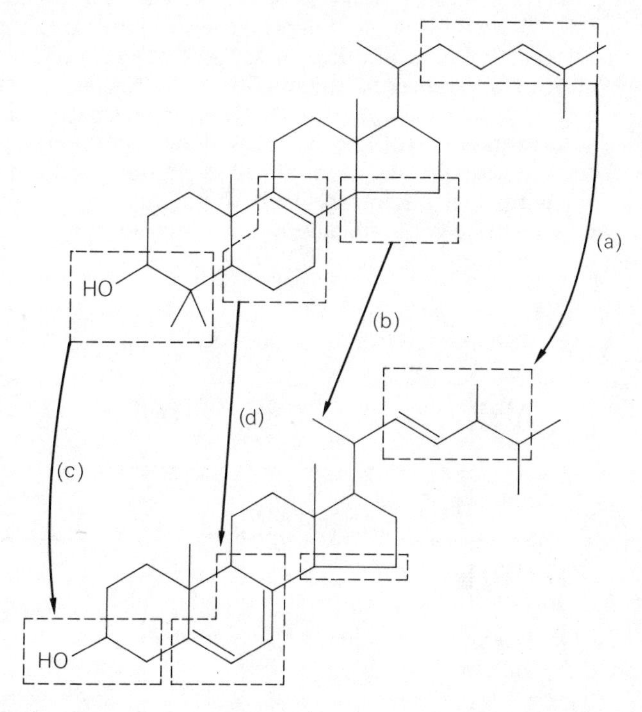

Fig. 3.4 Independent regions of transformation in the conversion of lanosterol into ergosterol.

The further transformations of 6-methylsalicylic acid by some of the fungi that produce it provide a further example of this 'metabolic grid' effect and illustrate how within such a grid of reactions both genotypic and phenotypic selectivity can arise, the former because some organisms only produce enzymes for some of the transformations, the second by suitable manipulations of the fermentation conditions (Bu'Lock, *et al.*, 1965; Bu'Lock, Shepherd & Winstanley, 1969; Forrester & Gaucher, 1972). Nearly all of the compounds shown in Fig. 3.5 have been isolated, as major or minor metabolites, from fungi of the *Penicillium urticae* (*P. patulum*, *P. griseofulvum*) series; not all of the reactions shown in Fig. 3.5 have been unequivocally demonstrated, but from the overall pattern of structures it is clear that a metabolic grid exists here. The full display of this pattern can be regulated environmentally; by altering the nutritional conditions of established cultures, a strain of *P. urticae* which normally produces patulin as its major metabolite has been made to produce, more selectively, (a) 6-methylsalicylic acid alone, (b) mainly 6-methylsalicylic acid and *m*-cresol, (c) mainly gentisyl alcohol, and (d) mainly gentisic acid. Other strains in the same group, and also strains of *Aspergillus fumigatus*, show similar specificities genotypically; in addition, *Penicillium islandicum* uses one part of this network to produce 3-hydroxyphthalic acid while some species of *Phoma* use another part of the sequence to make epoxydon, the epoxide of gentisyl alcohol (Turner, 1971). The cases of phenotypic selectivity depend on such features as (a) the fact that once formed, the 6-methylsalicylate synthetase complex is relatively stable and persists when formation of other enzymes in the sequence is blocked, either by inhibitors or by altering the nutritional status or (in a chemostat) the growth rate, (b) the cleavage enzyme which makes patulin has special trace metal requirements, (c) the hydroquinones in the series may be removed from the sequence as quinones in conditions of high oxidation potential, and so on. It is inherently likely that similar restrictions on the system should also arise genotypically in particular organisms.

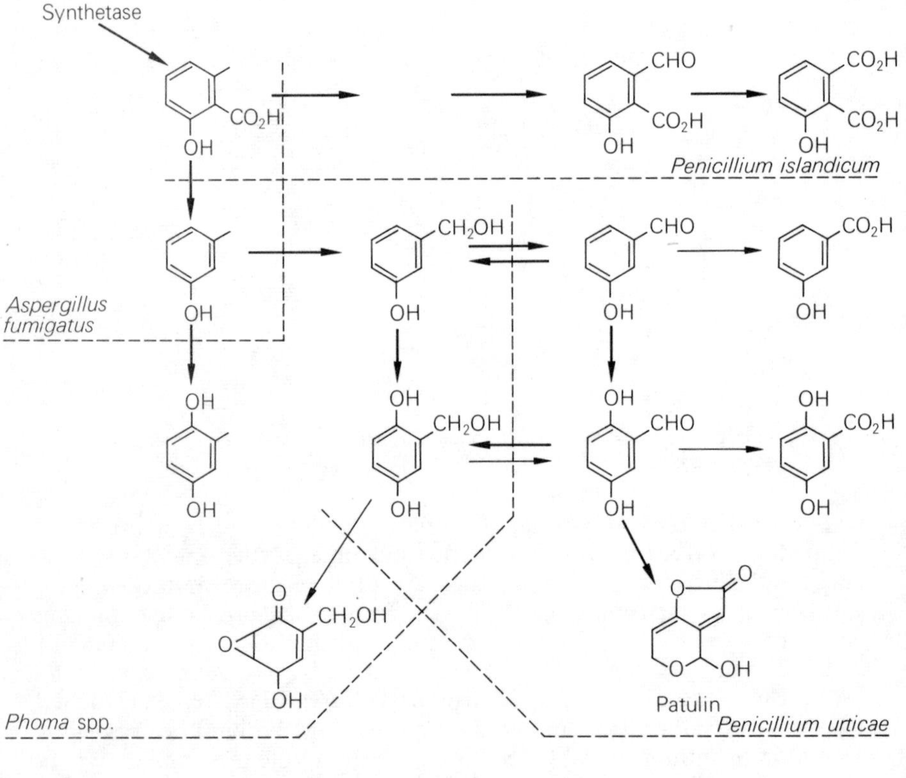

Fig. 3.5 Derivatives formed from 6-methylsalicylic acid in *Penicillium urticae* and other fungi. Arrows in heavy type show transformations which are directly demonstrable; those in lighter type are very probable; conjectural reactions and intermediates are omitted. All the quinols are in redox equilibrium with each other and with the corresponding quinones.

Quantitative and regulatory aspects of the enzymology of secondary biosynthesis are not well understood, largely because the variety of reactions to be studied (and the correspondingly slight importance of any single reaction) has often deterred skilled enzymologists from venturing into the field of secondary biosynthesis. One general observation is that whereas in many processes of primary metabolism the overall kinetics are normally governed by substrate levels, because the enzymes are usually present in excess (Polakis & Bartley, 1966; Srere, 1967; Wright, 1968), it is more common to find that secondary metabolism is largely enzyme-limited. This is in part a consequence of the position of secondary biosynthesis in the general economy of the cell, and a corollary of the manner in which secondary metabolic pathways branch off from the primary network. However it must be admitted that most experimental observations have been concerned with the initiation of secondary biosynthesis during particular phases of culture development: this is intrinsically an enzyme-limited phenomenon as we shall see. In fact there is surprisingly little direct experimental evidence for the generalization we have just offered, particularly for established 'production phase' systems in which secondary metabolism is fully operational. The most important supporting evidence is that there are so few recorded cases in which secondary metabolite production is appreciably increased as a direct result of increasing the supply of substrate. However, this clearly is the case for the supply of phenylacetic acid in the biosynthesis of benzylpenicillin, and other examples are not unknown. For example, in a Basidiomycete producing polyacetylene antibiotics (by a specialized dehydrogenation pathway from fatty acids) the rate of production was increased by supplying pregrown mycelium with ethanol in place of glucose; the percentage of acetyl-CoA incorporated into the polyacetylenes was unchanged by this treatment so that the increased biosynthesis must have been due to higher substrate levels rather than to changed enzyme proportions (Bu'Lock, Gregory & Hay, 1961). A quite different instance is provided by one particular reaction in the *Penicillium urticae* 'metabolic grid' (Fig. 5.5): here a key step is the conversion of gentisyl alcohol to the aldehyde, but the enzyme-catalysed equilibrium for this step actually lies in the reverse direction, so that the kinetics of the process in the overall system must be ascribed to substrate-level effects (Forrester & Gaucher, 1972).

Another type of control by substrate level is when the nutritional demands of secondary metabolism are more specialized. For example, if *Penicillium griseofulvum* is supplied only with the levels of chloride ion appropriate for normal growth the production of griseofulvin will clearly

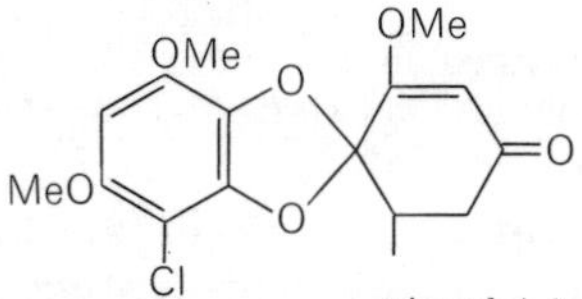

OMe OMe O O MeO O Cl

griseofulvin

be restricted; alternatively at higher substrate levels competition between chlorine and bromine can be demonstrated. Trace metal requirements for secondary biosynthesis are frequently more stringent than growth-requirements, though this is presumably accounted for by effects on the levels of active enzymes; the effect is very general, but hardly any detailed interpretations are available (Weinberg, 1970).

With regard to control by regulation of enzyme levels and enzyme activity most information relates to the relations between growth and specialized enzyme synthesis and this is discussed in the following section. However, we might also expect to find both allosteric effects on enzyme activity and operon-based derepression mechanisms in secondary biosynthetic systems, though once again direct evidence is rather rare (Demain, 1972). In so far as polyketide synthesis is dependent upon the formation of malonyl-CoA from acetyl-CoA, it will be affected by the same regulatory effects upon the carboxylase reaction as those which partly control fatty acid synthesis—for example, allosteric activation by citric acid, and feedback control by fatty acyl species (Wakil & Barnes, 1971). Supporting evidence for such effects on polyketide synthesis *in vivo* is sparse, but mechanisms of this kind would help to explain, for example, the adverse effects of fat-based antifoams on griseofulvin synthesis. Mechanisms of end-product inhibition have also been suspected in many secondary biosynthetic processes (Demain, 1972), but again none has been unequivocally demonstrated at an enzymological level, mainly because so little enzymology has been done in this field.

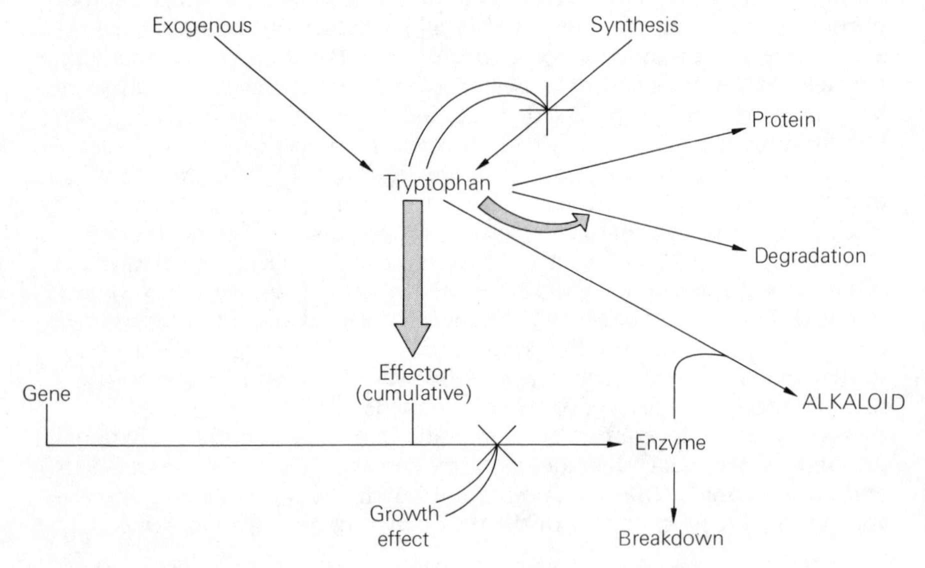

Fig. 3.6 Model for the regulatory system involving tryptophan induction and growth-linked suppression of ergot alkaloid synthesis in a *Claviceps purpurea* strain.

One of the best examples of a substrate-induced process in secondary biosynthesis is the case of ergot alkaloid synthesis in some strains of *Claviceps* sp. (Fig. 3.6), in which it appears that synthesis of a rate-limiting enzyme—apparently responsible for producing the basic dimethylallyl-tryptophan skeleton of the alkaloids—is partly controlled by inducer action of the substrate, tryptophan (Floss & Mothes, 1964; Bu'Lock & Barr,

1968; Vining, 1970; Řeháček, *et al.*, 1971). In support of this it can be shown that nonsubstrate analogues of tryptophan have a similar effect, and also that during the production phase there is a close relationship between the rate at which the rate-limiting step changes and the intramycelial tryptophan level. In this system, substrate-level induction apparently operates in conjunction with an overall control which is linked to growth rate. During rapid growth the inducer effect of tryptophan is cumulative but somehow 'latent' and no alkaloid is formed, whereas in the alkaloid-producing phase its inducer effect is directly apparent. In other strains of *Claviceps*, however, this overall regulation linked to growth-rate may be the only mechanism, since the tryptophan induction cannot always be demonstrated.

Control of processes by regulating the rates at which enzymes are synthesized is only economic when it is coupled with a provision for removing any unwanted excess of enzyme, *i.e.* by regulating synthesis against turnover. This is undoubtedly an important mechanism in secondary metabolism and there are several cases in which the effects of protein synthesis inhibitors on secondary biosynthesis have been interpreted in these terms. For example, the 'decay' of rate-limiting enzymes following inhibition of protein synthesis has been followed for ergot alkaloid synthesis in *Claviceps* (Bu'Lock & Barr, 1968) and for bikaverin synthesis in *Gibberella fujikuroi* (Bu'Lock, J.D. and Detroy, unpublished results). On the other hand not all enzymes of secondary metabolism are similarly labile, and in similar experiments with *Penicillium urticae* the 6-methylsalicylate synthetase seemed to be remarkably stable, once formed, though its further metabolism to patulin showed fairly rapid decay (Bu'Lock *et al.*, 1969).

3.3 Secondary Metabolism and Growth

We must now turn to the important feature of secondary metabolism mentioned in our introductory section, *viz.* that, irrespective of regulatory effects upon individual reaction steps, it is governed by an overall regulation which operates distinctively from that involved in co-ordinated growth. This is, in effect, the definition of secondary metabolism translated into terms of its regulatory mechanisms. The two activities, secondary biosynthesis and growth, are often complementary alternatives which may even be said to compete for key metabolic intermediates. This does not mean, as some have assumed, that growth and secondary metabolism are taken to be mutually exclusive and incompatible activities;* this would be very far from the truth, but before we can understand their interaction it will be necessary to consider more carefully what we usually mean by 'growth'.

For many microbiologists, and particularly for bacteriologists, the concept of growth relates directly to the self-reproducing properties of the living systems with which they deal. Bacteriologists *count* the cells in their

*Terms such as trophophase and idiophase, which were introduced to describe batch cultures with a well-marked division between growth-directed and secondary metabolic activities (Bu'Lock, 1965*b*) will become less useful when a better understanding of these activities is attained.

cultures; they use this as a measure of the multiplication of primary genetic materials and of the associated apparatus of biopolymer synthesis. The reservations which attach to the use of cell number of this manner are recognized and are seldom of criticial importance. On the other hand, mycologists who deal with filamentous multinucleate or even non-septate growths have no comparably convenient experimental measure which carries the same implications, and the changes which occur in their cultures are so obviously qualitative as well as quantitative that no single quantity can characterize their progress adequately. There is however an inevitable tendency to use an estimate of mycelial quantity—usually dry weight—as an all-purpose measure: to follow the progress of a culture, to compare one set of data with another, and quite simply 'to give us something to divide by'.

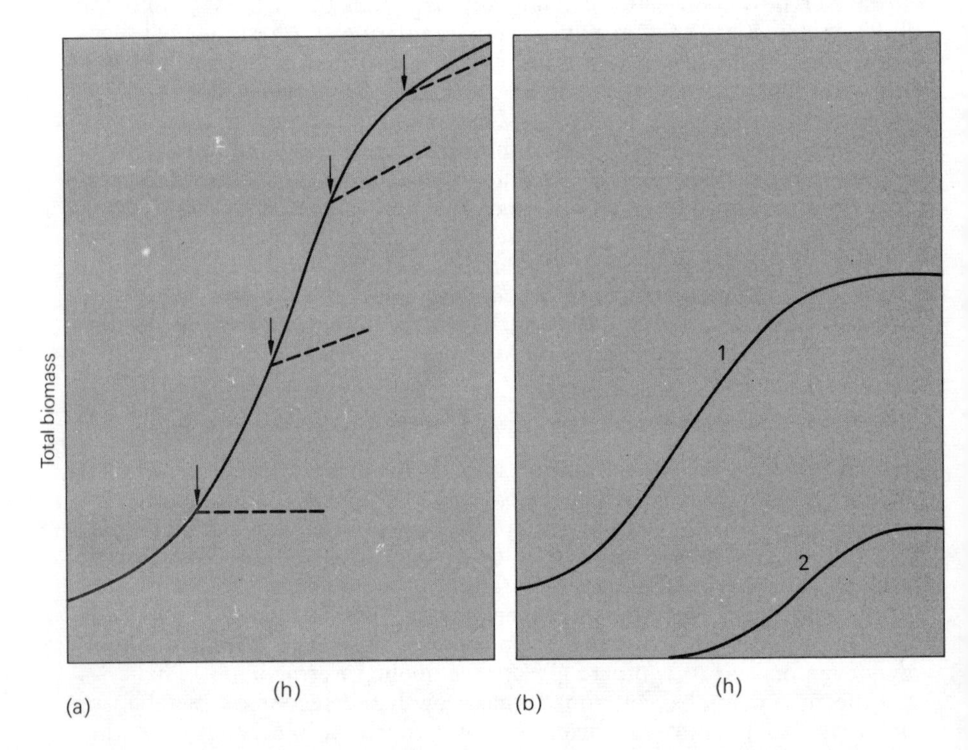

Fig. 3.7 (a) Total biomass *vs* time for nitrogen-limited *Gibberella fujikuroi*; full line, controls broken lines, after additions of cycloheximide. (b) Curves reconstructed from data of (a) to show (1) the cycloheximide-sensitive, and (2) cycloheximide-insensitive components in the total biomass.

Biomass represents the summation of all the processes which add to cell material, whether these are autocatalytic, as in the basic replicatory system of the nucleic acids and proteins required for nucleic acid and protein synthesis, or assimilatory (in the sense that both mycelium and people may become fatter and thicker). The two can be distinguished if we make more selective measurements—for example, using nucleic acid or protein content, or even 'total mycelial insolubles', as a better approximation for the autocatalytic component of biomass (Taber & Vining, 1963). Whenever this is done, the results emphasize that replicatory and assimilatory processes vary very considerably in their relative importance. Fig. 3.7(a) illustrates an experiment which shows how autocatalytic processes tend to diminish in importance, and assimilatory processes to increase, during a typical batch culture (Bu'Lock, J.D. & Detroy, unpublished results). *Gibberella fujikuroi* was grown on a low-nitrogen medium and gave the crude biomass curve shown as a full line. To a parallel series of cultures, at various times, sufficient cycloheximide was added to block the replicatory systems of DNA, RNA, and protein synthesis, and the subsequent biomass curves of the inhibited cultures are shown as broken lines. These latter represent those processes contributing to biomass production which are, to a first approximation, independent of the replicatory systems; they are negligibly slow at first, but they become faster as the culture develops. We can use the two sets of data to calculate, at least for illustrative purposes, the contributions of the cycloheximide-sensitive and -insensitive processes to the total biomass, and the result is shown in Fig. 3.7(b). The experiment (which can be repeated with many other fungi) tells us nothing about the nature of the cycloheximide-insensitive 'assimilatory' processes, although other data for *G. fujikuroi* suggest that the deposition of polysaccharide makes an important contribution, but their relationship to 'growth' clearly relates these processes to the type of biosynthesis we have called 'secondary'. The experiment gives some direct reality to the type of subdivision of the growth concept used in some of the more successful mathematical analyses of batch fermentation kinetics (Ramkrishna, Fredrikson & Tsuchiya, 1966). Very similar conclusions about the relative importance of 'multiplicatory' and 'development' activities in batch culture were reached by Dawson (1972) from work on synchronized 'chemostat' cultures of *Candida utilis*.

In Chapter 5 of this book, Righelato has presented a full account of the concepts of 'unlimited' and 'limited' growth, together with the complications which both concepts involve when filamentous fungi are being considered. The two states correspond to the 'enzyme-limited' and 'substrate-limited' extremes in enzyme kinetics. In 'unlimited' growth the rate of uptake of substrates into the mycelium, and the rate at which they are used for mycelial development, are governed by intrinsic properties of the mycelium and are therefore as high as they can possibly be for the system; in 'limited' growth these rates are partly or wholly governed by external parameters, such as the concentration of a particular nutrient, and are correspondingly sub-maximal. 'Unlimited' growth is intrinsically *balanced* in that the relative contributions of individual processes remain constant, and consequently it is truly replicatory. Batch cultures of fungi may or may not begin with a period of 'unlimited' growth, but the progressive consumption of substrates inevitably carries them into and through a spectrum of increasingly severely limited growth. Under these conditions the rapidity with which the growth rate declines is governed by the rate at which the concentration of limiting substrate is lowered (by being consumed) in comparison with the magnitude of the (Michaelis-type) 'affinity' constant of the mycelium for that substrate.* With a constantly-changing growth rate it no longer follows that growth is balanced, and mycelial differentiation can (and usually does) occur. *Secondary metabolism is an aspect of the differentiation which 'limited' growth usually implies.*

A large number of examples supporting this generalization could be enumerated, but for present purposes it is better to present just three of the cases for which the available data are more crucial. The first comes from the work of Taber on *Claviceps* species (Taber & Vining, 1963; Taber, Brar & Giam, 1968). To obtain a high level of ergot alkaloids synthesis with *C. purpurea* isolates the usual procedure is to grow the mycelium on a synthetic medium based on mannitol and ammonium succinate with a relatively low initial supply of phosphate. Batch cultures on such media show a period of very rapid growth with negligible alkaloid synthesis. When all the phosphate in the medium has been consumed, so that continuing metabolism necessitates turnover and redistribution of the phosphate in the mycelium, nett protein synthesis is drastically reduced, and even though the mycelial *weight* continues to increase quite rapidly it is clear that *growth* is quite severely limited. This is brought out when measurements of crude biomass are replaced by measures of 'total mycelial insolubles', even though the latter still include assimilatory polymeric material (polysaccharide accretion) as well as the autocatalytic macromolecules. Under these conditions a phase of spectacularly rapid alkaloid synthesis occurs. Having established this as the normal pattern, Taber found an alkaloid-producing strain of *C. paspali* which was more exacting in its nutritional requirements and would only show this 'phased' pattern of batch culture development on a complex medium of glucose and yeast extract. If this more demanding strain was grown on the normal synthetic medium, growth of the batch was much slower, both *ab initio* and throughout—*i.e.* always 'limited'; under these conditions alkaloid production occurred quite steadily throughout the fermentation.

A somewhat similar situation involving carbon nutrition is well-known and important in the production of penicillins by *Penicillium chrysogenum*. Here the traditional batch medium contains a proportion of carbon sources which are readily utilized (glucose and/or lactate) and a larger proportion of the much less readily utilized disaccharide, lactose. These carbon sources are consumed in a diauxic manner (*i.e.* successively) so that when only lactose remains the growth rate slows down markedly (see footnote, below). It is during this phase of very 'limited' growth that penicillins are formed, provided that other conditions are appropriate. Moreover, if a quantity of glucose is added during this production phase, penicillin production comes to a halt, and it is not resumed until the added glucose is nearly all consumed. Conversely, the requirement for lactose in the production phase can be quite satisfactorily replaced by a *slow* continuous feeding of glucose and this constitutes a very valuable working procedure for industrial

***i.e.* the constant K_s in the Monod equation $\mu/\mu_{max} = S/K_s + S$ (see p. 100). Note that when the relevant substrate is not a 'true' nutrient but a substance which the organism must itself convert into a usable form (*e.g.* a polysaccharide which must be hydrolysed, or a nitrate which must be reduced) it may be effectively 'limiting' at all times, and the Michaelis-Monod model is then quite misleading as a guide to the internal responses of the mycelium.

fermentations (Soltero & Johnson, 1953; Demain, 1968; Hockenhull & MacKenzie, 1968).

The third example comes from work on *Gibberella fujikuroi* and shows that the *type* of secondary metabolite produced may be quite closely controlled by the *degree* to which growth is limited. This fungus has been the subject of extensive studies showing the existence of an initial 'unlimited' or 'balanced' phase in batch cultures, leading to conditions of nutritional limitation the onset of which depends in a very regular manner upon the initial composition of the culture medium. This is because during balanced growth the different ingredients of the medium are consumed in constant mutual proportions, so that the initial composition determines which nutrient first becomes limiting. The work included detailed measurements of mycelial composition showing the effects of different successions of nutrient 'exhaustion' after the end of the balanced phase; for example, the accumulation of mycelial fat and carbohydrate under all conditions except those of carbon limitation. The detail and full extent of this work can only be appreciated by reference to the original papers (Borrow *et al.*, 1961; Borrow *et al.*, 1964). In our own continuation of work on *G. fujikuroi* (Bu'Lock, J.D., Hošťálek, and Al-Shakarchi, unpublished results) we studied more specific secondary metabolites under the particular condition of nitrogen limitation, especially the diterpenoid gibberellins (for which this fermentation is run commercially) and the characteristic polyketide pigments, bikaverins; we observed that as nitrogen limitation in batch culture became progressively more severe, the bikaverins were produced first and the main phase of gibberellin synthesis came later (Fig. 3.8). Nitrogen-limited growth was therefore examined under chemostat conditions, in which the overall growth rate and the prevailing level of the limiting nutrient are directly linked and under experimental control.

'Unlimited' growth is not strictly attainable in the chemostat but the equivalent maximum specific growth rate, μ_{max}, can be determined by extrapolation; for *Gibberella fujikuroi* with glycine as the N source it is *c* 0.18^{-1}. Growth rates of $0.10\ h^{-1}$ or less were sufficiently 'limited' to elicit some signs of 'unbalanced' growth, such as the accumulation of mycelial carbohydrate, but so long as the growth rate was held above $0.05\ h^{-1}$ no bikaverins were formed. Below this growth rate they were produced just as in batch culture. However, for gibberellin production to be significant, even lower growth rates, of $0.01\ h^{-1}$ or less, were required, corresponding to extremely low levels of the limiting substrate.

Such observations indicate that the pattern of secondary metabolism in a culture will in general be determined not merely by the *existence* of a limitation upon replicatory growth but more particularly by the *intensity* of that limitation. If we also recognize that different secondary biosynthetic pathways may also have their own special metabolic requirements (for example, that severe nitrogen-limitation may be very favourable for the synthesis of polyketides and terpenoids but can hardly be optimal for oligopeptide production) we see that the *character* of the growth limitation is also in question.

This same point was illustrated in a slightly different way from chemostat data by Dean (1972), who tabulated the effects of dilution rate on the activity of individual enzymes in a range of micro-organisms, though nearly all the data available at that time related to bacteria. In his tabulation were enzymes whose activity increased at higher dilution rates (*e.g.* glutamate dehydrogenase and hexokinase in glucose-limited *Aspergillus nidulans*), enzymes whose activity increased at lower dilution rates (*e.g.* glucose-6-phosphate dehydrogenase in sucrose-limited *Candida utilis*), and enzymes whose activity at a certain range of dilution rates was minimal (*e.g.* hexokinase in sucrose-limited *C. utilis*) or maximal. All the data listed by Dean pertain to enzymes functioning in the main pathways of metabolism and the measured enzyme levels must all represent a balance between synthesis and turnover, but they provide a background against which the data for enzymes of secondary biosynthesis can be viewed without serious inconsistencies.

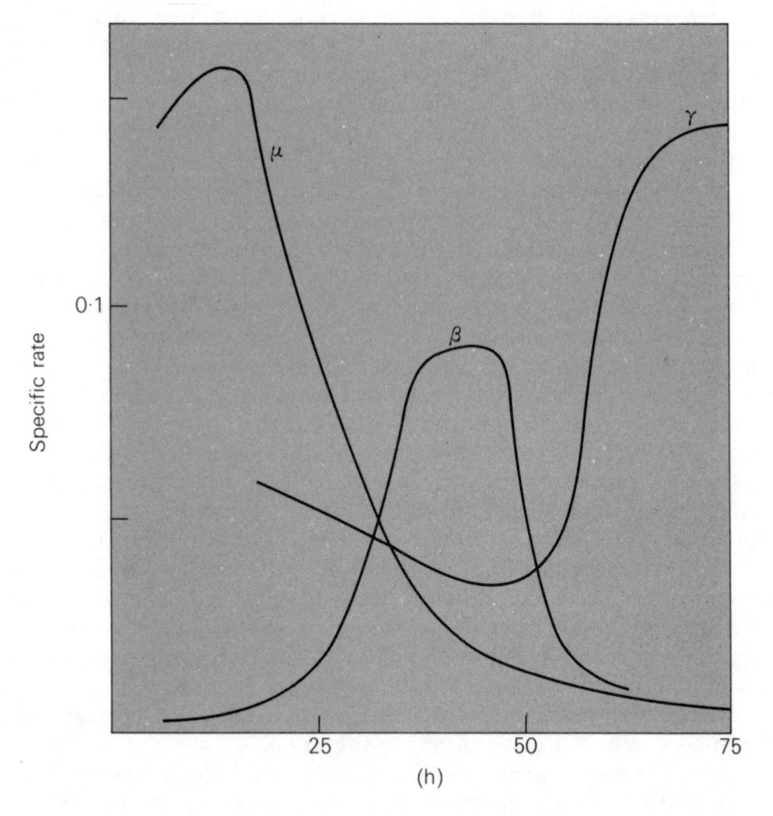

Fig. 3.8 Specific rates (per mycelial dry weight) for a nitrogen-limited batch culture of *Gibberella fujikuroi*: μ (biomass) in h^{-1}, β (bikaverins) in mg/g-h, γ (gibberellins) in mg/g-h.

So far as effects of growth limitation on primary metabolic pathways are concerned, it is not easy to judge from the large volume of data which can be collected (e.g. for fungi, Smith & Galbraith, 1971; Smith & Anderson, 1973) the extent to which the observed effects are (a) primary effects on levels of intermediates, (b) changed enzyme activity due to substrate effects on their activity or their synthesis, or (c) secondary effects on levels of intermediates following changes in enzyme activity, or (d) further consequences (*ad infinitum*). In batch cultures events occur in rapid and uncontrolled succession and do not necessarily equate with the more complete adjustments observed in the chemostat. Moreover, the events in any one system will depend to a marked degree upon features which may be specific to that system—the nature of particular endogenous reserve materials, and the particular ways in which the limiting nutrient can be utilized—as well as upon generalized response mechanisms. Consequently, only a few illustrative generalizations can be attempted. For example, since nitrogen is assimilated by the reductive amination of keto-acids, a falling nitrogen supply will be reflected first in increasing levels of tricarboxylic acid cycle intermediates, a lower demand for reduced coenzymes (the latter, *a fortiori*, if nitrate was being assimilated), and, until nett protein synthesis slows down correspondingly, a marked fall in the total amino acid pool. Between these primary effects and the actual adjustment of 'growth rate' there must be many interacting stages; one is normally the activation of proteases which through an increased turnover of amino acids permit new proteins to be synthesized. In comparison, the immediate effects of a reduction in glucose supply must be quite different even if they ultimately actuate the same controls on growth rate; for example, any turnover of reserve carbohydrates must be much more wasteful than the turnover of amino acids to alleviate nitrogen deficiency. To pursue all these effects, even for the range of fungi which have been studied, would be too arduous a task here, and in any event there are grounds for believing that so far as the development of secondary biosynthesis is concerned most of these changes in intermediary metabolism really constitute an 'interface' between the external nutritional conditions and the controls of new enzyme synthesis at gene level.

Mycelial age

In growing mycelium the hyphal tips are always 'young' and more distal portions of the hyphae are always progressively older. It is well-recognized that there are always very substantial structural and biochemical differences between the apical and more distal regions, even of aseptate hyphae (Zalokar, 1959; Bartnicki-Garcia, 1973). At any particular growth rate the overall distribution of this pattern of differentiation within the mycelium remains constant, because of the hyphal branching and colony fragmentation, but over a spectrum of decreasing growth rates the pattern must change and the proportions of 'older' type regions must increase; the changing growth rates are largely accommodated by decreasing the proportion of hyphal tips rather than by reducing their individual rates of extension (Trinci, 1969). The development of batch cultures can be correlated with such changes, as for example in the studies correlating classified morphological and histochemical observations with the phased development of several fermentations (Becker *et al.*, 1956; Becker *et al.*, 1965). Among more recent studies, Nash & Huber (1971) described a particularly clearcut analysis of morphological types in relation to the production of β-lactams by *Cephalosporium acremonium*.

The age-distribution of mycelium can be defined statistically and it then offers a parameter which is a function both of lapsed time and of the spectrum of growth-rates followed in a fermentation (Cook & James, 1964; Gottlieb & van Etten, 1965). Such a parameter can therefore be used to make significant correlations for the development of fermentations run under varied conditions, and as a computational description this method has had some success. However, this does not necessarily mean that all the characteristics of mycelium for which such correlations can be made are necessarily determined, in any meaningful way, by the kind of intrinsic and inevitable lapsed-time process which the word 'ageing' suggests to mortal men.

Fully-developed differentiation systems in micro-organisms do of course include intrinsic, and therefore not fully reversible, elements to which the broader concept of ageing might be applicable. Generally speaking, they occur when the environmental changes provoke internal responses which have the effect of 'insulating' the system in some respect, so that some further responses of the system are governed by internal factors. This is the phenomenon usually known as 'commitment', and seen in its simplest form in the sporulation sequence of bacilli. In this well-explored (but still mysterious) process (Szulmajster, 1973; Balassa, 1971; Sterlini & Mandelstam, 1969), the initial stimulus is an environmentally-determined reduction in growth rate, but among the responses which this evokes are changes which affect, for example, the permeability of the pre-spore compartment, and so make the further development of the spore increasingly irreversible. Thus in chemostat cultures at various growth rates, the rate at which sporulation is initiated in the population depends upon the degree of growth-limitation, but the rate at which bacteria committed to sporulation actually form spores is intrinsically determined, and is the same as in batch growth (Dawes, Kaye, & Mandelstam, 1969). Similarly, in chemostat studies of *Aspergillus niger*, Ng, Smith & McIntosh (1973) could show that although conidiation of this fungus is a process with marked commitment, the initiation of conidiation is determined by the degree and character of nutritional constraints upon growth.

The part played by compartmentalization responses of this kind in developmental processes in higher micro-organisms, in which commitment phenomena are equally well-marked, has been rightly noticed by Smith & Galbraith (1971) in their detailed review of fungal differentiation. Many mycologists would emphasize committed aspects of differentiation in such self-closing systems, and regard the environmentally-conditioned responses merely as a triggering mechanism; on the other hand workers in secondary metabolism tend to the view that 'the medium is the message' while admitting the possibility that some responses may involve commitment. The difference is merely one of emphasis and should not impede the unified view.

However, it is also possible to apply a statistically-based concept of 'ageing' to the development of secondary biosynthesis in ordinary batch culture systems, in which there may in fact be only minimal commitment, and I believe that however successful such an approach may be in terms of computational modelling it is biologically misleading. Once again, this can be illustrated from work on the *Gibberella fujikuroi* system in which bikaverin synthesis is evoked in nitrogen-limited growth (Bu'Lock, J.D. and Detroy, unpublished results). Mycelium removed from a batch culture before pigment synthesis has started, and transferred to a low-nitrogen medium, will respond by beginning to synthesize bikaverins, but only after varying periods of time. It can be shown that this delay period is the time needed for the mycelium to use up its internal pool of readily-available amino acid-nitrogen, and that this pool is larger in mycelium isolated earlier in the fermentation. To this extent the failure to produce pigment is intrinsically determined, but on the other hand in a normal batch culture the internal amino acid pool has more time to respond to the gradual depletion of the external source, so that when pigment production does begin it can be seen as a response to an external stimulus. More strikingly, once pigment production has begun it can be stopped, within a few hours, by adding external nitrogen; production of the rate-limiting synthetase is blocked just as effectively as it is by a general inhibition of protein synthesis, and bikaverin synthesis decays with a 1.5 hour half-life. This is entirely consistent with a theory which relates secondary metabolism to the concept of limited growth, but wholly inconsistent with its presentation in terms of mycelial age distribution. More generally, any transient features of a batch fermentation which can also be maintained indefinitely in a chemostat at an appropriate dilution rate do not, *a priori*, involve intrinsic or commitment effects.

Growth-linked regulation

We have seen that it is possible to define as 'secondary' all those processes which are intensified in 'limited' or sub-optimal growth; such a definition is found acceptably to include not only the production of structurally-distinct metabolites but also those 'abnormal' accumulations of primary metabolites which conform to similar regularity modes. It includes corresponding patterns of enzyme production, *a priori* for the enzymes effecting secondary processes, but also for secondary over-production of individual enzymes irrespective of their role. For any given system it embraces a whole complex of processes, some of which may in addition have their own more selective requirements, either for individual substrates or for specific induction mechanisms. All these processes are governed by an overall control mechanism which is linked to the degree (and character) of growth-limitation. In the first instance this mechanism is actuated in response to, or as part of, the mechanism by which external parameters bring about the limitation of growth, but in cases where the secondary response introduces irreversible changes into the system, aspects of its further development may be governed intrinsically instead of remaining accessible to external influences. It is apparent that in dealing with secondary metabolism in these extended and generalized terms we are looking at a wide spectrum of phenomena which are moreover of quite fundamental importance to the biology of micro-organisms. To explain it, we must enquire to mechanisms for the regulation of gene expression in relation to restraints upon replicatory processes.

Definitive evidence on the molecular biology of these phenomena is only available, so far, for their manifestations in bacteria, where they relate on the one hand to the well-defined sequence of bacterial sporulation (Szulmajster, 1973; Balassa, 1971) and on the other to the process known as catabolite repression. In using them as a guide to mechanisms in eukaryotes we must bear in mind the additional regulatory mechanisms which the structure of higher cells allows, but it is probably safe to assume that a view of the bacterial system will provide at least part of the model for which we are seeking. The sporulation process in, for example, *Bacillus subtilis*, is a clear example of a series of changes governed by successive initiations (and terminations) of specific enzyme syntheses, initially in response to a measurable degree of growth limitation but with an increasing element of commitment to an intrinsically-governed program. The process involves the selective and sequential regulation of a large number of genes which are neither adjacent nor correspondingly consecutive on the chromosome. Some of the regulated steps are indispensable to the further progress of the sequence, presumably because they are involved in the commitment process, while others are not so essential to sporulation though they fall under the same overall control; some have their own selective requirements in addition to the overall mechanism. The parallels to the various aspects of secondary metabolism which have already been discussed for fungi are very close, and indeed one of the first 'events' in the sporulation sequence is the production of typical secondary metabolites, the oligopeptide antibiotics, the appearance of which is governed by the simultaneous and co-ordinated synthesis (and subsequent breakdown) of the complex of enzymes specifically concerned with their biosynthesis, during the first 1.5–2 hours of the sporulation sequence (Tomino, Yamada, Itoh & Karahashi, 1967).

Although there are some experimental data which suggest that this pattern of gene expression is at least partly regulated at the ribosomal (translation) level, it is now believed that a major part of the general control in bacterial sporulation is exerted at the transcriptional level, and in particular through modifications of components in the DNA-dependent RNA polymerase which recognise specific sequences in the promoter regions of the relevant genes. This 'recognition' determines the initiation of m-RNA synthesis and it is connected with the presence of specific 'σ-factor' in the polymerase. The sub-unit pattern, and in particular the σ-factor, in the polymerase from sporulating cells is quite different from that in vegetative cells. There is evidence that the σ-factor interacts with a region of promoter gene, so that other transcriptional controls can still act on the operator gene within this more general mechanism. Such additional controls must be invoked in order to explain the full range of phenomena we have noted, such as the existence of specific induction mechanisms operating within the limits of the overall control and the manner in which some responses are sensitive to the *degree* of growth-limitation. The 'σ-factor mechanism' is valuable since it explains how a number of non-linked genes can be controlled simultaneously, but in turn it raises the question of how the production of one σ-factor rather than another is itself controlled; some further regulation mechanism, either for the genes which code for polymerase components or for genes controlling enzymes which modify the sub-units, is clearly implied (Szulmajster, 1973).

The second general control mechanism which has been studied at gene level in bacteria is that known as 'catabolite repression' (Paigen & Williams, 1970). This term has a long history and it has been widely applied, to relatively restricted effects due to high glucose levels (and perhaps mediated rather directly) on the one hand to broad-spectrum effects of growth-limitation, including controls of secondary metabolite production (Demain, 1968, 1972), on the other. Recently, the term has been increasingly used in the broader sense, and though the subject is still controversial one type of mechanism for catabolite repression seems to be gaining wider acceptance (Zubay, Schwarz, & Beckwith, 1970; Emmer, deCrombrugghe, Pastan & Perlman, 1970). As is well-known the *lac* operon in *Escherichia coli* cannot normally be de-repressed in the presence of sugars more readily utilized than galactose, other nutrients being non-limiting—*i.e.* in addition to its control by repressor protein there is an over-riding control by catabolite repression when growth is insufficiently 'limited'. It has been found that when exhaustion of nutrients provokes a decline in the growth rate of *E. coli* there is a corresponding increase in the intracellular level of cyclic adenine 3′,5′-monophosphate, cAMP, and this is an effector for the lifting of catabolite repression (Makman & Sutherland, 1965). It activates a specific receptor protein which controls the *lac* operon positively, it being suggested that when the activated protein is bound to the distal* part of the operator gene the operon is available for transcription (Riggs, Reiness and Zubay, 1971). Synthesis of the corresponding m-RNA will now proceed if the *lac*-repressor protein, which binds to proximal regions of the operator gene (Gilbert, 1972), is at the same time removed in the normal way by reaction with an inducer such as isopropyl thiogalactoside. There is little in the way of detailed metabolic understanding to explain why the level of cAMP rises when the growth rate is limited, though fairly direct couplings to either the rate of DNA replication or the rate of cell wall synthesis are not excluded; we also need more detailed studies of cAMP levels in a variety of systems with growth rates changing for a wider spectrum of reasons. Nevertheless, at the time of writing this cAMP-mediated 'pleiotropic positive control' is the most promising basis for a unified view of the generalized 'catabolite-repression' effect, not only for bacteria but also for higher micro-organisms. Indeed, a role for cAMP in fungal differentiation might provide an intriguing phylogenetic link between its function in bacteria and its very widespread importance as a mediator of hormone-controlled processes in higher organisms (which in fact pose cybernetic problems which are fundamentally similar to those raised by the theme of secondary metabolism in microbes).

Though the evidence is as yet indirect, the cAMP-mediated lifting of catabolite repression could itself be a mechanism controlling the nature of the specific sub-units in RNA polymerase, so providing a means for controlling the transcription of whole groups of unlinked genes in a co-ordinate manner. At the same time we might expect that different genes controlled by binding to the cAMP-activated protein would display quantitatively different responses in a spectrum of activator levels, which is one requirement for a generalized mechanism as we have seen (p. 47).

So far, supporting evidence for higher micro-organisms is very sparse and comes mainly from *ad hoc* experiments on the effects of added cAMP (or its less polar dibutyl derivative). In Myxomycetes, cAMP promotes

*More distant from the structural genes, *i.e.* earliest in the direction of transcription.

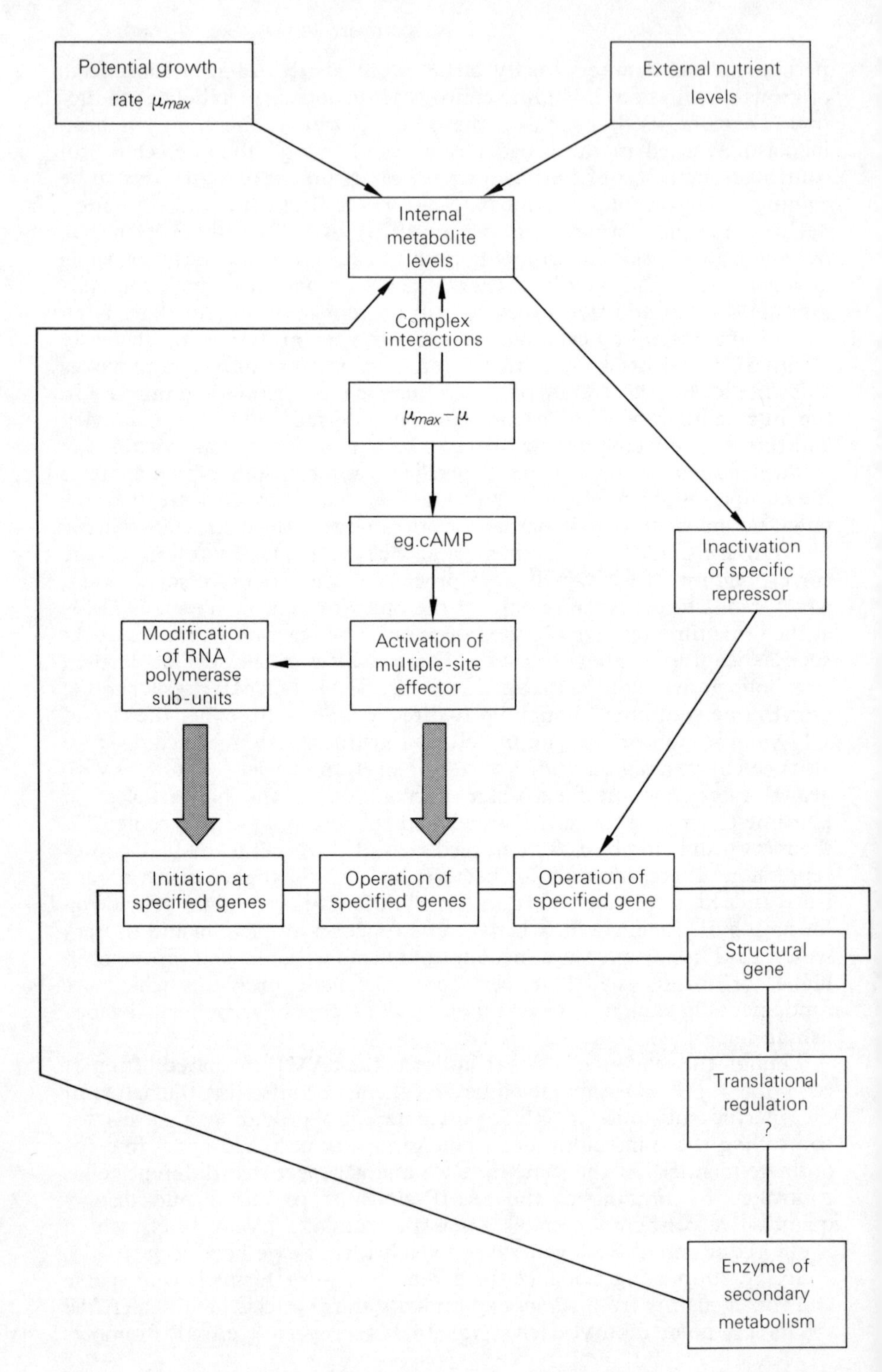

Fig. 3.9 A control mechanism for secondary metabolism. Broad arrows indicate controls acting at a multiplicity of gene sites.

aggregation at the end of the amoeboid phase, which is the beginning of a developmental sequence normally initiated by nutrient limitation and accompanied by sharp changes in enzyme levels and metabolic patterns (Bonner, 1971). In *Penicillium urticae*, added dibutyl-cAMP causes increased production of at least one enzyme in the sequence from 6-methylsalicylic acid to patulin, which is normally regulated by catabolite repression (Forrester & Gaucher, 1972); similarly in *Aspergillus niger* it promotes the complex switching of carbon utilization that leads to citric acid accumulation (Wold & Suzuki, 1973). In yeasts it prevents the nutrient inhibition of sporulation (Yanagishima, unpublished results). On the other hand, nutrient suppression of bikaverin synthesis in *Gibberella fujikuroi* is insensitive to added dibutyryl-cAMP, though permeability problems are not excluded as a explanation of this negative result (Bu'Lock, J.D. and Adams, unpublished results).

Doubtless the number of sporadically-found instances will increase considerably pending more systematic demonstrations, but in the meantime it is tempting to conclude by an extrapolation from the available evidence to attempt an outline scheme (Fig. 3.9) and a general statement about the different and concurrent controls for genes involved in secondary metabolism:

(a) Reductions in replicatory growth cause increased levels of cAMP (or some similar substance).
(b) A cAMP-activated protein promotes transcription of some genes directly; different genes may be differently sensitive.
(c) The cAMP-mediated process may cause different σ-factors to be produced for the control of m-RNA initiation so that a different range of genes becomes accessible for transcription.
(d) Genes controlled by mechanisms (b) and/or (c) may still require induction or the removal of specific repressors (by conventional mechanisms).
(e) The above mechanisms can also operate in the reverse sense causing some genes to cease transcription.
(f) Since recognition elements in DNA may be transcribed, similar mechanisms at post-transcriptional stages (e.g. ribosomal factors specific for certain m-RNA types) are not *a priori* excluded.
(g) Enzyme levels are governed by breakdown as well as synthesis.
(h) A graded and therefore sequential effect over a multiplicity of genes can arise either through a continuous decline in replicatory growth as nutrients are used up or through changes initiated by the above mechanisms which accentuate the initial effect internally.

To which series of rather final-looking summaries I should surely add:

(z) All the above statements are liable to disproof at any moment, and merely constitute an ill-informed attempt to solve a problem which is too big for most of us.

3.4 References

BALASSA, G. (1971). The genetic control of spore formation in Bacilli. *Current Topics in Microbiology and Immunology* **56**, 99–192.

BARTNICKI-GARCIA, S. (1973). Fundamental aspects of hyphal morphogenesis. In *Microbial Differentiation: Symposium of the Society for General Microbiology* **23**, 245–67.

BECKER, Z.E., DMITRIEVA, S.V., BORISOVA, T.G., TURKOVA, Z.A., LISINA, E.S. & CHAPLINA, L.B. (1965). Peculiarities of development of fungi producing various antibiotic and antiblastic substances. *Mikrobiologiya* **34**, 653–60.

BECKER, Z.E., OSTROGLOV, A.A., SMIRNOVA, A.D., KOTSCHELEVA, N.A. & FADEEVA, N.P. (1956). Developmental phenomena in submerged cultures of *Penicillium chrysogenum. Antibiotiki* **2**, 40–7.

BENTLEY, R. & CAMPBELL, I.M. (1968). Secondary metabolism of fungi. Edited by M. Florkin & E.H. Stotz. *Comprehensive Biochemistry* pp. 415–90. Amsterdam **20**: Elsevier.

BENTLEY, R. & THIESSEN, C.P. (1957). Biosynthesis of itaconic acid in *Aspergillus terreus*: properties and reaction mechanism of *cis*-aconitic acid decarboxylase. *Journal of Biological Chemistry* **226**, 703–20.

BONNER, J.T. (1971). Aggregation and differentiation in the cellular slime moulds. *Annual Review of Microbiology* **25**, 75–92.

BORROW, A., JEFFERYS, E.G., KESSELL, R.H.J., LLOYD, E.C., LLOYD, P.B. & NIXON, I.S. (1961). The metabolism of *Gibberella fujikuroi* in stirred culture. *Canadian Journal of Microbiology* **7**, 227–76.

BORROW, A., BROWN, S., JEFFERYS, E.G., KESSEL, R.H.J., LLOYD, E.C., LLOYD, P.B., ROTHWELL, A., ROTHWELL, B., & SWAIT, J.C. (1964). The kinetics of the metabolism of *Gibberella fujikuroi* in stirred culture. *Canadian Journal of Microbiology* **10**, 407–44.

BU'LOCK, J.D. (1961). Intermediary metabolism and antibiotic synthesis. *Advances in Applied Microbiology* **3**, 293–342.

BU'LOCK, J.D. (1965*a*). *The biosynthesis of natural products*: McGraw-Hill, London.

BU'LOCK, J.D. (1965*b*). Aspects of secondary metabolism in fungi. Edited by Z. Vaněk & Z. Hošťálek. *Biogenesis of Antibiotic Substances* pp. 61–71. New York: Academic Press and Academy of Sciences, Prague.

BU'LOCK, J.D. & BARR, J.G. (1968). A regulation mechanism linking tryptophan uptake and synthesis with ergot alkaloid synthesis in *Claviceps. Lloydia* **31**, 342–54.

BU'LOCK, J.D., GREGORY, H., & HAY, M. (1961). The rate of polyacetylene synthesis by Basidiomycete B-841. *Journal of the Chemical Society (London)* 1961, 3544–5.

BU'LOCK, J.D., HAMILTON, D., HULME, M.A., POWELL, A.J., SMALLEY, H.M., SHEPHERD, D. & SMITH, G.N. (1965). Metabolic development and secondary biosynthesis in *Penicillium urticae. Canadian Journal of Microbiology* **11**, 765–78.

BU'LOCK, J.D., SHEPHERD, D., & WINSTANLEY, D.J. (1969). Regulation of 6-methylsalicylate and patulin synthesis in *Penicillium urticae. Canadian Journal of Microbiology* **15**, 279–85.

COOK, J.R. & JAMES, T.W. (1964). Age distribution of cells in logarithmically growing cell populations. In *Synchrony in cell division and growth* pp. 485–95, Edited by E. Zeuthen. New York: Interscience.

DAWES, I.W., KAY, D. & MANDELSTAM, J. (1969). Sporulation in *Bacillus subtilis*: establishment of a time-scale for the morphological events. *Journal of General Microbiology* **56**, 171–9.

DAWSON, P.S.S. (1972). Continuously synchronized growth. *Journal of Applied Chemistry and Biotechnology* **22**, 79–103.

DEAN, A.C.R. (1972). Influence of environment on the control of enzyme synthesis. *Journal of Applied Chemistry and Biotechnology* **22**, 245–59.

DEMAIN, A.L. (1968). Regulatory mechanisms and the industrial production of microbial metabolites. *Lloydia* **31**, 395–418.

DEMAIN, A.L. (1972). Cellular and environmental factors affecting the synthesis and excretion of metabolites. *Journal of Applied Chemistry and Biotechnology* **22**, 345–62.

DIMROTH, P., WALTER, H. & LYNEN, F. (1970). Biosynthese von 6-methylsalicylsaure. *European Journal of Biochemistry* **13**, 98–110.

DIMROTH, P., GREULL, G., SEYFFERT, R. & LYNEN, F. (1972). 6-Methylsalicylic acid synthetase. *Hoppe-Seyler's Zeitschrift für Physiologische Chemie* **353**, 126.

EMMER, M., DECROMBRUGGHE, B., PASTAN, I. & PERLMAN, R.L. (1970). Cyclic AMP receptor protein of *E. Coli. Proceedings of the National Academy of Sciences, U.S.A.* **66**, 480–7.

FLOSS, H.G. & MOTHES, U. (1964). Uber den einfluss von tryptophan und analogen verbindungen auf die biosynthese von clavinalkaloiden in saprophytischer kultur. *Archiv für Mikrobiologie* **48**, 213–21.

FORRESTER, P.I. & GAUCHER, G.M. (1972). Conversion of 6-methyl-salicylic acid into patulin by *Penicillium urticae*; *m*-hydroxybenzyl alcohol dehydrogenase from *Penicillium urticae*. *Biochemistry* **11**, 1102–7; 1108–14.

GILBERT, W. (1972). The *lac* repressor and the *lac* operator. In *Polymerisation in biological systems: Ciba Foundation Symposium* 7, 245–56.

GOTTLIEB, D. & VAN ETTEN, J.L. (1965). Changes in fungi with age. *Journal of Bacteriology* **91**, 161–8.

HOCKENHULL, D.J.D. & MACKENZIE, R.M. (1968). Preset nutrient feeds for penicillin fermentation and defined media. *Chemistry and Industry (London)* 1968, 607–10.

MAKMAN, R.S. & SUTHERLAND, E.W. (1965). Cyclic adenosine 3′,5′-phosphate in *E. coli*. *Journal of Biological Chemistry* **240**, 1309–14.

MULHEIRN, L.J. & RAMM, P.J. (1972). The biosynthesis of sterols. *Chemical Society Reviews* **1**, 259–91.

NASH, C.H. & HUBER, F.M. (1971). Antibiotic synthesis and morphological differentiation of *Cephalosporium acremonium*. *Applied Microbiology* **22**, 6–10.

NG, A.M.L., SMITH, J.E. & MCINTOSH, A.F. Conidiation of *Aspergillus niger* in continuous culture. *Archiv für Mikrobiologie* **88**, 119–26.

PAIGEN, K. & WILLIAMS, B. (1970). Catabolite repression and other mechanisms in carbohydrate utilization. *Advances in Microbial Physiology* **4**, 251–324.

POLAKIS, E.S. & BARTLEY, W. (1966). Changes in the intracellular concentrations of adenosine phosphates and nicotinamide nucleotides during the aerobic growth cycle of yeast on different carbon sources. *Biochemical Journal* **99**, 521–33.

RAMKRISHNA, D., FREDRICKSON, A.G., & TSUCHIYA, H.M. (1967). Dynamics of microbial propagation: models considering inhibitors and variable cell composition. *Biotechnology and Bioengineering* **9**, 129–70.

ŘEHÁČEK, Z., KOSOVÁ, J., ŘIČICOVÁ, A., KAŠLÍK, J., SAJDL, P., ŠVARC, S., & BASAPPA, S.C. (1971). Role of endogenous tryptophan during submerged fermentation of ergot alkaloids. *Folia Microbiologica* **16**, 35–40.

RIGGS, A.D., REINESS, G., & ZUBAY, G. (1971). Purification and DNA-binding properties of the catabolite gene activator protein *Proceedings of the National Academy of Sciences, U.S.A.* **68**, 1222–5.

SJOLAND, S. & GATENBECK, S. (1966). Studies on the enzyme synthesizing the aromatic product alternariol. *Acta Chemica Scandinavica* **20**, 1053–9.

SMITH, J.E. & ANDERSON, J.G. (1973). Differentiation in the Aspergilli. In *Microbial Differentiation: Society for General Microbiology Symposium* **23**, 295–337.

SMITH, J.E. & GALBRAITH, J.C. (1971). Biochemical and physiological aspects of differentiation in the fungi. *Advances in Microbial Physiology* **5**, 45–134.

SOLTERO, F.V. & JOHNSON, M.J. (1953). The effect of the carbohydrate nutrition on penicillin production by *Penicillium chrysogenum* Q-176. *Applied Microbiology* **1**, 52–7.

SRERE, P.A. (1967). Enzyme concentrations in tissues. *Science* **158**, 936–7.

STERLINI, J.M. & MANDELSTAM, J. (1969). Commitment to sporulation in *Bacillus subtilis* and its relationship to the development of actinomycin resistance. *Biochemical Journal* **113**, 29–37.

SZULMAJSTER, J. (1973). Initiation of bacterial sporogenesis. In *Microbial Differentiation: Society for General Microbiology Symposium* **23**, 45–83.

TABER, W.A. & VINING, L.C. (1963). Physiology of alkaloid production by *Claviceps purpurea*: correlation with changes in mycelial polyol, carbohydrate, lipid, and phosphorus-containing compounds. *Canadian Journal of Microbiology* **9**, 1–14.

TABER, W.A., BRAR, S.S. & GIAM, C.S. (1968). Patterns of *in vitro* ergot alkaloid production by *Claviceps paspali* and their association with different growth rates. *Mycologia* **60**, 806–26.

TOMINO, S., YAMADA, M., ITOH, H. & KURAHASHI, K. (1967). Cell-free synthesis of gramicidin-S. *Biochemistry* **6**, 2552–60.

TRINCI, A.J.P. (1969). A kinetic study of the growth of *Aspergillus nidulans* and other fungi. *Journal of General Microbiology* **57**, 11–24.

TURNER, W.B. (1971). *Fungal metabolites*. Academic Press, London and New York.

VINING, L.C. (1970). Effect of tryptophan on alkaloid biosynthesis in cultures of a *Claviceps* species: *Canadian Journal of Microbiology* **16**, 473–80.

WAKIL, S.J. & BARNES, E.M. (1971). Fatty acid metabolism. In *Comprehensive Biochemistry*, pp. 57–104. Edited by M. Florkin & E.H. Stotz. Amsterdam: Elsevier.

WEINBERG, E.D. (1970). Biosynthesis of secondary metabolites: roles of trace metals. *Advances in Microbial Physiology* 4, 1–44.

WOLD, W.S.M. & SUZUKI, I. Cyclic AMP and citric acid accumulation by *Aspergillus niger*. *Biochemical and Biophysical Research Communications* **50**, 237–44.

WRIGHT, B.E. (1968). An analysis of metabolism underlying differentiation in *Dictyostelium discoideum*. *Journal of Cellular Physiology* 72, 145–60.

ZALOKAR, M. (1959). Growth and differentiation of *Neurospora* hyphae. *American Journal of Botany* **46**, 602–10.

ZUBAY, G., SCHWARTZ, D. & BECKWITH, J. (1970). Mechanism of activation of catabolite-sensitive genes: a positive control system. *Proceedings of the National Academy of Science, U.S.A.* **66**, 104–10.

Additional References

BU'LOCK, J.D., DETROY, R.W., HOŠŤÁLEK, Z. & MUNIM-AL-SHAKARCHI, A. (1974). Regulation of secondary biosynthesis in *Gibberella fukikuroi*. *Transactions of the British Mycological Society*, **62**, 377–89.

UNO, I. & ISHIKAWA, T. (1973). Metabolism of adenosine 3′,5-cyclic monophosphate and induction of fruiting-bodies on *Coprinus macrorhizus*. *Journal of Bacteriology*, **113**, 1249–55.

CHAPTER 4

Strain Improvement and Strain Stability in Filamentous Fungi

J. R. JOHNSTON

4.1 Introduction

This chapter concerns the genetical basis of a strain's characteristics, *i.e.* the genome underlying the phenotype. By way of introduction some aspects of the genetical methodology used with fungi have been included. This aspect of genetics has been comprehensively reviewed by Fincham & Day (1971). In particular, the emphasis in this chapter will be placed on the applications of microbial genetics to industrial mycology.

Recently, there has been an upsurge of interest in the area of applied microbial genetics. The last few years have seen the publication of Sermonti's book (1969), the staging in Prague in 1970, of *The First International Symposium on the Genetics of Industrial Micro-organisms*, and, in Vienna in 1971, of *The Symposium on Use of Radiation and Radioisotopes for Genetic Improvement of Industrial Micro-organisms*. The Prague papers have been published in book form (Vaněk, Hošťálek & Cudlín, 1973) and have been excellently summarized by Calam (1972). Much of this applied work is published in journals of Eastern European countries *e.g.* recent reviews in Russian and in Slovak by Alikhanian (1972) and Alačevíc (1972) respectively. Fortunately, most of this literature can be traced through *Chemical Abstracts, Genetical Abstracts* and *Plant Breeding Abstracts* (*Economic Lower Plants* Section).

4.2 Basic Genetics

Genetic analysis in higher plants and animals is based upon observation of the inherited characteristics of progeny of particular sexual crosses over at least one, but customarily two or more generations. Chromosomal (nuclear) genes are defined by the (Mendelian) segregation of their alternative forms (alleles) during meiosis and the dominance relationships between alleles by the phenotype of hybrid (heterozygous) individuals. That different genes are located on the same chromosome (gene linkage) can be inferred from the lack of independent assortment of both pairs of alleles into meiotic products. The frequency of genetic recombination in a dihybrid (*i.e.* the proportion of progeny lacking the original, or parental, combinations of genes) should be approximately 0.5 if the two genes are unlinked (located on different chromosomes, or far apart on the same chromosome). The recombination frequency between two linked genes however, will be significantly less than this and its level is used as a measure of the (genetic) distance between genes. For example, a recombination frequency of 0.3 may be used to place the two genes 30 units (centimorgans) apart on a particular chromosome. The extension of these observations in various dihybrid and trihybrid crosses permits the construction of chromosome maps illustrating the positions of linked genes in terms of their frequencies of recombination with each other. The basis of such studies rests upon (a) knowledge of the sexual mechanisms of the material under investigation, *e.g.* pollen-production, self-fertilization, and (b) use of suitable mutants either by selection of natural variants or production of such mutants by physical or chemical agents (mutagens).

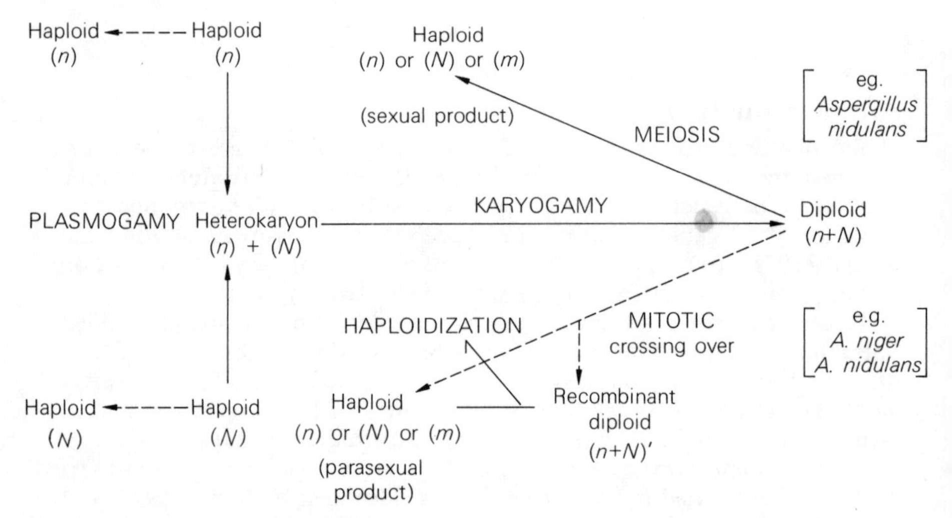

Fig. 4.1 Summary of the ploidy relationships and mechanisms of recombination in some filamentous fungi. (— — — — — denotes mitotic divisions). (Adapted, in part, from Ball, 1973*a*.)

The same two basic premises apply to the genetic analysis of micro-organisms, with the additional proviso that, in some, the mechanisms of recombination may be, solely or additionally, non-sexual. For example, genes may be transferred from one bacterium to another by viruses acting as the carriers, or recombinant progeny may be produced by some fungi without the advent of meiosis. The foundation of genetical analysis with the filamentous fungi is, then, to elucidate the ploidy relationships and mechanisms of recombination inherent in the life-cycles of the species. A summary of these is presented schematically in Fig. 4.1, and the full life-cycle of *Aspergillus nidulans* is illustrated in Fig. 4.2.

A major difference between the life-cycles of higher organisms and those of fungi is the prolonged, and generally predominant, haploid phase of filamentous fungi. As in higher organisms, genetic recombination can occur in certain fungi by meiotic segregation. Few of the industrially important filamentous fungi are, however, endowed with such a conventional sexual cycle. Fortunately, the filamentous fungi will often form heterokaryons and, from these, occasional diploid nuclei will result from

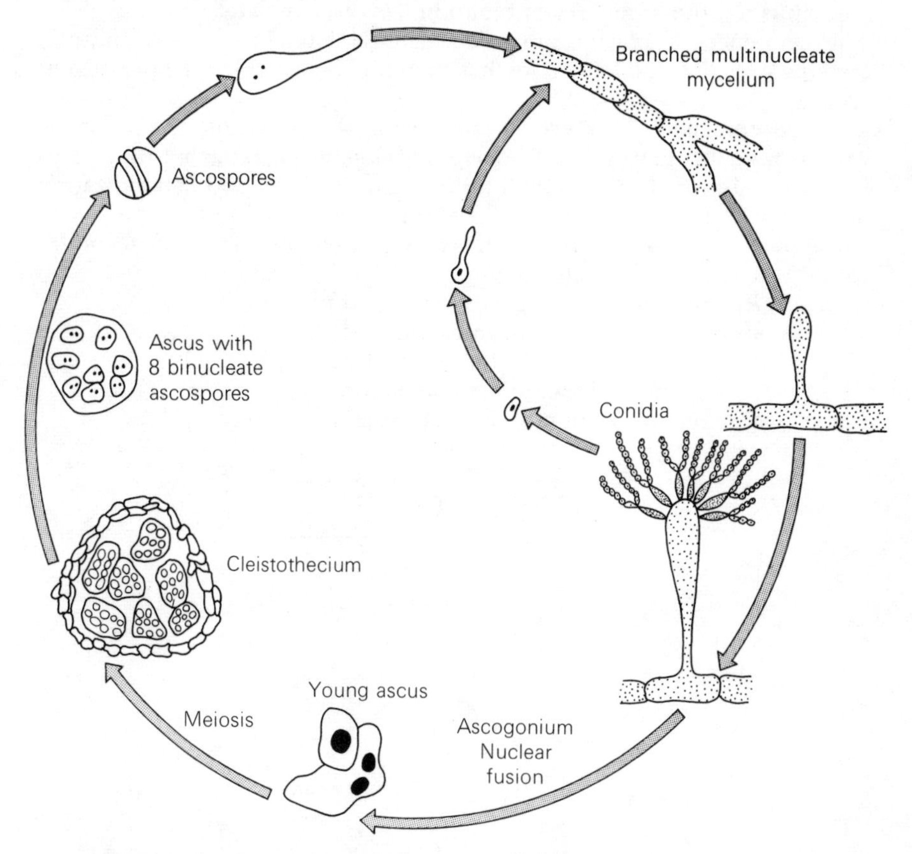

Fig. 4.2 Life cycle of *Aspergillus nidulans* (after Fincham & Day, 1971).

karyogamy. Although recombination is much less frequent, it can occur in diploids by the mechanisms of mitotic crossing over and mitotic nondisjunction, *i.e.* a parent chromosome and its newly-synthesized copy fail to disjoin and therefore both migrate to the same mitotic pole and are both included in the same daughter nucleus, the other daughter nucleus thus lacking a copy of this particular chromosome. Repeated non-disjunction for different chromosomes during a series of mitotic divisions can eventually lead to both parental and recombinant haploid nuclei, such haploidization being the equivalent of the meiotic independent-assortment of nonlinked genes. Considering genes of a particular chromosome which are present in the diploid in the heterozygous condition, mitotic non-disjunction results in a diploid nucleus in which all genes of this chromosome are now present in homozygous condition. On the other hand, as illustrated in Fig. 4.3, mitotic crossing over produces homozygosity only for those genes which are more distant from the centromere (distal) than the position of the cross-over. Subsequent haploidization from such a homozygous diploid yields haploids which are recombinants for linked genes. In this way, mitotic crossing over followed by haploidization accomplishes the same recombination as does meiotic crossing over. The sequence of plasmogamy, karyogamy, haploidization or mitotic crossing over plus haploidization has therefore been named the parasexual cycle.

The occurrence of the events comprising the parasexual cycle can be very infrequent, however. In *Aspergillus nidulans*, karyogamy may be as low as 1 in 10^6 or 10^7 nuclei, and mitotic crossing over and haploidization may occur only once in each 500 and 10^3 nuclear divisions respectively. The *mitotic recombination index* (Imi) may be calculated from the formula:

$$\text{Imi} = [E + (n - 1)h]d$$

where E = the number of cross-overs per diploid nucleus
n = the number of chromosome pairs
h = the proportion of diploid nuclei which undergo haploidization
d = the proportion of diploid nuclei per colony

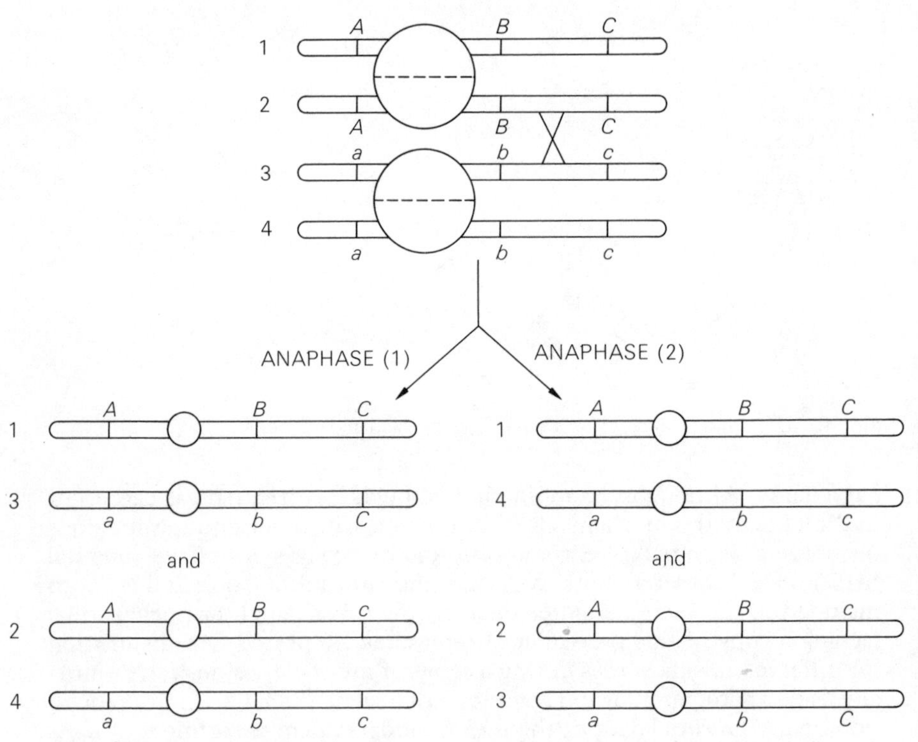

Fig. 4.3 Mitotic crossing-over leading to allelic segregation in diploid daughter cells. The two possible anaphase distributions should be equally probable and thus crossing over should result in segregation of the *C,c* alleles in 50% of cases. Segregation, and therefore homozygosity, will only occur for genes situated distal to the crossover position.

In *Aspergillus nidulans*, this index is of the order 10^{-5}. In *A. niger*, however, the rates of mitotic crossing over and haploidization are much higher and Lhoas (1967) has estimated the mitotic recombination index as 2×10^{-3}. He has suggested that the frequency of mitotic crossing over may be as high as 0.2 and, by parasexual analysis, has assigned genes to six different chromosomes. To ease genetical analysis, the frequency of haploidization can be increased considerably by exposure of diploids to the amino acid analogue, *p*-fluorophenylalanine, which probably affects the mitotic spindle. Several agents, such as ultraviolet light, formaldehyde and nitrogen mustard, increase the frequency of both haploidization and mitotic crossing over, while the latter is induced specifically by fluorouracil. Although mitotic recombination and the parasexual cycle have been investigated most extensively in *Aspergillus nidulans* and, to a lesser extent in several other *Aspergillus* species such as *A. niger*, *A. oryzae* and *A. fumigatus* as well as *Penicillium chrysogenum*, these processes apparently occur in a fairly wide range of other filamentous fungi such as *Fusarium oxysporium*, *Verticillium albo-atrum*, *Cephalosporium mycophilum*, *Cochliobolus sativus* and *Emericellopsis* spp.

In common with other micro-organisms, the mutants of filamentous fungi most extensively used in genetic analysis are auxotrophic, *i.e.* requiring for growth a nutrient over and above those required by the wild-type, parent organism. These auxotrophs most generally are unable to synthesize a particular amino acid, purine or pyrimidine, or vitamin. Other requirements, as in *Penicillium chrysogenum* for example, may be for reduced nitrogen as ammonium ion or reduced sulphur as thiosulphate. Drug-resistant mutants, which have been widely used in bacteria and yeast, have been more difficult to isolate in the mycelial fungi. Among the successes, however, are resistance to acriflavin in *Aspergillus* spp., and to 8-azaguanine in *Aspergillus* spp. and *Penicillium* spp. Some morphological mutants have also proved extremely useful, particularly those in which spore pigmentation is altered. Such mutants as *white* and *yellow* in *A. nidulans* and *brown*, *yellow* and *bright green* in *P. chrysogenum* are clearly distinguishable from the wild-type. Other common morphological mutants are affected in their growth rate or their degree of conidiation. Some mutants are pleiotropic, *i.e.* mutation of one gene results in the alteration of two or more characteristics. For example, in *P. chrysogenum*, an auxotrophic mutant may also be impaired in penicillin production or a mutant producing an increased titre of antibiotic may also display altered morphological characteristics.

Some fungal mutations are cytoplasmic or extranuclear *i.e.* their mode of inheritance is different from that expected of gene mutations, or is non-Mendelian. Certainly genetic material in structures similar to chromosomes exists outside of the nucleus in organelles such as mitochondria and chloroplasts. Several continuously-varying characteristics such as growth rate, spore germination and perithecial density have been shown to be under cytoplasmic control in *Aspergillus glaucus* (Jinks, 1964). One of the better-studied cytoplasmic mutants is *poky* (reduced growth rate) in *Neurospora crassa* but this mutation has not clearly been shown to be mitochondrial as has the extensively-studied, similar mutation, *petite*, in *Saccharomyces cerevisiae*. Recently, however, the phenomena of senescence in *Aspergillus amstelodami* and cold-sensitivity in *A. nidulans* have been shown to be associated with mitochondrial mutation (Handley & Caten, 1973; Waldron, 1973).

4.3 Strain Improvement

This term is applied to selection or production of an organism with an altered genotype which is considered better suited for *our* purposes than the original, parent strain. It should be emphasized that the altered strain is only improved with respect to its service to *us*. It is highly likely that the genetic change represents degeneration, for, in becoming more specialized, the organism has probably lost out in adaptability to environmental change and thus in its survival-fitness. We have therefore 'induced degeneration of the organism' (Pirt, 1969).

There are basically three ways by which strains may be improved: (1) by selection alone, (2) by mutation followed by selection and (3) by recombination and selection. There may be *practical* difficulties but, in general, there are no theoretical reasons why the same genetic principles which underlie many dramatic increases in the productivity levels of higher plants and animals should not be applied to microbial productivity. Indeed, in the cases of *Penicillium* mutants and bakers' yeast hybrids, to name two outstanding examples, great improvements in particular characteristics have already been achieved.

Selection

Significant improvements in various micro-organisms have been effected simply by testing various available strains and selecting the best, and thus capitalizing upon natural genetic variation. Some examples of these are contained in Table 4.1. This list is far from exhaustive but the examples chosen cover a wide range of products or desirable characteristics of filamentous fungi. Frequently, large numbers of strains are subjected to testing and a single strain eventually selected. Thus Mehrotra & Krishna (1966) determined the best producer of amylase among 875 strains of *Aspergillus niger*, 14 of *A. oryzae* and 8 of *A. flavus*.

Sometimes a particular strain has been found to comprise two or more genetic types and an improved performance has been obtained by using one of the components as a pure culture. In other cases, so-called training

Table 4.1 Some microbial characteristics improved by selection

Organism	Product	Characteristics	Reference
Aspergillus spp. *Rhizopus* spp.	proteases	yield others	Keay, 1971
A. oryzae	amylolytic enzymes	yield	Cojocaru & Milena, 1967
Claviceps spp.	lysergic acid alkaloids	yield	Kelleher, 1969
Coriolus versicolor	Basidio-laccase	phenol oxidase activity	Fukumoto & Schichiji, 1971
Fusarium spp.	biomass	mycelial morphology protein content nutritional value	Scammell, 1973

has been carried out by cycles of exposure to some selective agent. For example, by repeated cultures in the presence of sodium pentachlorophenate, Cojocaru & Milena (1967) selected a strain of *Aspergillus oryzae* with a 33% or 50% increase in amylolytic activity depending on whether growth was surface or submerged. In most cases, however, initial selection should be regarded as merely a first step in producing superior strains. Just as H.J. Müller's discovery of artificially-induced mutation in 1926, by providing the opportunity to obtain a vastly extended range of mutants, heralded a new era of genetics, *induced* mutants have extended the horizons of strain improvement.

Mutation and selection

The production of mutant strains for industrial use has become more widespread in recent years. Yet to date the achievements probably represent only the tip of the iceberg and future years should see much more extensive applications of genetic manipulation to industrial microorganisms. It is to be hoped that 'whenever a fermentation process is being used for production purposes the search for improved strains should be regarded as an important aspect of the development programme' (Calam, 1970).

The methodology of mutant induction and selection has been outlined by Adler (1971) and presented in more detail by Calam (1970). Besides use of the more traditional agents of ionizing—and ultraviolet—radiations, increasing attention is being paid to the newer chemical mutagens, such as nitrous acid, base analogues, and in particular alkylating agents. Substances such as N-methyl-N′-nitro-N-nitrosoguanidine (NTG) can be most effective mutagens. For example, increases in mutant frequencies of 10^3–10^4, unaccompanied by extensive killing, have been obtained by phenethyl nitrogen mustards and ethylene-imino pyrimidines in *Aspergillus nidulans* and use of NTG can result in as many as 10% auxotrophic survivors.

Occasionally, mutagen specificity for a particular desired effect has been observed, as in the case of the mutagen diepoxybutane and double auxotrophy for adenine and inositol in *Neurospora crassa.* More generally, however, mutagen specificity has been absent as in a comparison of the effects of ultraviolet light and diepoxybutane in inducing auxotrophs and spore colour mutants in *Penicillium chrysogenum* and *Aspergillus nidulans* respectively. Because various mutagens *can* produce different effects, such as gross chromosomal changes (aberrations) by ionizing radiations compared with small structural changes in genes (point mutations) by ultraviolet light, it may be advisable to employ a range of mutagens during an improvement programme. However, it should be noted that some chromosome aberrations produced by ionizing radiations (*e.g.* translocated parts of chromosomes) may lead to problems with the stability of the mutant finally selected for production. Mutagen treatment is generally applied to either resting or germinating spores.

Selection of improved mutants is usually much more laborious than in the case of auxotrophic, colour, or resistant mutants. Efficient screening procedures are the most important factor and statistical consideration may need to be applied for optimal testing methods. In industry, testing is usually carried out in shake culture, a batch of which should include controls with a previously-typed strain for detection of any batch variations in production. The selection procedure becomes more critical during later stages of a programme when improvements are usually small than in the early stage when high enhancement levels may be obtained. Typically, the first round of screening involves only single and not replicate flasks, but a high proportion of isolates are re-tested later. The idea is to test a large number of isolates for approximate production rather than a smaller number more accurately. Calam (1970) describes an example from a programme for improved penicillin titre based upon statistical principles. Since the proportion of isolates showing a 5% increase was much higher than those giving a 10% increase, it was more efficient to attempt a 10% increase by two successive 5% increases. It was also more economical to perform successive mutations and screen a smaller number of isolates (*e.g.* 200) than to screen a larger number from a single mutagen treatment. Re-testing of the most promising 50 isolates was then carried out using four flasks each and the best five isolates selected for the next cycle of mutation and selection. These initial lines are subjected to several such cycles and eventually a suitable superior mutant for production is isolated. Such multi-stage programmes are similar to those used in plant-breeding and crop-selection. There may, however, be problems in scale-up, such as the limitations of highly aerated 5 l pilot fermenters compared with shake flasks in assaying penicillin titre improvement in *Penicillium chrysogenum* (Elander *et al.*, 1973).

Industrial mutants fall into two broad categories. There are those selected by extensive mutation-selection programmes, in which the performance of large numbers of mutagen-surviving organisms are tested in more-or-less random fashion. Stable superior mutants can usually be found but rarely is the genetical basis or indeed are the underlying biochemical mechanisms, of the enhanced characteristic uncovered. More clearly defined biochemical mutants may also be used. These may carry identified auxotrophic mutations and/or regulator gene mutations which interfere with the organism's normal control over synthesis. In many cases the biochemistry of the resulting phenotypic change is at least partially understood. To date, this latter class of mutants comprises largely bacteria, but it appears likely that industrial fungal mutants will increasingly be of this type. As stated by Demain (1972*a*): 'Much basic work on catabolite regulation must be done in the applied area if we are to utilise the recent findings of molecular biology. Experimentation with mutants resistant to catabolite repression as fermentation organisms would appear to be of primary importance'.

A selection of examples of mutants used in various industrial processes is listed in Table 4.2. For convenience, the remainder of this section is sub-divided according to the type of fermentation product.

ANTIBIOTICS Besides the outstanding achievement with *Penicillium chrysogenum*—namely, the increase in penicillin activity from 250 units cm^{-3} in 1945 via the legendary Wisconsin strain, Q176, to around $10-13 \times 10^3$ units cm^{-3} under optimal conditions today (McCann & Calam, 1972)—yields of many other antibiotics have also been enhanced by use of selected mutants. The vast majority, however, are mutants of Actinomycetes. Elander *et al.* (1973) describe the lineage of improved Wyeth strains of *P. chrysogenum* derived from Wisconsin strain Q176 by successive mutagenesis with ultraviolet light, methyl-bis-(β-chloroethyl) amine (NM), and diepoxybutane. All these strains conidiated excellently and showed a progressive capacity to synthesize penicillin G (benzylpenicillin) and penicillin V (phenoxymethylpenicillin). An overall improvement in penicillin G production of approximately 25-fold that of strain Q176 was obtained. Of the other filamentous fungi, mutants of *Cephalosporium acremonium* producing increased titres of cephalosporin C have been reported (Dennen & Carver, 1969).

Table 4.2 Some improved characteristics of industrial mutants

Organism	Product	Characteristic	Reference
Aspergillus cinnamomeus	amylases	yield	Kurushima *et al.*, 1972
A. niger	citric acid	yield	Siechertova & Leopold, 1969
Cephalosporium acremonium	cephalosporin C	yield	Dennen & Carver, 1969
Eremothecium ashbyii	riboflavin	yield	Stepanov & Zhdanov, 1972
Fusarium moniliforme	gibberellins	yield	Jeffreys, 1970
Fusarium moniliforme	gibberellins	ratio of components	Erokhina, 1969
Penicillium chrysogenum	penicillin	yield	Elander *et al.*, 1973

Other mutants have been used in attempting to identify some of the genes involved in antibiotic synthesis. By analysing titres of conidial isolates from heterokaryons of *Aspergillus nidulans*, Holt & MacDonald (1968*a*) showed that penicillin production is under the control of nuclear genes and is not cytoplasmically inherited. Ball (1973b) has isolated single-step, increased titre mutants which have been designated determinants (probably genes) *t1*, *t2*, *t5*. Some of these have been mapped by rigorous parasexual analysis, *t2* having been allocated to haploidization group (chromosome?) I, and *t5* tentatively to haploidization group III.

Only a few studies of the use of regulatory mutants in antibiotic biosynthesis have been reported. Goulden & Chattaway (1969) showed that valine feedback inhibition of acetohydroxy acid synthetase is reduced in an improved mutant of *Penicillium chrysogenum.* The intermediate acetolactate is apparently common to valine and penicillin syntheses. Similarly a mutation which prevented lysine inhibition of the precursor α-aminoadipic acid (Kinoshita, 1971) might well lead to increased penicillin synthesis. In *Cephalosporium acremonium*, increased cephalosporin titre in a synthetic medium has been obtained using a mutant incapable of assimilating sulphate (Nüesch, Treichler & Liersch, 1973). Methionine stimulates synthesis of cephalosporin and whereas growth of the wild-type strain was inhibited by high concentrations of DL-methionine, the sulphate mutant was not. It has been proposed that sulphite arising from conversion of sulphate inhibits formation of a methionine permease and that this inhibition is therefore absent in the mutant strain. Many antibiotics may in fact regulate their own synthesis by feedback mechanisms. If so, mutants in which feedback inhibition or repression is absent or reduced could produce higher yields. It may also prove possible to isolate mutants with enhanced abilities to carry out structural transformations of antibiotics (Perlman, 1971).

VITAMINS, AMINO-ACIDS AND EDIBLE PROTEINS Riboflavin for animal feeds is produced by fermentations using the ascomycetes *Eremothecium ashbyii* and *Ashbya gossypii*. The isolation of induced superior mutants of *E. ashbyii* has recently been reported by Soviet workers (Stepanov & Zhdanov, 1972). Enhanced yields have also been obtained in Japan using a caffeine-resistant mutant of *Eremothecium* (Yamada, Nakahara & Fukui, 1971). This increase is thought to be due to a reduction of inhibition of some intermediate of the riboflavin pathway by xanthine, of which caffeine is an analogue. Improved mutants of *A. gossypii* are reported by Demain (1972*b*). In his review of 'over-production' of riboflavin by these yeast-like moulds, he suggests that they might be genetically constitutive for pathway enzymes or that, more likely, synthesis of the riboflavin repressor is inhibited under the fermentation conditions practised.

Compared with bacteria, fungi are little used commercially for production of amino-acids. Yet it is in this area that auxotrophic, analogue-resistant, and regulatory mutants really come into their own. Many examples of amino-acid overproducing mutants which are derepressed, or have feedback-resistant enzymes, or may be prototrophic revertants (having only wild-type requirements) are discussed by Demain (1972*a*). In fungi, there are only examples from yeasts to be cited. Ethionine-resistant or N-acetyl-1-norleucine-resistant mutants of *Hanseniaspora valbyenis* accumulate methionine (Kinoshita, 1971) and mutants of *Hansenula anomala* are used for industrial production of tryptophan (Hütter, 1971). The latter are resistant to high levels of the precursor anthranilic acid which can then be fed in high concentrations since feedback inhibition by tryptophan acts upon anthranilate synthetase.

Increasing attention is being given to the production of edible protein by micro-organisms. Filamentous fungi such as *Fusarium* sp. can be grown on waste-products of the food carbohydrate industries when supplemented with inorganic nitrogen and phosphates and can provide protein of suitable amino-acid composition (Courts, 1973). Strains of *Aspergillus, Penicillium* and *Rhizopus* have also been cited as suitable for protein production (Kihlberg, 1972), and their protein has the advantage of adequate methionine content. Superior mutant strains apparently have not yet been developed. However, Heden & Molin (1971) have suggested that 'production of genetically improved starter organisms for use both in the food industry and for distribution in developing countries will probably become a new task for the development work in some pilot plants'. And among criteria for selecting a micro-organism for food protein, Bhattacharjee (1970) lists 'known genetic and physiological properties and ability to improve genetically'.

Certainly there should be considerable scope for manipulating the amino-acid profiles of proteins by mutation. This principle can be illustrated by an example from yeast. A mutant strain of *Candida utilis* in which protein methionine content is increased by 24% has been isolated by Okanishi & Gregory (1970). It has been suggested that these mutants may be derepressed for synthesis of a methionine-rich protein. It should also be possible to isolate mutants which show optimal utilization of the substrate in question and maximal production of the desired protein composition. Continuous culture, with its ability to select mutants with the most efficient substrate-utilization, should play an important part in such developments. Mutants reduced in the proportion of non-proteinous material, such as cell wall carbohydrate, should also prove worthy of search.

ORGANIC ACIDS AND ENZYMES The organic acid produced in greatest amounts by fungal fermentations is citric acid, when the most frequently used organism is *Aspergillus niger*. Seichertová & Leopold (1969*a*) present results of test fermentations with the 45 most active mutants induced by ultraviolet irradiation of four production strains of *A. niger*. The maximum improvements in yields were 11% in one strain and 18% in another.

There have been several earlier reports of enhanced yields of amylases by mutants of *Aspergillus* species, notably *A. oryzae*, although some more recent attempts have been less successful (Meyrath, Bahn & Han, 1971). Although screening tests following exposure to ultraviolet or ionizing radiation indicated many potential improved mutants, production tests failed to show more than small increases in yields of α-amylase. At least one more cycle of mutation and selection might have led to more significant improvements in this instance. By using four cycles of ultraviolet-induced mutation and selection, Kurushima, Fujii & Kitahara (1972) obtained a mutant of *Aspergillus cinnamomeus* exhibiting a six-fold increase in production of acid-stable α-amylase. The industrial production of fungal enzymes would seem to be an area ripe for exploitation by the use of regulatory mutants such as those constitutive for the enzyme in question. Such a development ought to see an important role for mutant strains in the production of fungal amylases, hydrolases and proteases.

MISCELLANEOUS PRODUCTS AND ECOLOGICAL ASPECTS Induced mutants have been used to enhance yields of various gibberellins produced by strains of *Fusarium moniliforme* (*Gibberella fujikuroi*. Breakdown of the higher mutant activities into gibberellin A_3 (gibberellic acid) and gibberellin A_{13} is given by Jefferys (1970) in his review of gibberellin fermentations. Besides these mutants providing an overall increase in yield, other induced mutants have been selected for their production of component gibberellins in different proportion to the parent strain (Erokhina, 1969). Some mutants for instance produce only gibberellins A_1 and A_3. Induced mutants of *Claviceps paspali* enhanced in ergot alkaloid production have been reported by Kelleher (1969).

The future may see other areas, some ecological, where fungal mutants prove superior to existing strains. For instance, fungi such as *Rhizopus* and *Fusarium* can be used in pollutant fermentations with the double purpose of utilizing undesirable effluent substances and producing some useful product (McLoughlin, 1972). An example of this is production of fumaric acid from spent sulphite liquor by *Rhizopus* sp. Perhaps utilization of mutant strains could even spread to fungal control of insect pests. *Beauvaria bassinia* and *Metarrhizium anisophiae* are used as spore powders or dusts but their main disadvantages are their slow invasion of the insect body and their occasional cause of allergic reaction in man. Conceivably it should be possible to isolate mutants of these fungi which grow faster in insects and to which man might possibly be less allergic.

Recombination and selection

Mutation itself represents only the foundation stone of genetic variability. The myriad of different genetic structures (genotypes) which may be built upon this foundation is the result of genetic recombination. The number of possible genotypes from different combinations of a large number of alternative genes (alleles) is very great indeed. Natural selection can of course operate upon new genomes produced by recombination as well as by mutation.

The application of hybridization and recombination techniques to industrial strains of filamentous fungi has proved considerably more difficult than anticipated. Unfortunately, information on results obtained in industrial laboratories is not abundant. Pathak & Elander (1971) report the construction of a vigorous diploid strain (WC-9) of *Penicillium chrysogenum*. This diploid effectively represents genome-duplication of the haploid production strain rather than a true hybridization of different types. Nevertheless, it displays significantly higher glucose-utilization, respiration, and enzyme activity than the parent haploid. Other true, parasexual diploids are efficient producers of penicillin G and penicillin V (Elander *et al.*, 1973).

Table 4.3 Some organisms and products for which yield improvement has been attempted or realized by recombination and selection

Organism	Product	Reference
Aspergillus niger	citric acid	Seichertova & Leopold, 1971
A. oryzae	proteases	Ikeda *et al.*, 1957
Cephalosporium acremonium	cephalosporin	Nüesch *et al.*, 1973
Emericellopsis terricola	synnematin B	Fantani, 1962
Fusarium moniliforme	gibberellins	Calam *et al.*, 1973
Penicillium chrysogenum	penicillin	Ball, 1973*b*; Elander *et al.*, 1973
P. patulum	griseofulvin	Calam *et al.*, 1973

Although potential improvement by parasexual breeding of *Penicillium chrysogenum* is simple in theory, difficulties have been encountered in practice. MacDonald and co-workers (for references, see MacDonald, 1968) found that haploid segregants from a diploid cross were predominantly of parental genotype, thus hindering isolation of haploid recombinant strains. This result was probably due to structural differences in the chromosomes of the two remotely related parent strains used. However, the use of more closely related parent strains, and of haploidization induced by *p*-fluorophenylalanine (Ball, 1971) help to solve this problem and Ball (1973*b*) has succeeded in obtaining haploid recombinants producing increased titres of penicillin (Table 4.3).

Formal genetic analysis has also been undertaken in *Penicillium chrysogenum* (see page 67). The genes *t2* and *t5*, which increase titre in a parental strain yielding 3000 units cm^{-3}, recombine (*i.e.* giving strain *t2*, *t5*)

to produce 8000 units cm^{-3}. Genetic analysis of genes involved in penicillin synthesis in *Aspergillus nidulans* has also been performed. The latter has the advantages of a great deal of background genetic knowledge being available and of easier genetic analysis whether by meiotic or parasexual means, than with *P. chrysogenum*. Penicillin titres, however, are very much lower in *A. nidulans*, amounting to only 6 units cm^{-3} in wild-type strains. Several mutant genes which confer higher titres have been identified by MacDonald and co-workers. For example, the recessive gene *pen-01*, which has been mapped on chromosome 8 (*ribo 2* linkage group), increases penicillin titre to 20 units cm^{-3} (Holt & MacDonald, 1968*b*). A few recombinant strains from crosses of *pen-01* to wild-type have given titres up to 40 units cm^{-3}. Recently, mutant gene *pen-02* has been shown to be dominant over its wild-type allele and gene *pen-03* semi-dominant (MacDonald, Holt & Ditchburn, 1973). The former gene has been allocated to chromosome 4 (*pyro 4* linkage group) and the latter may be linked to *gal 1* on chromosome 3. Recombinant strains carrying different pair combinations of the genes *pen-01*, *pen-02* and *pen-03*, do not, however, produce higher titres than does the *pen-01* mutant itself (20 units cm^{-3}). The action of these enhanced-titre genes therefore does not appear to be additive.

Following upon this work, Merrick & Caten (1973) are applying some of the techniques used when breeding for quantitative characteristics in higher organisms. The inheritance of penicillin titre in *Aspergillus nidulans* shows continuous variation typical of polygenic control, and genetically different selection lines have been established. As suspected, many genes contribute to antibiotic synthesis by a particular strain and in this case, apparently most of these polygenes act in an additive fashion.

Genetic analysis of citric acid synthesis by *Aspergillus niger* has been conducted in Poland (Ilczuk, 1971). This has involved isolation of prototrophic heterokaryons from auxotrophic mutants of production strains and further selection of diploids from these heterokaryons. Although heterokaryons showed no increases in yield over haploid production strains, all diploids derived from high yield parents gave enhanced activities ranging from 8% to 18% in excess of the parent production strains.

In some respects, these results contrast with those obtained with *Aspergillus oryzae* in attempts to improve kojic acid production and protease activity (Ikeda *et al.*, 1957). In this instance, some heterokaryon strains gave high yields of kojic acid yet all diploid strains derived from heterokaryons were weak-producers. However, one presumed-haploid segregant of a diploid strain produced 40% more kojic acid than the higher yielding of the two parent strains. In these experiments, strains thought to be polyploid were also constructed by hybridization. With respect to protease activity, diploid strains were on this occasion superior to their haploid parents and an assumed-tetraploid hybrid exhibited the highest protease activity of all strains tested.

Heterokaryons have been difficult to synthesize in *Cephalosporium acremonium* and those occasionally obtained are usually unstable (Nüesch *et al.*, 1973). Stable diploid strains have also been difficult to isolate and often these conidiate either poorly or not at all. However, mitotic segregation readily occurs in the diploids obtained and can also be induced by *p*-fluorophenylalanine.

Some *Emericellopsis* species which produce the penicillin, synnematin B, have a sexual stage. In *E. terricola* var. *glabra*, Fantani (1962) isolated some recombinants from sexual crosses with improvement in activity of up to 50%. In *E. salmosynnemata*, heterokaryons and parasexual diploids have been made and some parasexual recombinants produce enhanced yields.

Recombination experiments with *Penicillium patulum* and *Fusarium moniliforme* have produced results often difficult to interpret (Calam, Daglish & Gaitskell, 1973). Production cultures of *P. patulum* can apparently hybridize to provide presumptive diploids but it is not clear that these are the result of an orthodox parasexual process. Although stable, their spore size is smaller than that expected of diploids, and often may be no greater than the haploid parents. More conventional, presumptive diploids are obtained with *F. moniliforme* but, here again, other diploids producing small spores may arise. Some of these gave enhanced yields of gibberellic acid compared with already high-yielding parent strains.

In contrast with some bacteria and Actinomycetes, recombination by means of transformation (transfer of naked genes without cell fusion) and transduction (transfer of genes by viruses acting as agents) has not yet been found in the fungi. Virus-like particles have, however, been observed in *P. chrysogenum* (Banks *et al.*, 1970) and these may be transferable to other fungi (Lhoas, 1972). We may yet therefore see virus-mediated recombination contributing to the construction of improved strains of industrial fungi.

4.4 Strain Stability

Esser (1971) makes the following statement: 'From a geneticist's point of view one of the main problems of research involved in fungal fermentation seems to be the breeding of suitable strains with optimal effectiveness and the possibilities to keep these characters constant over a long range of vegetative propagation (*i.e.* stable strains)'. Having discussed the topic of strain improvement in the previous section, attention is now turned to the subject matter of the latter part of this statement.

It should be emphasized at this point that the genetic stability of a strain is determined not just by its genotype. Like most other phenotypic characteristics, the degree of genotypic expression is dependent upon environment. Thus the genetic stability of a strain may vary with the particular environmental conditions involved. For instance, the level of stability may be temperature-dependent or growth-rate mutations may be expressed only in a stress environment (Wills, 1968). In fact, from the view point of selective fitness, it can be argued that too low a level of mutability is a disadvantage to a strain and that selection favours a constant level of mutability (Zamenhof, Heldenmuth & Zamenhof, 1966). It should also be stressed that genetic stability is a *quantitative* characteristic, the inheritance of which will often be polygenic. For convenience, the remainder of this section will be sub-divided into (1) strain maintenance, (2) strain degeneration, and (3) population dynamics. These topics are of course closely interrelated, and to some extent the terms themselves are interchangeable.

Strain maintenance

In using this term we are usually referring to *storage* of organisms and, more precisely, to preservation of some of their particular characteristics which are considered important. As well as the more obvious hazards of death and contamination, there is genetic change during storage to be considered. There are many varied ways of maintaining strains but probably the best method depends upon both the organism and particular characteristics being considered. For example, recently reported methods include the use of glass beads upon which the organisms are deposited before freeze-drying (Nagel & Kunz, 1972) and storage at room temperature in the dark, thereby eliminating the need for cold storage or freeze-drying (Antheunisse, 1972). For some characteristics, avoidance of these latter procedures, often thought of as precautions, may in fact be essential. An illustration is the case of some nystatin-resistant mutants of yeast which die rapidly in the refrigerator but store well at room temperature (Karunakaran & Johnston, 1974). Storage of cultures of a high-yielding mutant of *Penicillium chrysogenum* at −196°C for 3–5 years was characterized by higher survival and less yield decay compared with storage at 4°C over the same period (MacDonald, 1972).

Strain degeneration

In this section, deleterious changes during *use* of a strain, but without population analysis being performed, will be considered. In mentioning the problem of strain instability in continuous culture, Righelato & Elsworth (1970) remark that 'indeed degeneration has proved a problem in antibiotic and vaccine-producing cultures'. As Sermonti (1969) points out, however, the word degeneration is used with different meanings, some of which can be contradictory. This is because, as with the word improvement, it depends whether degeneration of the strain *itself* or of its performance from *our* point of view is being considered. The *strain* degenerates when it suffers genetic changes which negatively affect its growth or sporulation. Sermonti (1969) suggests the term yield decay for decreasing product-formation by a strain, particularly when related to antibiotic product. For, in fact, this deterioration from *our* point of view may concur with an *increase* in the strain's vigour (growth and/or sporulating ability). Thus the vulnerability of many an improved mutant (*e.g.* for enhanced yields) is its pleiotropism for reduced vigour which can result in lower-producing revertants having a selective advantage during culture.

Sermonti (1969) gives examples of both strain degeneration during selection of yield-improvement mutants and yield decay in strains of *Penicillium chrysogenum*. Since selected mutants often have reduced vigour expressed in their lower linear growth rate and poorer conidiation, titre-reduction is best explained by the appearance and selective advantage of more vigorous back-mutants with inferior penicillin synthesis. To counteract the latter process, the original high-producing strain must be frequently re-isolated. Sometimes, however, even freshly isolated high-activity cultures progressively deteriorate and no preventive measures can be taken.

Such cases have generally been attributed to senescence. Difficulties in maintaining levels of synnematin with strains of *Emericellopsis* (Fantani, 1962) lysergic acid alkaloid production with strains of *Claviceps* (Kelleher, 1969), and riboflavin with a strain of *Ashbya gossypii* (Obermann, H., personal communication) have also been reported.

The term senescence describes progressive strain deterioration, affecting characteristics such as growth, conidiation, and perithecial formation, which eventually culminates in death. The phenomenon has been studied principally in *Aspergillus glaucus*, and *Podospora anserina* by Jinks and others (see Jinks, 1963). Although some forms of degeneration are reversible, true senescence, characterized by the spread of degeneration through hyphae as if by infection, is irreversible. Such a trait is strain-dependent and apparently under genetic control. When exhibited, its appearance may initially be due to a gene mutation but its inheritance is extrachromosomal and dominant. In most instances, only microconidia, which have a minimal amount of cytoplasm, do not transmit the condition. To explain the infective nature of senescence in filamentous fungi, it has been suggested that the inherited determinant is an episome or a latent virus which can switch to a vegetative state and initiate the ageing process. Similar ideas have been proposed to explain the strain degeneration often accompanying selected yield-improvement in *Penicillium* spp. (Sermonti, 1969).

A different type of strain instability may be due to an alteration in chromosome number or chromosome structure. Thus Roper (1971) has studied the chromosomal mechanisms which result in formation of vegetative variants in strains of *Aspergillus nidulans* carrying duplications and/or translocations. He proposes that an imbalance among chromosome segments can affect the fidelity of chromosome replication so that a daughter nucleus may differ in genotype from the parent nucleus. The term mitotic non-conformity has been suggested to describe this form of instability due to chromosome aberrations.

Ball (1973*a*) discusses a similar phenomenon in production strains of *Penicillium chrysogenum*. A chromosomal determinant of instability has been identified and allocated to a particular linkage group. Part of a chromosome present in duplicate in this essentially haploid strain (*i.e.* partial chromosome duplication) may cause the unstable condition. In crosses to morphologically stable haploids, this type of instability is dominant but stable diploids can be recovered as colony sectors. Such instability would seem to be a limitation in attempting to increase yields by constructing unbalanced genome strains, *i.e.* carrying several copies of increased-titre genes in an otherwise haploid genome. Even with 'balanced-genome' strains, filamentous fungi may show greater instability in the diploid or triploid than in the haploid condition (Ball, 1973*a*). Spontaneous frequencies of aneuploidy from diploid, and particularly from triploid, strains are often much higher than in the case of hybrid yeast strains.

Population dynamics

This area is, in effect, the study of microbial population genetics. To date, however, there have been few instances of microbial geneticists and population geneticists joining forces to tackle problems in this area which, after all, should be a natural meeting ground for them. Population geneticists have rarely turned their attention to micro-organisms and, with the exception of pioneering work during the Fifties, it has not been fashionable for microbiologists, in particular microbial geneticists, to analyse culture populations.

Such efforts as have been made have involved primarily bacteria. Thus, shortly after inventing the chemostat, Novick, Szilard and collaborators began applying it to the frequency of mutants in cultures and the measurement of bacterial mutation rates (for example, Novick, 1955). These early studies were extended by Ryan and co-workers to include analysis of population kinetics, with the introduction of such concepts as periodic selection. Other aspects of microbial evolution, namely the dependence of bacterial selection kinetics upon environmental factors, and the long-term genetic interactions between *Escherichia coli* and T phages have also been investigated (Zamenhof & Eichhon, 1967; Horne, 1970). Some features of bacterial mixed-culture dynamics, in particular competition between two organisms for a single growth-limiting substrate have recently been discussed by Veldkamp & Jannasch (1972).

Although the analysis of population changes could be important in any experiments involving considerable amounts of microbial growth, such as fermentations where successive inoculations are by serial transfer, it is particularly applicable to processes including *continuous* cultivation. Certainly it has been argued that establishment of mutants in most continuous cultures is unlikely and some experiments with bacteria have failed to detect any differences in phenotypes even after months of culture. While a high degree of stability may generally be true for wild-type bacterial cultures, however, a quite different state of affairs may exist for cultures of improved mutants, or of fungi. Unfortunately, although there has been a steady increase in the use of continuous culture techniques in microbial research, the organisms used only rarely include the filamentous fungi. The theory of continuous cultivation has been developed for unicellular organisms and, in practice, continuous culture of mycelial fungi may be handicapped or thwarted by blocked tubes or valves, growth on the vessel wall and other surfaces, and a lack of homogeneity throughout the culture (*e.g.* because of pellet growth). Nevertheless, short periods of continuous cultivation of *Penicillium chrysogenum* (for references see Pirt & Righelato, 1967) and *Aspergillus nidulans* (Bainbridge *et al.*, 1971) have been achieved. Technical improvements for the continuous growth of *P. chrysogenum* have recently been developed and, with these, steady states lasting up to three weeks obtained (Brunner & Röhr, 1972).

Although no analysis of population changes has been reported for the filamentous fungi, results obtained with yeast seem pertinent to any extended application of continuous culture techniques to the mycelial fungi. In using certain mutant strains of *Saccharomyces cerevisiae*, wild-type revertant cells are strongly selected for, and replace the original mutant cells within a few weeks of continuous culture. This result has obvious implications for mutant production strains of inferior vigour to the wild-type. A less predictable result is the similar, if slower, selection of homozygous mitotic recombinants over the parental heterozygous diploid cells. Such selection may restrict or prohibit the use of filamentous diploid strains in prolonged continuous processes.

4.5 References

ADLER, H.I. (1971). Techniques for the development of novel micro-organisms. In *Radiation and radioisotopes for industrial micro-organisms*, pp. 241–9. International Atomic Energy Agency, Vienna.

ALAČEVÍC, M. (1972). Importance of microbial genetics in the improvement of the productivity of micro-organisms in industry. *Acta Biologica Jugoslavia Ser Genetica* **4**, 119–27.

ALIKHANIAN, S.I. (1972). Applied problems of micro-organism genetics. *Isitologia & Genetika* **6** (1), 3–12.

ANTHEUNISSE, J. (1972). Preservation of micro-organisms. *Antonie van Leeuwenhoek* **38**, 617–22.

BAINBRIDGE, B.W., BULL, A.T., PIRT, S.J., ROWLEY, B.I. & TRINCI, A.P.J. (1971). Biochemical and structural changes in non-growing maintained and autolysing cultures of *Aspergillus nidulans*. *Transactions of the British Mycological Society* **56**, 371–85.

BALL, C. (1971). Haploidization analysis in *Penicillium chrysogenum*. *Journal of General Microbiology* **66**, 63–9.

BALL, C. (1973*a*). The genetics of *Penicillium chrysogenum*. *Progress in Industrial Microbiology* **11**, 12, 47–72.

BALL, C. (1973*b*). Improvement of penicillin productivity in *Penicillium chrysogenum* by recombination. In *Genetics of industrial micro-organisms*, pp. 227–37. Edited by Vaněk, Z., Hošťálek, Z. & Cudlín, J. Prague: Academia.

BANKS, G.T., BUCK, K.W., CHAIN, E.B., DARBYSHIRE, J.E., HEMMELWEIT, F., RATTI, G., SHARPE, T.J. & PLANTEROSE, D.N. (1970). Antiviral activity of double stranded RNA from a virus isolated from *Aspergillus foetidus*. *Nature, London* **227** 505–7.

BHATTARCHARJEE, J.K. (1970). Microorganisms as potential sources of food. *Advances in Applied Microbiology* **13**, 139–61.

BRUNNER, H. & RÖHR, M. (1972). Novel system for improved control of filamentous micro-organisms in continuous culture. *Applied Microbiology* **24**, 521–3.

CALAM, C.T. (1970). Improvement of microorganisms by mutation, hybridization and selection. In *Methods in microbiology*, Volume 3A, pp. 435–59. London & New York, Academic Press.

CALAM, C.T. (1972). Genetics and development of industrial strains. *Process Biochemistry* **7** (7), 29–31.

CALAM, C.T., DAGLISH, L.B. & GAITSKELL, W.S. (1973). Hybridization experiments with *Penicillium patulum* and *Fusarium moniliforme*. In *Genetics of industrial microorganisms*, pp. 265–82. Edited by Vaněk, Z., Hošťálek, Z., & Cudlín, J. Prague: Academia.

COJOCARU, C. & MILENA, S. (1967). The use of moulds for obtaining amylolytic enzymatic preparations used in the spirit industry. *Lucrarile Institutului de Cercetari Alimentare* **8**, 249–61.

COURTS, A. (1973). Recent advances in protein production. *Process Biochemistry* **8** (2), 31–3.

DEMAIN, A.L. (1972*a*). Cellular and environmental factors affecting synthesis and excretion of metabolites. *Journal of Applied Chemistry and Biotechnology* **22**, 345–62.

DEMAIN, A.L. (1972*b*). Riboflavin oversynthesis. *Annual Review of Microbiology* **26**, 369–88.

DENNEN, D.W. & CARVER, D.D. (1969). Sulfatase regulation and antibiotic synthesis in *Cephalosporium acremonium*. *Canadian Journal of Microbiology* **15**, 175–81.

ELANDER, R.P., ESPENSHADE, M.A., PATHAK, S.C. & PAN, C.H (1973). The use of parasexual genetics in an industrial strain improvement programme with *Penicillium chrysogenum*. In *Genetics of industrial micro-organisms*, pp. 239–53. Edited by Vaněk, Z., Hošťálek, Z. & Cudlín, J. Prague: Academia.

EROKHINA, L.I. (1969). Some characteristics of mutants of *Fusarium moniliforme*. *Genetika* **5** (11), 143–7.

ESSER, K. (1971). Application and importance of fungal genetics for industrial research. In *Radiation and radioisotopes for industrial micro-organisms*, pp. 83–91. International Atomic Energy Agency, Vienna.

FANTANI, A.A. (1962). Genetics and antibiotic production of *Emericellopsis* species. *Genetics* **47**, 161–77.

FINCHAM, J.R.S. & DAY, P.R. (1971). *Fungal genetics*. Oxford and Edinburgh: Blackwell Scientific Publications.

FUKUMOTO, J. & SHICHIJI, S. (1971). Industrial production of enzymes. In *Biochemical and industrial aspects of fermentation*, pp. 91–117. Edited by Sakaguchi, K., Uemura, T. & Kinoshita, S. Tokyo: Kodansha Ltd.

GOULDEN, S.A. & CHATTAWAY, F.W. (1969). End-product control of acetohydroxy acid synthetase by valine in *Penicillium chrysogenum* Q176 and a high penicillin-yielding mutant. *Journal of General

Microbiology **59**, 111–18.

HANDLEY, L. & CATEN, C.E. (1973). Vegetative death: a mitochondrial mutation in *Aspergillus amsteladami* (abstract). *Heredity* **31**, 136.

HEDÉN, C.G. & MOLIN, N. (1971). The productivity of micro-organisms—a catalytic factor for research and transdisciplinary co-operation. *Symposium Society for General Microbiology* **22**, 1–14.

HOLT, G. & MACDONALD, K.D. (1968*a*). Penicillin production and its mode of inheritance in *Aspergillus nidulans*. *Antonie van Leeuwenhoek* **34**, 409–16.

HOLT, G. & MACDONALD, K.D. (1968*b*). Isolation of strains with increased penicillin yield after hybridization in *Aspergillus nidulans*. *Nature, London* **219**, 636–7.

HORNE, M.T. (1970). Co-evolution of *Escherichia coli* and bacteriophages in chemostat culture. *Science* **168**, 992–3.

HÜTTER, R. (1971). Regulation and amino-acid biosynthesis and industrial production of amino acids. In *Radiation and radioisotopes for industrial micro-organisms*, pp. 169–79. International Atomic Energy Agency, Vienna.

IKEDA, Y., NAKAMURA, K., UCHIDA, K. & ISHITANI, C. (1957). Two attempts upon improving an industrial strain of *Aspergillus oryzae* through somatic recombination and polyploidization. *Journal of General and Applied Microbiology* **3**, 93–101.

ILCZUK, Z. (1971). Genetics of citric acid producing strains of *Aspergillus niger*. IV. Citric acid synthesis by heterozygous diploids of *Aspergillus niger*. *Nahrung Chemistry, Biochemistry, Mikrobiology, and Technology* **15**, 381–8.

JEFFERYS, E.G. (1970). The gibberellin fermentation. *Advances in Applied Microbiology* **13**, 283–316.

JINKS, J.L. (1963). Cytoplasmic inheritance in fungi. In *Methodology in Basic Genetics*, pp. 325–54. Edited by Burdett, W.J. San Francisco: Holden-Day.

JINKS, J.L. (1964). *Extrachromosomal inheritance*. New Jersey: Prentice-Hall Incorporated.

KEAY, L. (1971). Microbial proteases. *Process Biochemistry* **6** (8), 17–21.

KELLEHER, W.J. (1969). Ergot alkaloid fermentations. *Advances in Applied Microbiology* **11**, 211–44.

KIHLBERG, R. (1972). The microbe as a source of food. *Annual Review of Microbiology* **26**, 427–66.

KINOSHITA, S. (1971). Metabolic control in the modern fermentation industry. In *Biochemical and industrial aspects of fermentation*, pp. 9–35. Edited by Sakaguchi, K., Uemura, T. & Kinoshita, S., Tokyo: Kodansha Limited.

KURUSHIMA, M., FUJII, T. & KITAHARA, K. (1972). Enzymatical and morphological variations by ultra-violet irradiation (studies on the fungal amylases Part II) (abstract). *Agricultural and Biological Chemistry* **36**, A26.

LHOAS, P. (1967). Genetic analysis by means of the parasexual cycle in *Aspergillus niger*. *Genetical Research* **10**, 45–61.

LHOAS, P. (1972). Mating pairs of *Saccharomyces cerevisiae* infected with double stranded RNA viruses from *Aspergillus niger*. *Nature, London* **236** 86–7.

MCCANN, E.P. & CALAM, C.T. (1972). The metabolism of *Penicillium chrysogenum* and the production of penicillin using a high yielding strain at different temperatures. *Journal of Applied Chemistry and Biotechnology* **22**, 1201–8.

MACDONALD, K.D. (1968). The persistence of parental genome segregation in *Penicillium chrysogenum* after nitrogen mustard treatment. *Mutation Research* **5**, 302–5.

MACDONALD, K.D. (1972). Storage of conidia of *Penicillium chrysogenum* in liquid nitrogen. *Applied Microbiology* **23**, 990–3.

MACDONALD, K.D., HOLT, G. & DITCHBURN, P. (1973). The genetics of penicillin production in *Aspergillus nidulans* (abstract). *Heredity*, **31**, 131–2.

MCLOUGHLIN, A.J. (1972). Fermentation of pollutants. *Process Biochemistry* **7** (1), 27–39.

MEHROTRA, B.S. & KRISHNA, N. (1966). Amylase from Indian strains of moulds. In *Annual National Academy of Sciences of India*, pp. 86 (abstract).

MERRICK, M. & CATEN, C.E. (1973). Hybridization and selection for penicillin production in *Aspergillus nidulans*—a biometrical approach. *Journal of Applied Chemistry and Biotechnology* **23**, 705–6.

MEYRATH, J., BAHN, M. & HAN, H.E. (1971). Induction of amylase-producing mutants in *Aspergillus oryzae* by different irradiations. In *Radiation and radioisotopes for industrial micro-organisms*, pp. 137–55. International Atomic Energy Agency, Vienna.

NAGEL, J.G. & KUNZ, L.J. (1972). Simplified storage and retrieval of stock cultures. *Applied Microbiology* **23**, 837–8.

NOVICK, A. (1955). Growth of bacteria. *Annual Review of Microbiology* **9**, 97–110.

NÜESCH, J., TREICHLER, H.J. & LIERSCH, M. (1973). The biosynthesis of cephalosporin C. In *Genetics of industrial micro-organisms*, pp. 309–34. Edited by Vaněk,

Z., Hošťálek, Z. and Cudlín, J. Prague: Academia.

OKANISHI, M. & GREGORY, K.F. (1970). Isolation of mutants of *Candida tropicalis* with increased methionine content. *Canadian Journal of Microbiology* **16**, 1139–43.

PATHAK, S.G. & ELANDER, R.P. (1971). Biochemical properties of haploid and diploid strains of *Penicillium chrysogenum. Applied Microbiology* **22**, 366–71.

PERLMAN, D. (1971). Microbial transformations of antibiotics. *Process Biochemistry* **6** (7), 13–14.

PIRT, S.J. (1969). Microbial growth and product formation. *Symposium Society for General Microbiology* **19**, 199–221.

PIRT, S.J. & RIGHELATO, R.C. (1967). Effect of growth rate on the synthesis of penicillin by *Penicillium chrysogenum* in batch and chemostat cultures. *Applied Microbiology* **15**, 1284–90.

RIGHELATO, R.C. & ELSWORTH, R. (1970). Industrial applications of continuous culture: pharmaceutical products and other products and processes. *Advances in Applied Microbiology* **13**, 399–417.

ROPER, J.A. (1971). Vegetative instability in fungi. The role of chromosome aberrations. In *Radiation and radioisotopes for industrial micro-organisms*, pp. 113–21. International Atomic Energy Agency, Vienna.

SCAMMELL, G. (1973). Selection problems in the biomass industry. *Journal of Applied Chemistry and Biotechnology*, **23**, 710.

SEICHERTOVÁ, O. & LEOPOLD, H. (1969*a*). Die Aktivierung von Stammen des *Aspergillus niger* I. Die Bildung von Mutanten durch wiederholte Bestrahlung der Stamme mittels UV-Licht. *Zentralblatt für Bakteriologie, Parasitenkunde, Infektionskrankheiten und Hygiene, Abt 11* **123**, 558–63.

SEICHERTOVÁ, O. & LEOPOLD, H. (1969*b*). Die Aktivierung von Stammen des *Aspergillus niger* II. Die Benutzung der parasexuellen Hybridisation. *Zentralblatt für Bakteriologie, Parasitenkunde, Infektionskrankheiten und Hygiene, Abt 11* **123**, 564–70.

SERMONTI, G. (1969). *Genetics of antibiotic-producing micro-organisms.* London and New York: Wiley-Interscience.

STEPANOV, A.I. & ZHDANOV, V.G. (1972). Use of mutagenic agents for breeding of *Eremothecium ashbyii*, the producer of riboflavin. *Genetika* **8** (6), 86–91.

VANĚK, Z., HOŠŤÁLEK, Z. & CUDLÍN, J. (Eds.) (1973). *Genetics of industrial micro-organisms. Actinomycetes and fungi.* Prague: Academia.

VELDKAMP, V. & JANNASCH, H.W. (1972). Mixed culture studies with the chemostat. *Journal of Applied Chemistry and Biotechnology* **22**, 105–23.

WALDRON, C. (1973). Cold-sensitive mutants of *Aspergillus nidulans* (abstract). *Heredity*, **31**, 135.

WILLS, C. (1968). Yeast partial dominance: effect of environment and background genotype. *Science* **160**, 549–50.

YAMADA, K., NAKAHARA, T. & FUKUI, S. (1971). Petroleum microbiology and vitamin production. In *Biochemical and industrial aspects of fermentation*, pp. 61–90. Edited by Sakaguchi, K., Uemura, T. & Kinoshita, S. Tokyo: Kodansha Ltd.

ZAMENHOF, S. & EICHHORN, H.H. (1967). Study of microbial evolution through loss of biosynthetic functions: establishment of 'defective' mutants. *Nature, London* **216**, 456–8.

ZAMENHOF, S., HELDENMUTH, L.H. & ZAMENHOF, P.J. (1966). Studies on mechanisms for the maintenance of constant mutability: mutability and the resistance to mutagens. *Proceedings of National Academy of Sciences, U.S.A.* **55**, 50–8.

Additional References

KARUNAKARAN, V. & JOHNSTON, J.R. (1974). Death of nystatin-resistant mutant of *Saccharomyces cerevisiae* during refrigeration. *Journal of General Microbiology*, **81**, 255–6.

CHAPTER 5

Growth Kinetics of Mycelial Fungi

R. C. RIGHELATO

5.1 Introduction

Studies of the growth physiology of micro-organisms have, for the most part, been concerned with bacteria and to a lesser extent yeasts. Mycelial fungi, for sound practical reasons, have been relatively neglected. Fungi grow slowly by comparison with many bacteria so experiments may take a long time to complete; the morphology of their growth form, usually as long filaments, as mats or as pellets, makes them rather difficult to handle and introduces further problems of heterogeneity. Moreover, they are prone to differentiate from the vegetative form to sporing forms which sometimes involve multicellular structures of considerable complexity. However, the economic and social importance of fungi as sources of food and biologically active metabolites has stimulated considerable interest in their growth physiology. A pre-requisite of most quantitative studies is an understanding of the way in which new mycelium is produced, at least insofar as this enables the experimenter to describe and predict changes in the type and concentration of the experimental organism. The kinetics dealt with here are intended primarily as an aid in the design of experiments that involve the growth of fungi, and in the interpretation of results. They can also be used to develop models describing the behaviour of growing systems which may then be used to predict the effects of changes in the models' parameters.

There is a substantial body of literature describing the growth kinetics of bacteria much of which is backed up by sound experimental evidence (Monod, 1942; Dean & Hinshelwood, 1966; Fencl, 1966). Similar principles can be applied to studies of mycelial fungi, with appropriate modifications for the differences in their morphology; the main difference is, of course, growth in multicellular or coenocytic units of considerably variable size instead of discrete single cells. In this chapter, the growth kinetics generally applicable to microbes will be outlined and a more detailed consideration will be given to those aspects peculiar to mycelial fungi. Much of what is said is to be equally applicable to those actinomycetes with a filamentous habit.

In considering the growth of fungi I have assumed that, as with bacteria, their growth is fundamentally an autocatalytic process, giving exponential growth in an unrestricted system. It should be pointed out, however, that several authors do not support this view, and Mandels (1965) in an authorative treatise on fungi stated that '. . . growth is not autocatalytic in filamentous fungi'. Here examples of exponential growth will be given and deviations from the exponential growth pattern are explained without rejecting my original premise.

5.2 Unrestricted Growth

Exponential growth of cell mass

Most, if not all, microbes grow autocatalytically when placed in an environment with an excess of all substrates necessary for growth and in the absence of inhibitors; the increase in mass is, therefore, exponential. The specific growth rate, the amount of organism produced by a unit amount of organism in unit time, is a constant characteristic of the strain and environment (equation *1*):

Rate of growth = constant × organism concentration

$$\frac{dx}{dt} = \mu x \qquad (1)$$

Equation (*1*) can be simply integrated to give the concentration of organism at any time (t) after the onset of exponential growth:

$$x_t = x_0 \, e^{\mu t} \qquad (2)$$

and

$$\ln x_t = \mu t + \ln x_0 \qquad (3)$$

Growth of this type is most conveniently represented in a graphical form using the natural logarithm of organism concentration against time. A straight line whose slope is equal to the specific growth rate is obtained (equation *3*). Using simple media exponential growth is often seen (Zalokar, 1959; Pirt & Callow, 1960; Trinci, 1969). It is usually preceded by a lag phase and followed by a deceleration or stationary phase (Fig. 5.1). The media used in industrial cultures are frequently complex, containing a variety of carbon and nitrogen sources. In such media substrates may be used sequentially with concomitant sequential changes in growth rate. Because of this an exponential growth phase cannot always be discerned.

The specific growth rate observed in unrestricted growth is the maximum the organism is capable of under the particular conditions of pH, medium, temperature *etc.*, since, as defined, all substrates are present in excess. Maximum specific growth rates of filamentous fungi in the order of 0.3 h^{-1}, a doubling time of about two hours, are not uncommon, though the variation between species is very large and the type of medium has a significant effect. Media with complex carbon and nitrogen sources usually support higher growth rates than do simple media. Maynard-Smith (1969) considered a number of mechanisms which could determine the maximum growth rate: congestion of the cell with the biosynthetic machinery, spatial organization within the cell resulting in diffusion times becoming limiting and a limitation on the maximum rate of macromolecular syntheses. For bacteria he favoured the last possibility, suggesting that RNA synthesis was rate-limiting probably at the messenger transcription stage. Also with bacteria, there is some evidence that the maximum rate of energy generation determines the maximum growth rate

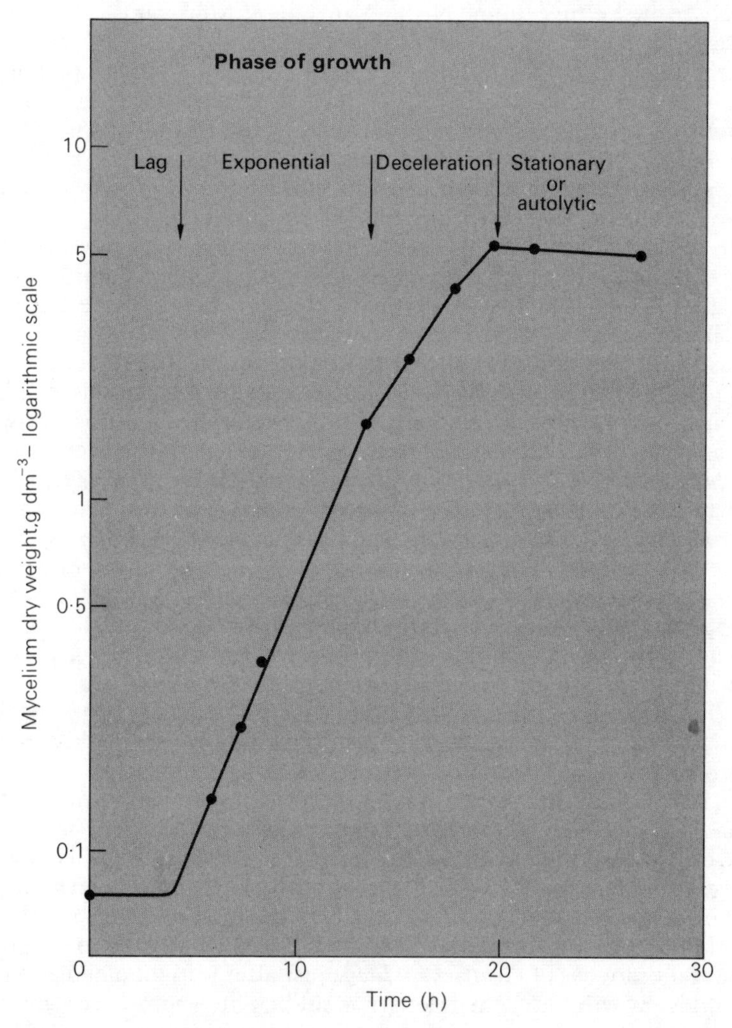

Fig. 5.1 Kinetics of growth of *Geotrichum lactis* on a medium containing inorganic salts, vitamins and 10 g dm^{-3} glucose in a shake flask at 25°C, redrawn from Trinci (1971).

(Harvey, 1970). Whatever the determinant of the maximum growth rate, under fixed conditions it has a characteristic value for each organism.

Exponential growth and branching

Increase in mass of bacteria and yeasts is usually accompanied by increase in length or diameter followed by division, so that the volume of a cell varies only over an approximately two-fold range. With mycelial fungi, growth involves an increase in hyphal length over a very much larger range; but exponential growth implies that all or a constant proportion of the hyphal length contributes to new growth. This can be attained through branching.

Proteins and nucleic acids are distributed fairly evenly throughout actively growing mycelium (Nishi, Yanagita & Maruyama, 1968; Fencl, Machek, Novak & Seichert, 1969), and electron microscopy of growing cultures shows that most cells have an even distribution of mitochondria, ribosomes and nuclei (Righelato, Trinci, Pirt & Peat, 1968). However, hyphal wall growth appears to occur only at the tips of hyphae (Katz & Rosenberger, 1971). Macromolecules or their monomers synthesized further back along the hyphae could contribute to growth of the tip thereby allowing the extension rate to be proportional to the overall length of the hyphae. Trinci (1969) showed that the germ tubes emerging from conidia of *Aspergillus nidulans* grew exponentially until 120 μm, long after which the specific hyphal extension rate became linear and branching ensued. Prolonged exponential growth of hyphal systems occurs because hyphae branch, producing more tips. Smith (1924) found that the total length of a hyphal system, *i.e.* the sum of the lengths of all the branches, increased exponentially. Thus, even when all tips grow at a constant linear rate, exponential increase in total length can occur *provided branching occurred at a rate proportional to the total hyphal length.* Katz, Goldstein & Rosenberger (1972) proposed a model for exponential growth and branch initiation based on observations on *A. nidulans* grown in shake flask culture. They concluded that (i) *the rate of extension at the growing points of short hyphae is proportional to both the hyphal length and the specific growth rate*, *i.e.* short hyphae grow exponentially, (ii) *each growing point can extend at a rate between a low value and a maximum characteristic of the organism and independent of the mass specific growth rate*, (iii) *a new growing point is formed when the capacity of the hypha to extend exceeds that of the existing points.* It follows from this that the branching frequency would be proportional to specific growth rate. Katz, *et al.* (1972) found that this was true of *A. nidulans* when grown on three media each of which supported a different growth rate.

Careful measurement of hyphal length at the margin of colonies of *Aspergillus nidulans* showed that the hyphal growth rate decreased prior to apical branching and increased again to the original rate a short while after the branch was formed (Trinci, 1970). Trinci (1970) also found that the lateral branches of *Geotrichum lactis* just after initiation had a remarkably high growth rate, their length doubling in about seven minutes; subsequently they slowed to a rate somewhat lower than that of the leading hyphae. Though it is clear from these examples that hyphal extension rates may vary around the time of branching, perhaps due to reorganization of the cell wall at the branching point, the overall picture of exponential increase in cell mass and total hyphal length, through branching, seems well established for homogeneous cultures.

5.3 Restricted Growth

The exponential, or unrestricted, growth phase in batch cultures is usually short. It is often preceded by a lag phase and followed by a phase of decelerating growth rate. The term restricted growth is used here loosely, to cover any growth condition in which the maximum growth rate cannot be expressed.

The lag phase

The lag phase is generally thought to be a period of adaptation to the conditions imposed by a change of environment. Kjeldgaard, Maaløe & Schaechter (1958) showed that a period of slow growth of *Salmonella typhimurium* preceded exponential growth at the maximum rate after transfer to a simple medium from one rich in organic nitrogen. The lag phase presumably involved induction of enzymes required for the new medium and an increase in the protein-synthesizing machinery. They also found that growth rate and RNA content were closely related. Fig. 5.2 shows how increasing the dilution rate of a sucrose-limited chemostat culture of *Penicillium chrysogenum* affects RNA and growth rate. There

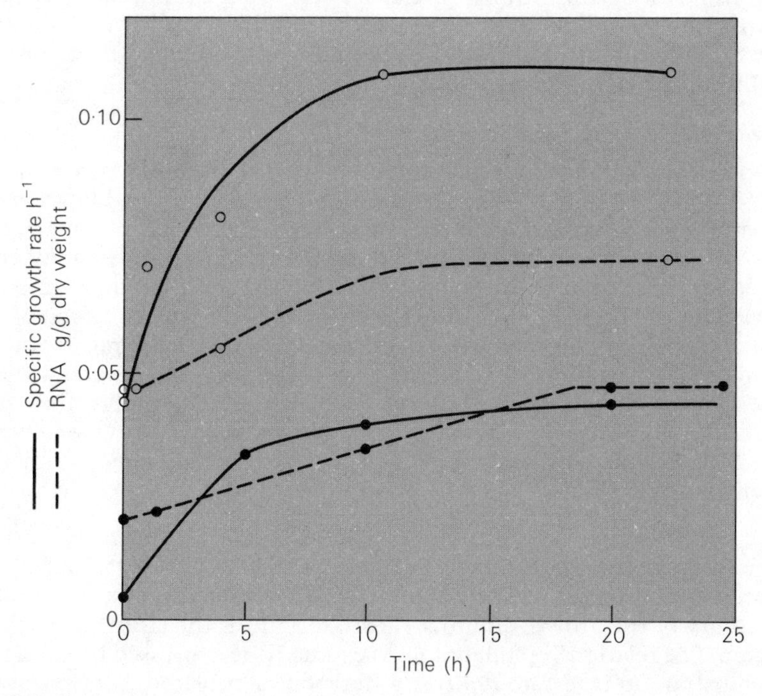

Fig. 5.2 Changes in growth rate and RNA content of *Penicillium chrysogenum* after increasing the dilution rate of a glucose-limited chemostat culture (O) from 0.43 h^{-1} to 0.11 h^{-1}, the maximum specific growth rate; (●) from 0.003 h^{-1} to 0.043 h^{-1}.

was a short lag period during which the RNA content increased before growth at the higher rates ensued.

Sometimes the lag phase may be due to modification of the environment by the organism. For instance, there is evidence that high oxygen concentrations inhibit the growth of some organisms (Hughes & Wimpenny, 1969); hence growth at the maximum rate is preceded by a period of slow growth during which the microbes' consumption of oxygen reduces the concentration in the medium to a sub-inhibitory level. Other examples of substrate inhibition are the substrate alcohols with methanol and ethanol oxidizing organisms and some biodegradable toxic wastes (Edwards, 1970). Sometimes products of metabolism are necessary in low concentrations for growth to occur at the maximum rate. Carbon dioxide, for instance, which is required for some anaplerotic reactions, has been reported to stimulate the growth of several fungi (Hartman, Keen & Long, 1972); hence a period of slower growth while the required metabolite concentrations build up.

For each of the cases mentioned above it would be necessary to modify equation (*1*) so as to express μ as a function of the concentration of the toxic or required substrate, which is itself a function of the organism concentration or organism concentration × time. Thus μ becomes a function of x and t up to a maximum value of μ. As the precise kinetics depend on the type of inhibition exerted by the substrate or product these will not be considered further. Simple cases of first order inhibition of specific growth rate were dealt with by Leudeking & Piret (1959) and Zines & Rogers (1970); more complex relationships can be derived from enzyme inhibition kinetics (Webb, 1963).

Deceleration and stationary phases

A simple calculation using equation (*2*) shows that one milligram of mould with a doubling time of three hours would produce a few kilograms in three days and several million tons in a week. In practice, of course, this does not happen; exponential growth at the maximum rate rarely continues for more than a day or two. Probably the most common cause of the slowing of specific growth rate is a decrease in the concentration of one or other substrate to a *growth limiting concentration.* Growth rate is usually considered to be a function of substrate concentration and several workers have presented data from bacteria and yeasts that support this view (Monod, 1942; Novick & Szilard, 1950; Moser, 1958; Van Uden, 1969). The Monod equation relating growth rate to substrate concentration is usually written:

$$\frac{dx}{dt} = \mu_{max}\left(\frac{s}{k_s+s}\right)x \qquad (4)$$

where μ_{max} is the maximum growth rate constant, characteristic of the organism, medium and culture conditions, k_s is the half saturation constant for uptake of the growth-limiting substrate and s is the substrate concentration. The relationship might more accurately be expressed by considering substrate uptake rate ds/dt as a function of substrate concentration. Growth rate dx/dt is then the uptake rate multiplied by a yield constant (Y):

$$\frac{ds}{dt} = \frac{1}{Y}\frac{dx}{dt} = q_{max}\left(\frac{s}{k_s+s}\right)x \qquad (5)$$

Here it is assumed that Y, q_{max} and k_s are constants though it will be shown that for some substrates Y is a function of specific growth rate (see page 91); and the induction of high affinity substrate assimilation pathways at low substrate concentrations is now well-documented (Tempest, Meers & Brown, 1970; Neville, Suskind & Roseman, 1971; Medveczky & Rosenberg, 1971). If, however, x and s are the only variables in equation (5), at least for some sets of conditions:

$$\mu_{max} = q_{max} Y$$

In batch cultures k_s is usually very small compared with the initial concentration of $s(s_R)$, so growth occurs at μ_{max} until s approaches k_s and the growth rate falls rapidly to zero as the substrate becomes exhausted. Then approximately:

$$x = s_R Y$$

The relationship between specific growth rate and substrate concentration according to equation (5) is shown in Fig. 5.3.

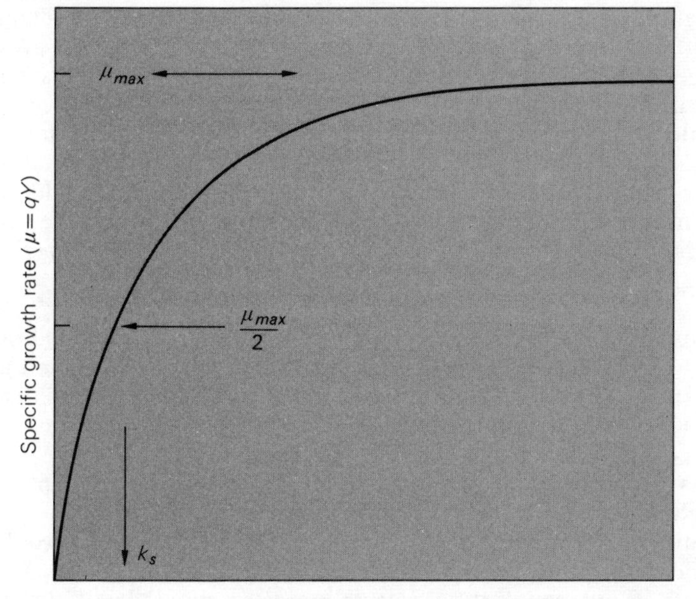

Fig. 5.3 The relationship between substrate concentration and growth rate.

Sometimes, particularly in industrial fermentations, substrates are supplied continuously to a culture. In this case exponential growth ceases when the exponentially increasing demand for the substrate depletes its concentration in the medium to a value in the order of the k_s and the rate of supply of the substrate to the culture determines the growth rate. The growth curve then takes the form dictated by the substrate feed rate.

i.e.

$$\frac{dx}{dt} = Y \frac{ds}{dt} \tag{6}$$

The supply of oxygen frequently becomes growth-limiting in fungal cultures due to its low solubility in aqueous media and the problems of obtaining high transfer rates from the gas phase to the aqueous even in well-shaken or stirred cultures. Another substrate often fed continuously at growth-limiting rates to batch cultures is the carbon source. By limiting growth in this way it is possible to greatly extend the period of productivity of batch fermentations (Hockenhull & Mackenzie, 1968; Edwards *et al.*, 1970). If the feed rate is more or less constant the growth rate (dx/dt) would also be constant but the specific growth rate $1/x \times dx/dt$ would fall as the organism concentration increased. Sometimes the specific growth rate needs to be kept constant in which case an exponentially increasing feed rate can be used (Martin & Felsenfeld, 1964). Usually, however, a more practical method of maintaining a constant specific growth rate is to use a continuous culture technique (see p. 88).

The pellet growth form

So far it has been assumed that the 'natural' law of growth is the exponential law and that deviations from it are due to limitations on the availability of substrates necessary for growth, or to inhibition of the mechanisms of growth. However, several workers have reported that fungal growth in homogeneous cultures is best fitted to a cube root law (Emerson, 1950; Marshall & Alexander, 1960). Marshall & Alexander (1960) used respiration rate as a precise measure of the growth of several fungi and actinomycetes which grew in a more or less pelleted form. Their method allowed them to make frequent readings over an approximately fifty-fold range of cell concentration. They showed that the linear fit to the cube root plot (equation 7) was significantly better than to the logarithmic plot (Fig. 5.4).

$$x^{1/3} = x_0{}^{1/3} + kt \qquad (7)$$

This apparent conflict was resolved by Pirt (1966) who pointed out that reports of 'cube root' growth were associated with the pelleted growth form. In such cases, he argued, when a pellet exceeds a certain diameter, substrates would not diffuse inwards fast enough to maintain unrestricted growth of the whole pellet mass. This is likely to occur because the pellet mass is proportional to the cube of its radius, whilst the surface area through which substrates must diffuse increases only with the square of the radius. After a while only a peripheral shell of the pellet would have sufficient substrate to maintain exponential growth; moreover, since the substrate affinity constant (k_s) is usually small compared with the external substrate concentration, the zone of the pellet exhibiting growth rates between zero and the maximum would be quite small. We can approximate and say that the pellet inside the exponentially growing shell is not growing at all, in which case:

$$\frac{dx}{dt} = \mu x_p \qquad (8)$$

where x_p is the mass of the peripheral growth zone of the pellet, and:

$$\frac{dr}{dt} = \mu\omega$$

or:

$$r = \mu\omega t + r_0 \qquad (9)$$

where r is the radius of the pellet and ω is the depth of the peripheral growth zone and is a constant for each growth-limiting substrate concentration. The mass of a pellet of density d is given by:

$$x = \tfrac{4}{3} r^3 \pi d \qquad (10)$$

Substituting for r in equation (9) gives:

$$x^{1/3} = \left(\frac{3}{4\pi d}\right)^{-1/3} \mu\omega t + x_0^{1/3} \qquad (11)$$

$(3/4\pi d)^{-1/3}$ is a constant numerically equal to 0.74 if the pellet density is assumed to be 0.1 g dry weight cm^{-3}. Thus the growth of a single pellet, or of a number of pellets of the same size, is predicted to follow a cube root growth pattern once r is greater than ω. While r is smaller than ω exponential growth would occur.

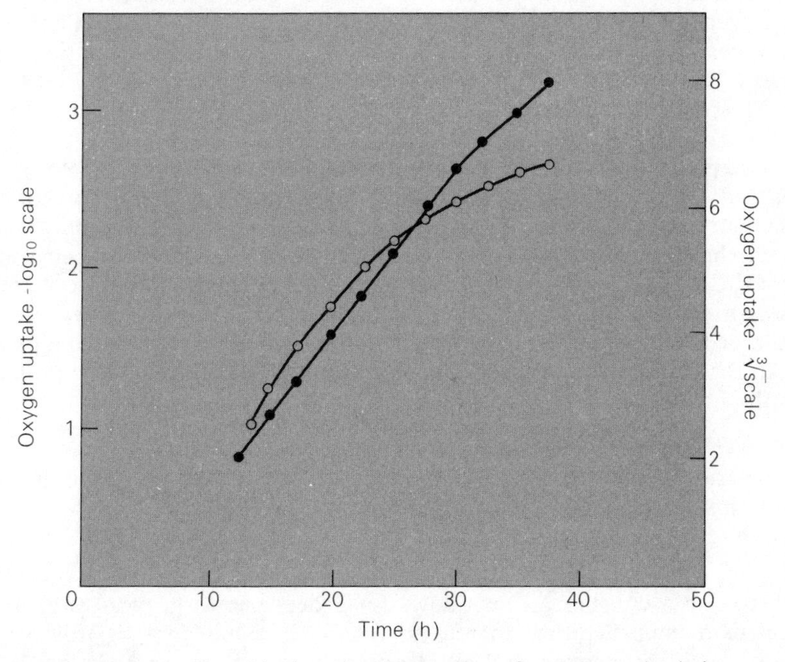

Fig. 5.4 Comparison of logarithmic and cube root plots of oxygen uptake during growth of *Nocardia* sp. Redrawn from Marshall & Alexander (1960).

Pirt (1966) went on to consider the size of the peripheral growth zone assuming that substrate entered the pellets only by diffusion. He concluded that ω would be determined by the rate of utilization of the growth-limiting substrate and the concentration outside the pellet. For obligatory aerobic growth, oxygen is probably the most common limiting substrate in this context because its solubility in water is very low and it is required in substantial quantities (Table 5.1). In such cases Pirt (1966) calculated the depth of the peripheral growth zone to be in the order of the outer 0.077 mm.

Clearly, these kinetics imply that there would be substantial differences in metabolism between the peripheral growth zone and the interior of the pellet. Microscopic examination of dense pellets shows that their interior may be dead or lysing (*e.g.* Camici, Sermonti & Chain, 1952). Obviously then, when the pellet radius is much greater than the peripheral growth zone the culture will be physiologically heterogeneous. It is desirable, therefore, to avoid pelleted growth wherever possible for most metabolic studies.

Microscopic observations show that pellet formation is due to close branching of hyphae. If hyphal breakage is proportional to hyphal length, the longer the internode the more separate hyphal units and the less the likelihood of a pellet being formed. Also, the shorter the internode length, *i.e.* the greater the branching frequency, the denser the pellet. If the branching frequency increases with growth rate (Katz *et al.*, 1971), pellet formation would be more likely at higher growth rates, as has been observed with *Penicillium chrysogenum* (Righelato *et al.*, 1968). There are, however, several other factors known to affect pellet formation: pH (Pirt & Callow, 1959), metal ion concentrations (Choudhary & Pirt, 1965) and complex or synthetic media (Pirt & Callow, 1959; Whitaker & Long, 1973).

The pellet growth form has an important effect on broth rheology; the viscosity of the culture broth is lower than for an equal concentration of filamentous mould, allowing better gross mixing and faster oxygen transfer from the gas phase into the broth (Taguchi, 1971). Perhaps because of this the pelleted growth form is essential for efficient operation of tower fermenters (Ross & Wilkin, 1968). However, the diffusion paths from the bulk of the medium into the hyphae are considerably greater for pellets than for filaments, so higher concentrations of oxygen would be required in the broth.

The logistic growth curve

Decelerating growth rate is sometimes fitted to the so-called logistic growth law, the general expression for which is:

$$\frac{dx}{dt} = \mu x(1 - cx) \qquad (12)$$

where c is a constant. This allows for a decrease in growth rate as the organism concentration increases. Although some data fit this model (Hockenhull & Mackenzie, 1968; Constantinides, Spencer & Gaden, 1970) it is of limited value because as such it does not test any hypotheses concerning the cause of the slowing growth rate.

5.4 Continuous Culture Kinetics

So far I have dealt with growth in batch cultures; in these most substrates are added at the start of the culture, though usually oxygen and sometimes other substrates are added during cultivation. Usually the only products that are removed from batch cultures during their operation are heat and carbon dioxide. Because of this the concentrations of most substrates, organisms and products change continuously, so a precise description and control of the environment of the microbes is almost impossible. Moreover, in practice it is found that the exponential growth phase is usually only maintained for three to six generations, thus restricting the number and type of observations that can be made.

For many experimental purposes continuous culture techniques are preferable to batch cultures because growth rate, organism concentration and the concentrations of all substrates and products can be maintained constant for indefinite lengths of time. Moreover, growth rate can be controlled at any value between the maximum of which the organism is capable and values approaching zero. The interested reader can consult reviews of the role of continuous culture in microbiological and biochemical research (Tempest, 1970) and its role in industry (Righelato & Elsworth, 1970).

In a continuous culture, medium is fed at a rate (F) into the culture vessel whose volume (V) is held constant. The ratio F/V is known as the dilution rate (D). There are several types of continuous fermenter that have been used with fungi and actinomycetes. The best known is the chemostat, in which the flow rate of medium is controlled such that the dilution rate has any value below the maximum specific growth rate (μ_{max}) and one or more substrates in the medium is growth-limiting. The turbidostat operates at the maximum growth rate by adjusting the flow rate to maintain the organism concentration constant at a level at which its growth is not substrate-limited. The microbe concentration cannot always be measured by culture turbidity, particularly with filamentous organisms, so other parameters directly related to organism concentration, for instance respiration rate, can be used (Zines & Rogers, 1970).

There is a wide range of heterogeneous continuous culture systems but they have been used experimentally much less than homogeneous cultures. They are, nonetheless, of considerable significance; many natural environments probably correspond to plug flow reactors with varying degrees of back-mixing and retardation of the organism, and there has been some noteworthy industrial development of continuous tower fermenters (Purssell & Smith, 1968; Ross & Wilkin, 1968). The kinetics of the heterogeneous reactors are difficult to ascertain accurately and will not be dealt with here. A number of homogeneous tanks in series approximates to a plug flow reactor and may be more useful experimentally since efficient aeration and control of such things as pH can be applied more easily. The multistage tower fermenters, that have been developed in recent years, have the construction advantages of the simple tower fermenter with the advantage of a series of homogeneous fermenters (Kitai, Tone & Ozaki, 1969; Falch & Gaden, 1969; Prokop *et al.*, 1969).

The morphology of fungi in submerged culture is not well adapted to continuous submerged cultivation. Masses of hyphae tend to accumulate on fermenter walls and block overflow lines, so that the type of fermenters used for bacteria were enlarged and modified for experimental chemostat cultures of *Penicillium chrysogenum* (Pirt & Callow, 1960; Righelato & Pirt, 1967). A horizontal multistage vessel was designed by Means, Savage, Reusser & Koepsell (1962) for filamentous organisms. Unless new pellets are generated in some way, pelleted organisms cannot be used in continuous cultures. The kinetics given below have only been tested with filamentous moulds and yeasts in homogeneous cultures; little information is available on pelleted and heterogeneous cultures.

The kinetics of growth in homogeneous continuous cultures have been described and tested with bacteria by many workers (Monod, 1950; Herbert, Elsworth & Telling, 1956; Pirt, 1965) and it has been found that these can be applied to the growth of filamentous moulds (Pirt & Callow, 1960; Righelato *et al.*, 1968). The organism mass balance is given thus:

$$\frac{dx}{dt} = \mu x - Dx \qquad (13)$$

In the steady state $(dx/dt) = 0$ and $\mu = D$. The substrate mass balance can be written similarly:

$$\frac{ds}{dt} = D(s_R - s) - qx$$

or:

$$\frac{ds}{dt} = D(s_R - s) - q_{\max}\left(\frac{s}{k_s + s}\right)x \qquad (14)$$

Provided D and s_R are held constant the system will tend to a steady state in which $ds/dt = 1/Y\,(dx/dt) = 0$. The steady state values of s and x are then:

$$s = k_s\left(\frac{D}{q_{\max}\;Y - D}\right) \qquad (15)$$

$$x = Y(s_R - s) \qquad (16)$$

If the dilution rate is much less than the maximum specific growth rate $(= q_{\max} Y)$ the concentration of residual substrate is usually small because k_s is usually very small compared with s_R; so the steady state organism concentration approximates to Ys_R. The chemostat gives unique steady states at any dilution rate below that equal to the maximum specific growth rate and allows independent control of organism concentration and growth rate by control of s_R and D. The turbidostat, on the other hand, can only be operated effectively at dilution rates close to $\mu_{\max}$ since it is only in that range that x varies significantly with D (Fig. 5.5). At dilution rates above $\mu_{\max}$ the organisms are washed out at a rate equal to $(D - \mu_{\max})$ unless a continuous inoculation or organism recycle system is used. The kinetics of a number of developments of the simple chemostat, multistage cultures and cultures with feedback of organism can be developed from the basic relationships given above (Fencl, 1966; Pirt & Kurowski, 1970).

Because they run under a fixed set of conditions for indefinite periods continuous cultures select in favour of mutants or contaminants better able to use the growth-limiting substrates; the kinetics of selection are dealt with in detail by Powell (1958) and Kubitschek (1970). Inspection of equation (*14*) shows that mutants with a higher yield constant (Y), a higher maximum substrate assimilation rate ($q_{\max}$) or a lower half-saturation constant (k_s) would have a higher specific growth rate and hence outgrow the parent. At growth rates much below $\mu_{\max}$ the lower k_s would be the most advantageous mutation. Mutant selection is often a problem in continuous cultures (Pirt & Callow, 1961; Reusser, 1961; Slezak & Sikyta, 1964), though it may be turned to use provided the character to be selected confers a growth advantage.

5.5 Maintenance, Endogenous Metabolism, Turnover and Death

When a culture becomes exhausted of the carbon and energy source, instead of growth simply stopping as predicted by equation (*6*), lysis occurs and the cell concentration decreases (Trinci & Righelato, 1970). Net lysis can be prevented by feeding the carbon source at a low rate, termed the maintenance rate (Righelato *et al.*, 1968); thus there is a requirement for carbon source to maintain a growth rate of zero. Growing cultures also have a requirement for carbon and energy sources in addition to that used for growth; when the specific substrate utilization rate (q) is plotted against the specific growth rate (μ) a straight line with a positive intercept on the q axis is obtained (Tempest & Herbert, 1965; Righelato *et al.*, 1968; Von Meyenburg, 1969). To fit this, equation (*6*) must be modified:

$$\frac{ds}{dt} = \frac{1}{Y_G}\frac{dx}{dt} + mx$$

or:

$$q = \frac{\mu}{Y_G} + m \qquad (17)$$

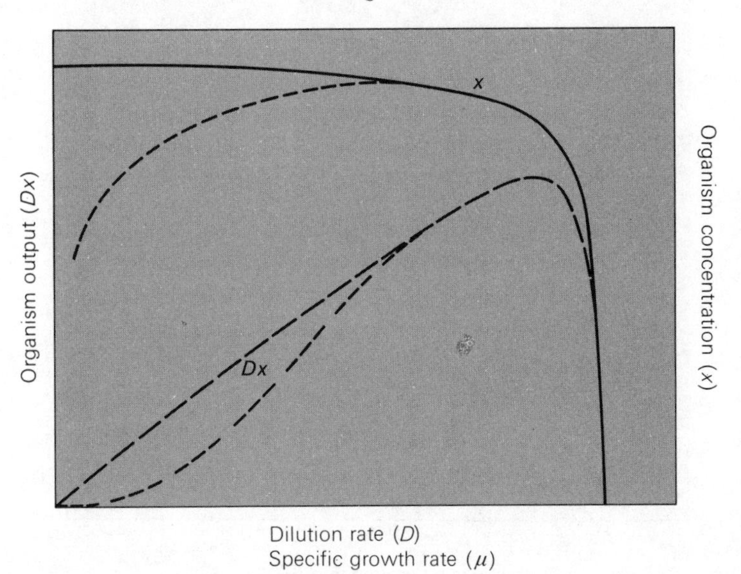

Fig. 5.5 The relationship between organism concentration, output rate and dilution rate in steady state chemostat cultures. The organism concentration is represented by continuous line when the growth-limiting substrate is used according to equation (*14*) and by dotted lines when part of the limiting substrate is used for cell maintenance (equation *17*).

It was found with *Penicillium chrysogenum* that the value of m obtained from equation (17) was equal to the requirement for glucose and oxygen to maintain a glucose-limited culture without growth (Righelato *et al.*, 1968). The glucose used was almost entirely oxidized, presumably providing energy for growth to replace lysed cells which were present in the cultures, *i.e.* cell turnover, and energy for intracellular turnover of proteins and nucleic acids, the maintenance of osmotic barriers and other aspects of endogenous metabolism. All of these processes could be independent of growth rate, in line with the fit of the data to equation (*17*) (Fig. 5.6). It is obvious from equation (*17*) that as the growth rate is decreased the significance of the maintenance coefficient increases; since fungi grow relatively slowly a substantial part of the energy source is used for maintenance. With *Penicillium chrysogenum*, for instance, the maintenance rate was 10% of the rate of glucose utilization at the maximum growth rate; and at a specific growth rate of 0.005 h^{-1}, such as is held in penicillin fermentations, it represents 70% of the glucose used (Fig. 5.6).

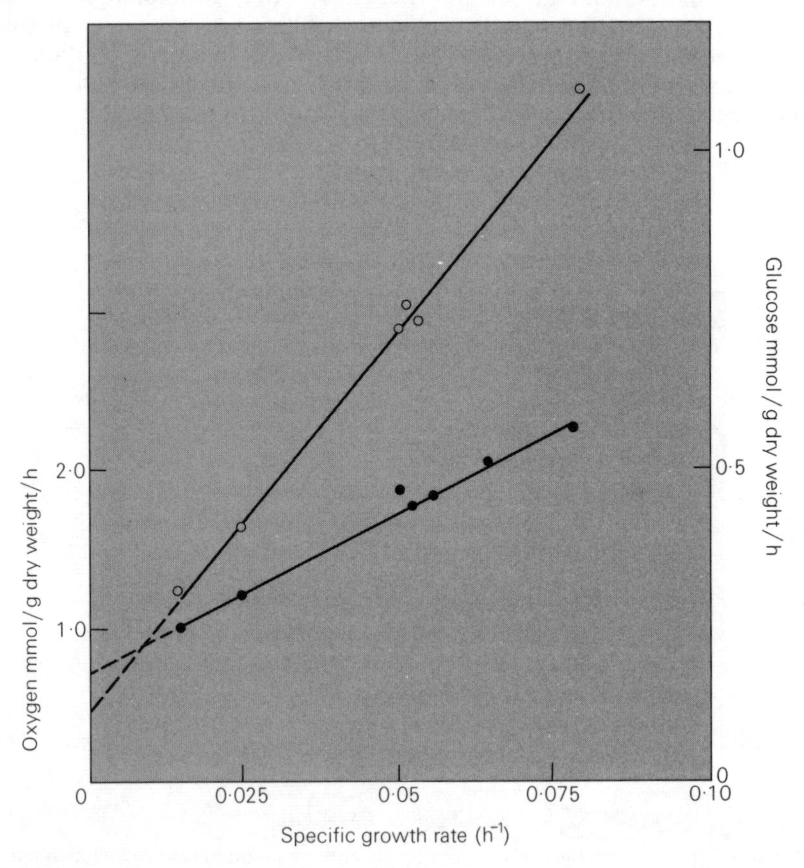

Fig. 5.6 The relationship between growth rate and the rates of oxygen (●–●) and glucose (○–○) utilization in glucose-limited chemostat cultures of *Penicillium chrysogenum* (Righelato *et al.* 1968).

Coupled with the concept of maintenance is that of cell death. It has been found that actively growing cultures of bacteria contain a proportion of dead cells, the viability being lower at lower growth rates (Tempest, Herbert & Phipps, 1967). Sinclair & Topiwala (1970) found that the number of dead cells could be fitted to a model which assumed a first order death rate:

$$\frac{dx}{dt} = \mu x - \gamma x - Dx \qquad (18)$$

where γ is the death rate constant. In the steady state the viability $x/(x+y)$, where x is the number of viable cells and y the number of dead cells, is given by:

$$\frac{x}{x+y} = \frac{D}{D+\gamma}$$

Provided the viability can be measured, the death rate can be calculated. With mycelial fungi this is usually not possible because often both apparently viable and apparently dead cells are found in the same hyphae (Trinci & Righelato, 1970). Thus most estimates of mycelial concentration include both live and dead cells, and the measured growth rate (μ) would be underestimates of the growth rate of the viable population ($\mu+\gamma$). The term m in equation (*17*) contains γ/Y, the substrate requirement for replacing dead cells, as well as the substrate requirement for the maintenance of viable cells.

5.6 Growth-limiting Substrates and Yield Constants

The carbon and energy source

Growth-limitation by the carbon and energy source is met more often than other substrate limitations. Since this is usually the most expensive medium component, efficient utilization is important, and usually when a substrate supply is growth-limiting a maximum efficiency of utilization is obtained. Secondly, many biosyntheses are catabolite-repressed and restriction of the carbon source, by feeding or by using slowly catabolized substrates, generally derepresses. For aerobic organisms, oxygen must be considered alongside the carbon source since stoichiometric equivalents are required; thus the weight of oxygen used by a fermentation is usually similar to the weight of carbohydrate. Although air is cheap, its addition to a culture is costly; the solubility of oxygen in aqueous media is *c*.9 mg dm^{-3} at atmospheric pressure and since a hundred or more times this amount may be required each hour, an efficient method of dissolving oxygen into the culture medium is required. This requirement for aeration accounts for much of the capital involved in building fermenters and for substantial energy costs for stirring and air compression. Usually fermentations are operated so as to have a small excess of oxygen to obtain a maximum efficiency of utilization of carbon source without spending unnecessarily on aeration, Better efficiencies still may possibly be obtained through co-limitation by carbon source and oxygen (Ryder & Sinclair, 1972).

As we have seen, the utilization of carbon and energy sources can be split into two components (equation *17*), a growth yield and a maintenance component. Using a graphical method (Fig. 5.6) the constants were found for several organisms in which virtually all the carbon source and oxygen consumed was recovered as cell mass and carbon dioxide (Table 5.1). The values of the growth yield and maintenance constants are remarkably similar for the four fungi in carbon source limited cultures. When the growth of *Penicillium chrysogenum* was limited by the nitrogen source or phosphate, the yield from glucose depended on the amount of glucose supplied in excess of the minimum requirement given in Table 5.1. The whole excess, up to three times the minimum requirement, was metabolized to glycerol, gluconic, malic and other organic acids (Mason, H.R.S. and Righelato, R.C., unpublished results).

Table 5.1 Growth yields and maintenance constants for glucose and oxygen in chemostat cultures limited by the carbon source

	Carbon source		Oxygen		
Organism	*Y* g cell/ g glucose	*m* g glucose g cell/h	*Y* G cell g O_2	*m* g O_2/ g cell/h	Reference
Aspergillus nidulans	0.45	0.029	—	—	Bainbridge *et al.* (1971)
Penicillium chrysogenum	0.45	0.022	1·56	0.024	Righelato *et al.* (1968)
*Saccharomyces cerevisiae**	0.51	—	0.94	0.019	Von Meyenberg (1969)
Torula utilis†	—	—	1.00	0.021	Tempest & Herbert (1965)

* Fully aerobic growth
† Growth on ethanol

The similarity of the growth yield constants under carbon-limited conditions suggests a maximum efficiency common to aerobic microbial systems. In the examples quoted, glucose and oxygen or ethanol and oxygen supplied all of the carbon and energy required for growth. Taking the yield constants 0.45 g cell/g glucose and 1.0 cell/g oxygen gives the approximate stoichiometry:

$$1 \text{ g atom glucose C} + 0.45 \text{ moles } 0_2 \rightarrow 0.55 \text{ g atoms cell C} + 0.45 \text{ moles } CO_2$$

where the cell contains 48% w/w carbon by dry weight and the respiratory quotient is 1. Thus, using glucose, nearly half of the sugar is oxidized providing energy. The quantity of energy-yielding substrate consumed is often used to assess the efficiency of the mechanisms of energy generation and its utilization for growth. The review by Payne (1970) describes several methods of estimating the efficiency of growth and gives data for a wide range of organisms and substrates.

The best known energy yield calculation is that of Y_{ATP}, the weight of cells produced per mole of ATP utilized. On the basis that anaerobic growth of bacteria and yeasts on substrates gives precisely known amounts of ATP, a value of 10.5 g dry weight/mole ATP was obtained (Bauchop & Elsden, 1960). The calculation is more difficult with aerobically-growing organisms since the P:O ratio is difficult to establish and usually has to be assumed. Stouthamer (1962) and Chen (1964) avoided this problem and obtained Y_{ATP} values of 10–11 for aerobically growing bacteria and a yeast.

They based their calculations on the total amount of substrate consumed. When the substrate used for maintenance is extracted values of *c.* 28g cells/mole ATP are obtained (Stouthamer & Bettenhausen, 1973).

ATP required = weight of cells/ Y_{ATP}

ATP produced = (oxygen consumed × P:O) + (ATP from substrate level phosphorylation)

If ATP required = ATP produced, the P:O ratio can be calculated by assuming a value for Y_{ATP}. Assuming $Y_{ATP} = 28$ and using the growth yield data in table 1, the P:O ratio for *Torula utilis* growing on ethanol was 0.6 and for *Penicillium chrysogenum* it was *c* 0.3. Since the maximum P:O ratio that can be expected is 3.0 for NADH-linked oxidations, a measurement of efficiency of energy conservation can be made. This type of calculation of Y_{ATP} and P:O ratio is open to many criticisms because of the number of assumptions it implies about the energy generating and utilizing mechanisms.

Another measurement of efficiency of growth and energy generation, equally open to criticism is $Y_{av\,e^-}$, the molar growth yield (g cell/mole substrate) divided by the number of available electrons in a molecule of the substrate. Available electrons are defined as those that can reduce oxygen during the complete biological oxidation of the substrate. On this basis glucose has 24 available electrons. The growth yield of 0.45 g cell/g glucose in Table 5.1 is equivalent to 81 g/mole and to a $Y_{av\,e^-}$ of 3.40. The mean for a wide range of fungi and bacteria (Payne, 1970) was 3.07 and the maximum recorded was 3.75. $Y_{av\,e^-}$ is an easy measurement to make for aerobically growing microbes; it requires only precise measurement of cell mass produced and substrate consumed and the knowledge that the sole products are cells and carbon dioxide.

Other substrates

There is relatively little information on the use of substrates other than the carbon and energy source to limit growth. Accumulation of carbohydrate storage products occurs when an excess of the carbohydrate substrate is present and another substrate limits growth (Herbert, 1958).

Table 5.2 Yield constants of *Penicillium chrysogenum* in chemostat culture

Growth-limiting substrate	Yield constant (Y) g dry weight/g substrate
Glucose	0.28*
$(NH_4)_2SO_4$ (nitrogen)	1.94
Na_2SO_4 (sulphur)	100.
KH_2PO_4 (phosphate)	19.6

Dilution rate was 0.02 h^{-1}

* This value is lower than the growth yield (Y_G) quoted in Table 5.1 due to the incorporation of the maintenance component:

$$Y = \frac{Y_G}{Y_G + (m/\mu)}$$

Growth limitation by a substrate other than the carbon source results in synthesis of secondary metabolites derived from acetyl-CoA such as polyketides and isoprenoids, in some fungi (Bu'Lock, 1965; Gatenbeck, 1965). In *Penicillium chrysogenum* the excess carbohydrate present during ammonium and phosphate-limited growth is converted to organic acids mostly malic and gluconic which accumulate in the medium (Mason, H.R.S. and Righelato, R.C., unpublished results). Yield constants for growth on several substrates in chemostat cultures of *P. chrysogenum* are given in Table 5.2; as these were obtained when the substrates were growth-limiting they are probably close to the maximum efficiency of utilization.

5.7 Surface Growth of Fungi

Growth of fungi on flat surfaces, usually agar, is a common feature of strain maintenance and strain selection procedures though it is rarely used now for production of metabolites. In recent years quantitative studies of growth on agar, stimulated by a desire to correlate characters observable during colony growth with characters expressed in submerged culture, have shown surface growth to be as lawful as homogeneous growth in fermenters.

Linear growth of colony diameter is a well-known feature of growth on agar; the problem has been relating this character to the exponential growth observed in submerged culture. Using similar reasoning to that applied to the growth of pellets in submerged culture, Pirt (1967) developed a general model for colony growth and applied it to bacterial colonies. This was refined by Trinci (1971) for growth of filamentous fungi. It is assumed that initially the growth of the mass of a colony is exponential *i.e.*

$$x_t = x_0 e^{\mu t} = H\pi d r_t^2$$

$$= (\pi H d r_0^2) e^{\mu t} \qquad (19)$$

This assumes that the colony grows as a disc radius r, height H and density d. Taking logarithms equation (*19*) becomes:

$$\ln x_t = \ln x_0 + \mu t = \ln(r_t^2 \pi H d) = \ln(r_0^2 \pi H d) + \mu t$$

or:

$$\ln r_t = \ln r_0 + \frac{\mu t}{2} \qquad (20)$$

Once a colony reaches such a size that the diffusion of nutrient into its centre becomes smaller than the demand for the nutrient, exponential growth would cease at the centre. The mycelium in the annulus outside the substrate-limited centre would continue to grow at the maximum specific growth rate but inside the annulus the transition to zero growth rate is likely to be rapid because small changes in the substrate concentration at low growth-limiting levels have a large effect on growth rate (Fig. 5.3). Thus growth of the colony can be regarded as being solely due to exponential growth of the mycelium in the outer annulus. Let the annulus have a depth a which is constant and a mass x_a:

$$\frac{dx}{dt} = \mu x_a$$

x_a approximates to $2\pi rHad$ where a is small compared with r and:

$$\frac{dr}{dt} = \mu a \tag{21}$$

$$r_t = a\mu t + r_0$$

As a and μ are constants, equation (*21*) predicts linear growth of the colony. These kinetics are in agreement with the observations that the diameters of colonies of *Chaetomium* sp. (Plomley, 1959) and *Aspergillus nidulans* (Trinci, 1971) initially grew exponentially (equation 20) and then became linear at about 0.2 mm (equation *21*) (Fig. 5.7).

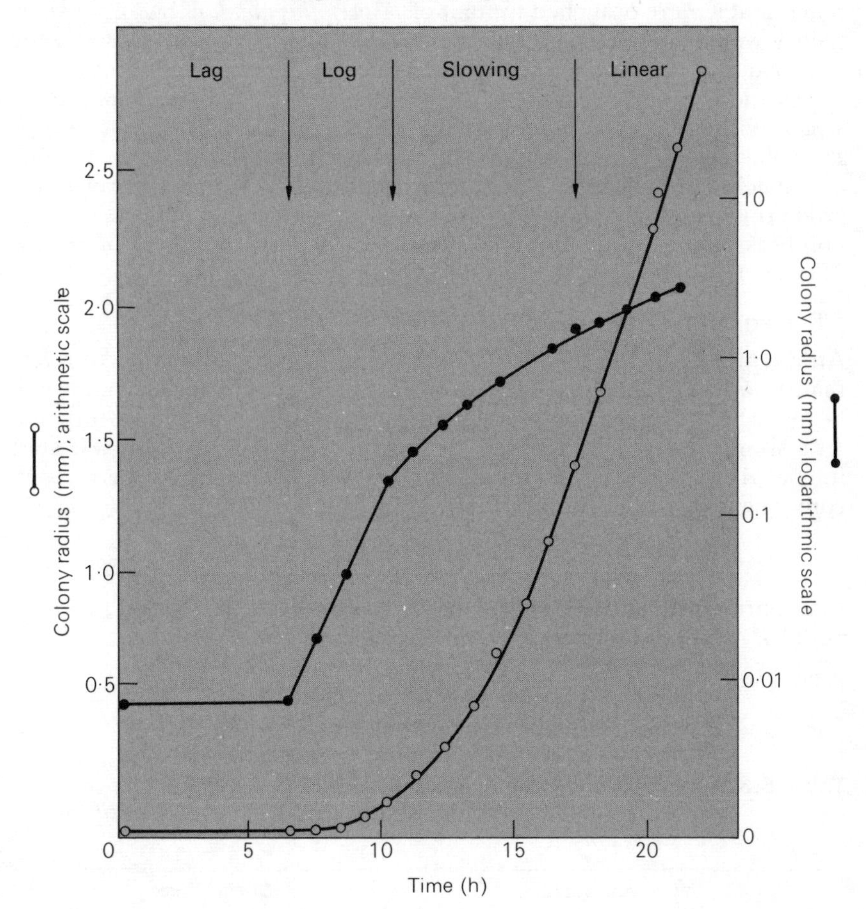

Fig. 5.7 Growth of colonies of *Aspergillus nidulans* on a glucose-salts agar, at 37°C (After Trinci, 1969).

In order to calculate the specific growth rate (μ), a must be measured as well as the increase in colony diameter. This can be done by cutting the colony along a chord and observing the growth rate of the colony edge on the outside of the cut (Trinci, 1971). Where the distance between the cut and the colony edge is less than a, the radial growth rate would decrease; where the cut is deeper into the colony than a there would be no effect on the radial growth rate. On the basis of this measurement of a, the peripheral growth zone, Trinci (1971) calculated the specific growth rate of nine species of fungi and found that the values obtained were very close to the specific growth rate in submerged culture on the same medium. He also observed a substantial variation of a between strains even when the specific growth rates were similar. This may be explained by variations in hyphal diameter or degree of branching, both of which would affect the distance covered by growing hyphae. Clearly the less the branching and the narrower the diameter, the faster would be the apical extension rate for a fixed mass growth rate. For instance, Bainbridge & Trinci (1970) noted that a more branched mutant of *Aspergillus nidulans* had a substantially lower colony radial growth rate and a smaller peripheral growth zone but an almost identical specific growth rate in submerged culture. Similar observations have been made with a series of strains of *Penicillium chrysogenum* with different branching frequencies (Table 5.3). These results show that colony growth rate measurements can be related to morphology and specific growth rate in submerged culture. Although still relatively crude this technique could be very useful in scanning the effects of large numbers of conditions or some characters of large numbers of strains (Trinci, 1970; Trinci & Gull, 1970; Okanishi & Gregory, 1970).

5.8 Product Formation

An attempt to fit the diverse kinetics of product accumulation to a few mathematical models would be doomed to failure. No such attempt will be made here. Instead, just a few elementary relationships will be outlined. In a simple analysis of a closed system product (P) formation was split into a growth-associated component (α) and a non-growth associated component (β) (Leudeking & Piret, 1959):

$$\frac{dP}{dt} = \alpha\mu x + \beta x \tag{22}$$

In a continuous culture steady state $dP/dt = 0$ and $\mu = D$, the product output (DP) and product concentration (P) are:

$$DP = \alpha D x + \beta x$$

$$P = \alpha x + \frac{\beta x}{D}$$

Table 5.3 A comparison of colony radial growth rates on agar with specific growth rate and branching in submerged cultures of *Penicillium chrysogenum*

	Agar surface		Shake flask	
Strain	Radial growth rate µm/h	Peripheral growth zone µm	Specific growth rate h^{-1}	Total hyphal length / Number of growing points µm
1	38	500	0.126	25
2	96	800	0.348	44
3	110	1200	0.232	54
4	140	—	0.116	101

If β is very small compared with α, the maximum output is obtained at high dilution rates; the prime example of this is cell mass as the product, when $\alpha=1$ and $\beta=0$. In such cases the output increases as D increases and the product concentration is independent of dilution rate, up to just below the maximum specific growth rate at which point the cell concentration decreases (Fig. 5.6).

If α is small compared with β, the output is independent of dilution rate and the product concentration is inversely proportional to dilution rate; penicillin synthesis by *Penicillium chrysogenum* falls into this category (Pirt & Righelato, 1968).

For simple batch cultures, integration of equation (*22*) gives

$$P = \left(\alpha + \frac{\beta}{\mu}\right) x_0 \, e^{\mu t}$$

If β/μ is small compared with α, product concentration and output rate are maximized by using as high a specific growth rate and as high an inoculum concentration as possible. Continuous cultures are clearly preferable to batch cultures in this situation since they represent the extreme case of the input concentration (x_0) being equal to the output concentration of cells.

If α is small compared with β/μ, product concentration is maximized by maximizing x_0 and minimizing μ. In practice this is done by having a two-step growth curve; initially a high growth rate is used to produce a high organism concentration followed by a long period of slow growth (Gaden, 1959; Deindoerfor, 1960; Hockenhull & Mackenzie, 1968).

Other, more sophisticated, models have been developed which for instance relate product formation to cell age (Shu, 1961; Brown & Vass, 1973) or to differentiation (Megee, Kinoshita, Fredericksen & Tsuchiya, 1970). Megee *et al.* (1970) based their model on observed functional differentiation of hyphae behind the extending tip on agar surfaces; by adjusting a series of cell-age related constants they fitted data from several fermentations to their model. One criticism of their model is that the differentiation they observed could have been due to differences in the environment around the immobilized hyphae in different parts of the colony. Thus the differentiation may not occur in homogeneous submerged cultures.

Many fermentation products are unstable in fermentation broths and a decay term must be added to equation (*22*) so the product concentrations are reduced. For continuous cultures in the steady state:

$$P = (\alpha D x + \beta x)\left(\frac{1}{D+b}\right)$$

and for batch cultures:

$$P = (\alpha\mu + \beta)\left(\frac{1}{\mu+b}\right) x_0 \, e^{\mu t}$$

Often in industrial practice, fermentations do not fit the simple kinetics outlined here; growth curves are rarely exponential for long, lags occur in induction of product formation and specific rates of synthesis change with cell-age. So it is often difficult to devise and to test models with predictive value and to discern effective control parameters. Nonetheless, much of the practical value of studying growth kinetics lies in the process optimization and control that stems from fermentation modelling.

5.9 Table of Symbols

a depth of peripheral growth zone of a surface colony
b first order decay constant (h^{-1})
c empirical constant
d density of organism in a pellet or colony
D dilution rate
F flow rate of medium
H height of colony
k_s half-saturation constant for limiting substrate
k cube root growth rate constant
m specific rate of utilization of limiting substrate at zero specific growth rate, the maintenance requirement
q specific rate of utilization of limiting substrate
r radius of colony
s concentration of limiting substrate in the culture
s_R concentration of limiting substrate in influent medium or uninoculated medium
t time
V volume of culture
x mass of organism
x_a mass of organism in peripheral growth zone of colony
x_p mass of organism in peripheral growth zone of pellet
Y mass of organism/mass of substrate used
Y_G mass of organism/mass of substrate used after extraction of the maintenance usage of substrate
α growth associated constant
β non-growth associated constant
γ death rate (h^{-1})
μ specific growth rate (h^{-1})
ω depth of peripheral growth zone of pellet

5.10 References

BAINBRIDGE, B.W. & TRINCI, A.P.J. (1970). Colony and specific growth rates of normal and mutant strains of *Aspergillus nidulans*. *Transactions of the British Mycological Society* **53**, 473–5.

BAINBRIDGE, B.W., BULL, A.T., PIRT, S.J., ROWLEY, B.I. & TRINCI, A.P.J. (1971). Biochemical and structural changes in non-growing maintained and autolysing cultures of *Aspergillus nidulans*. *Transactions of the British Mycological Society* **56**, 371–85.

BAUCHOP, T. & ELSDEN, S.R. (1960). The growth of micro-organisms in relation to their energy supply. *Journal of General Microbiology* **23**, 457–69.

BROWN, D.E. & VASS, R.C. (1973). Maturity and product formation in cultures of micro-organisms. *Biotechnology and Bioengineering*, in the press.

BU'LOCK, J.D. (1965). Aspects of secondary metabolism in fungi. In *Biogenesis of antibiotic substances*, Ed. by Vaněk, Z. and Hošťálek, Z. New York: Academic Press.

CAMICI, L., SERMONTI, G. & CHAIN, E.B. (1952). Observations on *Penicillium chrysogenum* in submerged culture. I. Mycelial growth and autolysis. *Bulletin of the World Health Organisation* **6**, 265–276.

CHEN, S.L. (1964). Energy requirement for

microbial growth. *Nature, London* **202**, 1135–6.

CHOUDHARY, A.Q. & PIRT, S.J. (1965). Metal complexing agents as metal buffers in media for the growth of *Aspergillus niger*. *Journal of General Microbiology* **41**, 99–107.

CONSTANTINIDES, A., SPENCER, J.L. & GADEN, E.L. (1970). Optimisation of batch fermentation processes. I. Development of mathematical models for batch penicillin fermentations. *Biotechnology and Bioengineering* **12**, 803–30.

DEAN, A.C.R. & HINSHELWOOD, C. (1966). *Growth function and regulation in bacterial cells*. Oxford: Clarendon Press.

DEINDOERFER, F.H. (1960). Fermentation kinetics and model processes. *Advances in Applied Microbiology* **2**, 321–34.

EDWARDS, V.H. (1970). The influence of high substrate concentrations on microbial kinetics. *Biotechnology and Bioengineering* **12**, 679–712.

EDWARDS, V.H., GOTTSCHALK, M.J., NOOJIN III, A.Y., TUTHILL, L.B. & TANNAHILL, A.L. (1970). Extended culture: the growth of *Candida utilis* at controlled acetate concentrations. *Biotechnology and Bioengineering* **12**, 975–99.

EMERSON, S. (1950). The growth phase in *Neurospora* corresponding to the logarithmic phase in unicellular organisms. *Journal of Bacteriology* **60**, 221–3.

FALCH, E.A. & GADEN, E.L. (1969). A continuous multistage tower fermenter, I. Design and performance tests. *Biotechnology and Bioengineering* **11**, 927–43.

FENCL, Z. (1966). A theoretical analysis of continuous culture systems. In *Theoretical and methodological basis of continuous culture of micro-organisms*. Ed. by Malek, I. and Fencl, Z. New York: Academic Press.

FENCL, Z., MACHEK, F., NOVAK, M. & SEICHERT, L. (1969). Control of culture activity as a function of growth rate in continuous cultivation. In *Continuous culture of micro-organisms* 173–87, Ed. by Málek, I., Beran, K., Fencl, Z., Ričica, J. and Smrčková, H. New York: Academic Press.

GADEN Jr., E.L. (1959). Fermentation process kinetics. *Biotechnology and Bioengineering* **1**, 413–29.

GATENBECK, S. (1965). Studies on the basic metabolism determining the biosynthesis of malonate-derived compounds in *Penicillium islandicum*. In *Biogenesis of antibiotic substances*. Ed. Vaněk, Z. and Hoštálek, Z. New York: Academic Press.

HARTMAN, R.E., KEEN, N.T. & LONG, M. (1972). Carbon dioxide fixation by *Verticillium albo-atrum*. *Journal of General Microbiology* **73**, 29–34.

HARVEY, R.J. (1970). Metabolic regulation in glucose-limited chemostat cultures of *Escherichia coli*. *Journal of Bacteriology* **104**, 698–706.

HERBERT, D. (1958). Some principles of continuous culture. In *Recent Progress in Microbiology. VII International Congress of Microbiology*, p. 381.

HERBERT, D., ELSWORTH, R. & TELLING, R.C. (1956). The continuous culture of bacteria; a theoretical and experimental study. *Journal of General Microbiology* **14**, 601–22.

HOCKENHULL, D.J.D. & MACKENZIE, R.M. (1968). Preset nutrient feeds for penicillin fermentation on defined media. *Chemistry and Industry*, 607–10.

HUGHES, D.E. & WIMPENNY, J.W.T. (1969). *Advances in Microbial Physiology* **3**, 197–232.

KATZ, D. & ROSENBERGER, R.F. (1971). Hyphal wall synthesis in *Aspergillus nidulans*: effect of protein synthesis inhibition and osmotic shock on chitin insertion and morphogenesis. *Journal of Bacteriology* **108**, 184–90.

KATZ, D., GOLDSTEIN, D. & ROSENBERGER, R.F. (1972). Model for branch initiation in *Aspergillus nidulans* based on measurements of growth parameters. *Journal of Bacteriology* **109**, 1097–1100.

KITAI, A., TONE, H. & OZAKI, A. (1969). Performance of a perforated plate column as a multistage continuous fermenter. *Biotechnology and Bioengineering* **11**, 911–26.

KJELDGAARD, N.O., MAALØE, O. & SCHAECHTER, M. (1958). The transition between different physiological states during balanced growth of *Salmonella typhimurium*. *Journal of General Microbiology* **19**, 607–16.

KUBITSCHEK, H.E. (1970). *Introduction to research with continuous culture*. New York: Prentice Hall.

LUEDEKING, R. & PIRET, E.L. (1959). Transient and steady states in continuous fermentation. Theory and experiment. *Biotechnology and Bioengineering* **1**, 431–59.

MANDELS, G.R. (1965). Kinetics of fungal growth. In *The Fungi*, Vol. 1, p. 599, London: Academic Press.

MARSHALL, K.C. & ALEXANDER, M. (1960). Growth characteristics of fungi and actinomycetes. *Journal of Bacteriology* **80**, 412–16.

MARTIN, R.G. & FELSENFELD, G. (1964). A new device for controlling the growth rate of micro-organisms: the exponential gradient generator. *Analytical Biochemistry* **8**, 43–53.

MAYNARD-SMITH, J. (1969). Limitations on growth rate. In *Microbial Growth 19th Symposium of the Society for General Microbiology*. Ed. Meadow, P. and Pirt, S.J. London: Cambridge University Press.

MEANS, C.W., SAVAGE, G.M., REUSSER, F. & KOEPSELL, A.J. (1962). Design and operation of a pilot plant fermenter for the continuous propagation of filamentous micro-organisms. *Biotechnology and Bioengineering* **4**, 5–16.

MEDVECZKY, N. & ROSENBERG, H. (1971). Phosphate transport in *Escherichia coli*. *Biochimica et Biophysica Acta* **241**, 494–506.

MEGEE III, R.D., KINOSHITA, S., FREDRICKSON, A.G. & TSUCHIYA, H.M. (1970). Differentiation and product formation in moulds. *Biotechnology and Bioengineering* **12**, 771–801.

MONOD, J. (1942). *Recherches sur la croissance des cultures bacteriennes*. Paris Hermann & Cie.

MONOD, J. (1950). La technique de culture continue. Théorie et applications. *Annales de l'Institut Pasteur* **T79**, 390–410.

MOSER, H. (1958). *The dynamics of bacterial populations maintained in the chemostat*. Publication No. 614. Washington: Carnegie Institution.

NEVILLE, M.M., SUSKIND, R. & ROSEMAN, S. (1971). A derepressible active transport system for glucose in *Neurospora crassa*. *Journal of Biological Chemistry* **246**, 1294–1301.

NISHI, A., YANAGITA, T. & MARUYAMA, Y. (1968). Cellular events occurring in growing hyphae of *Aspergillus oryzae* as studied by autoradiography. *Journal of General and Applied Microbiology* **14**, 171–82.

NOVICK, A. & SZILARD, L. (1950). Experiments with the chemostat on spontaneous mutations of bacteria. *Proceedings of the National Academy of Sciences (Washington)* **36**, 708–19.

OKANISHI, M. & GREGORY, K.F. (1970). Isolation of mutants of *Candida tropicalis* with increased methionine content. *Canadian Journal of Microbiology* **16**, 1139–43.

PAYNE, W.J. (1970). Energy yields and growth of heterotrophs. *Annual Reviews of Microbiology* **24**, 17–52.

PLOMLEY, N.J.B. (1959). Formation of the colony in the fungus *Chaetomium*. *Australian Journal of Biological Sciences* **12**, 53–64.

PIRT, S.J. (1965). The maintenance energy of bacteria in growing cultures. *Proceedings of the Royal Society, B* **163**, 224–31.

PIRT, S.J. (1966). A theory of the mode of growth of fungi in the form of pellets in submerged culture. *Proceedings of the Royal Society, B* **166**, 369–73.

PIRT, S.J. (1967). A kinetic study of the mode of growth of surface colonies of bacteria and fungi. *Journal of General Microbiology* **47**, 181–97.

PIRT, S.J. & CALLOW, D.S. (1959). Continuous flow culture of the filamentous mould *Penicillium chrysogenum* and the control of its morphology. *Nature, London* **184**, 307–10.

PIRT, S.J. & CALLOW, D.S. (1960). Studies on the growth of *Penicillium chrysogenum* in continuous flow culture with reference to penicillin production. *Journal of Applied Bacteriology* **23**, 87–98.

PIRT, S.J. & CALLOW, D.S. (1961). The production of penicillin by continuous-flow fermentation. *Scientific Reports of the Instituto Superiore di Sanita* (English edition) **1**, 250–9.

PIRT, S.J. & KUROWSKI, W.M. (1970). An extension of the theory of the chemostat with feedback of organisms, and its experimental realisation with a yeast culture. *Journal of General Microbiology* **63**, 357–66.

PIRT, S.J. & RIGHELATO, R.C. (1967). Effect of growth rate on the synthesis of penicillin by *Penicillium chrysogenum* in batch and chemostat cultures. *Applied Microbiology* **15**, 1284–90.

POWELL, E.O. (1958). Criteria for the growth of contaminants and mutants in continuous culture. *Journal of General Microbiology* **18**, 259–68.

PROKOP, A., ERICKSON, L.E., FERNANDEZ, J. & HUMPHREY, A.E. (1969). Design and physical characteristics of a multistage continuous tower fermenter. *Biotechnology and Bioengineering* **11**, 945–66.

PURSSELL, A.J.R. & SMITH, M.J. (1968). *Proceedings of the European Brewing Convention*, 1967, pp. 155–65. Amsterdam: Elsevier.

REUSSER, F. (1961). Continuous fermentation of novobiocin. *Applied Microbiology* **9**, 366–70.

RIGHELATO, R.C. & ELSWORTH, R. (1970). Industrial applications of continuous culture: pharmaceutical products and other products of processes. *Advances in Applied Microbiology* **13**, 399–417.

RIGHELATO, R.C. & PIRT, S.J. (1967). Improved control of organism concentration in continuous culture of filamentous micro-organisms. *Journal of Applied Bacteriology* **30**, 246–50.

RIGHELATO, R.C., TRINCI, A.P.J., PIRT, S.J. & PEAT, A. (1968). The influence of maintenance energy and growth rate on the metabolic activity, morphology and conidiation of *Penicillium chrysogenum*. *Journal of General Microbiology* **50**, 399–412.

ROSS, N.G. & WILKIN, G.D. (1968). Continuous microbiological processes involving filamentous micro-organisms. *British Patent* 1133875.

RYDER, D.N. & SINCLAIR, C.G. (1972). Model for the growth of aerobic micro-organisms under oxygen-limiting conditions. *Biotechnology and Bioengineering* **14**, 787–98.

SHU, P. (1961). Mathematical models for the product accumulation in microbiological processes. *Biotechnology and Bioengineering* **3**, 93–109.

SINCLAIR, C.G. & TOPIWALA, H.H. (1970). Model for continuous culture which considers the viability concept. *Biotechnology and Bioengineering* **12**, 1069–79.

SLÉZAK, J. & SIKYTA, B. (1964). Chlortetracycline and pigment formation by *Streptomyces aureofaciens* in continuous culture. In *Continuous culture of micro-organisms*. Publishing House of the Czechoslovak Academy of Sciences. Prague.

SMITH, J.H. (1924). On the early growth rate of the individual fungus hypha. *New Phytology* **23**, 65–78.

STOUTHAMER, A.H. (1962). Energy production in *Gluconobacter liquefaciens*. *Biochimica et Biophysica Acta* **56**, 19–32.

TAGUCHI, H. (1971). The nature of fermentation fluids. *Advances in Biochemical Engineering* **1**, 1–30.

TEMPEST, D.W. (1970). The place of continuous culture in microbiological research. *Advances in Microbial Physiology* **4**, 223–50.

TEMPEST, D.W. & HERBERT, D. (1965). Effect of dilution rate and growth-limiting substrate on the metabolic activity of *Torula utilis* cultures. *Journal of General Microbiology* **41**, 143–50.

TEMPEST, D.W., HERBERT, D. & PHIPPS, P.J. (1967). Studies on the growth of *Aerobacter aerogenes* at low dilution rates in a chemostat. In *Microbial physiology and continuous culture*. Ed. by Powell, E.O., Evans, C.G.T., Strange, R.E. and Tempest, D.W., Her Majesty's Stationery Office, London.

TEMPEST, D.W., MEERS, J.L. & BROWN, C.M. (1970). Synthesis of glutamate in *Aerobacter aerogenes* by a hitherto unknown route. *Biochemical Journal* **117**, 405–7.

TRINCI, A.P.J. (1969). A kinetic study of the growth of *Aspergillus nidulans* and other fungi. *Journal of General Microbiology* **57**, 11–24.

TRINCI, A.P.J. (1970). Kinetics of apical and lateral branching in *Aspergillus nidulans* and *Geotrichum lactis*. *Transactions of the British Mycological Society* **55**, 17–28.

TRINCI, A.P.J. (1971). Influence of the width of the peripheral growth zone on the radial growth rate of fungal colonies on solid media. *Journal of General Microbiology* **67**, 325–44.

TRINCI, A.P.J. & GULL, K. (1970). Effect of actidione, griseofulvin and triphenyltin acetate on the kinetics of fungal growth. *Journal of General Microbiology* **60**, 287–92.

TRINCI, A.P.J. & RIGHELATO, R.C. (1970). Changes in constituents and ultrastructure of hyphal compartments during autolysis of glucose-starved *Penicillium chrysogenum*. *Journal of General Microbiology* **60**, 239–49.

VAN UDEN, N. (1969). Kinetics of nutrient-limited growth. *Annual Reviews of Microbiology* **23**, 473–86.

VON MEYENBURG, H.K. (1969). Energetics of the budding cycle of *Saccharomyces cerevisiae* during glucose limited aerobic growth. *Archiv für Mikrobiologie* **66**, 289–303.

WEBB, J.L. (1963). *Enzymes and metabolic inhibitors*. Vol. 1. London and New York: Academic Press.

ZALOKAR, M. (1959). Enzyme activity and cell differentiation in *Neurospora*. *American Journal of Botany* **46**, 555–6.

ZINES, D.O. & ROGERS, P.L. (1970). The effect of ethanol on continuous culture stability. *Biotechnology and Bioengineering* **12**, 561–75.

Additional References

STOUTHAMER, A.H. & BETTENHAUSEN, C. (1973). Utilization of energy for growth and maintenance in continuous and batch cultures of micro-organisms. *Biochimica et Biophysica Acta* **301**, 53–70.

WHITAKER, A. & LONG, P.A. (1973). Fungal pelleting. *Process Biochemistry* **8**, 27–31.

CHAPTER 6

Historical Development of the Fungal Fermentation Industry

L. M. MIALL

6.1 Introduction

Some people seem to consider that the mould fermentation industry was born fully fledged with the advent of antibiotics. This is far from being the case. Admittedly, the development of penicillin, followed soon after by that of streptomycin, chloramphenicol, the tetracyclines, and many other antibiotics, caused a vast expansion in fermentation technology. However, there had been much work prior to the antibiotic era to build a base on which the industry grew. It is impossible to separate at this stage bacterial and fungal fermentations; firstly the techniques used are very similar and secondly it was not so very long ago that people were arguing whether the *Streptomyces*, the group of micro-organisms that make the majority of antibiotics, should properly be classified as bacteria or fungi.

The purpose of this chapter is, therefore, first to discuss developments in the industry before the antibiotic era. Then will be mentioned the other developments in the background sciences that enabled the industry to develop so rapidly, once it had been established that penicillin was the answer to the old hope of a compound that killed bacteria without injuring the host. Finally, some post-war developments will be listed, with particular reference to those not mentioned elsewhere in the book.

6.2 Fungal Fermentation on Solid Substrate

Cheese production

The origins of the industrial uses of mycelial fungi are lost in the mists of antiquity. Records of the production of Roquefort cheese, made in caves in the south of France from sheep's milk, go back about a thousand years. Other types of 'blue' cheese also have long histories, though of course it is only comparatively recently that the part played by the essential mould has been properly understood. It was only in the present century that Thom (1930) showed that the moulds that grew in all types of blue cheese were essentially strains of the same organism, *Penicillium roqueforti*, which is one of the few moulds that can grow under the conditions of oxygen limitation in the narrow air spaces of the curd. The other types of cheese which depend on the growth of moulds for flavour are exemplified by

Camembert, Brie, and similar varieties. Here the moulds concerned, *Penicillium camemberti* and *Penicillium caseicolum*, are surface growers.

Edible fungi

The use of the larger fungi as food rather than as flavouring agents no doubt goes back even further into the past, and here also the deliberate cultivation of the fungi has a very long history. The Chinese and Japanese grow an organism they call 'shi-ta-ké' on specially prepared wooden logs, which are inoculated with spores and left to incubate for many months before the sporophores are harvested for eating. This process may be at least twenty centuries old. *Agaricus campestris,* the cultivated mushroom best known in Europe, has been deliberately grown in caves in France since at least the eighteenth century. The techniques of cultivation are only applicable to the growth of the particular fungus and have been very thoroughly described elsewhere (Worgan, 1968; see Chapter 11).

Koji process

Of greater historical relevance to modern technology is the Japanese 'koji' process. Koji is the name given to a preparation of mouldy bran grown in semi-solid culture in trays; the cereal preparation having been inoculated with a culture now known to be *Aspergillus oryzae.* This again is a process with a very long history. It is the basis for the manufacture of saké, the most traditional alcoholic drink in Japan, and saké is believed to have been made at least since the eighth century. Its production involves fermenting a mixture of rice koji, steamed rice and water with yeast. Many other traditional Japanese foods are also based on koji, for example soy sauce and miso, a semi-solid cheese-like food based on soya beans, salt and rice koji (see Chapter 13). It is now obvious that growing a koji is essentially making a preparation of amylases and other enzymes, and that these saccharify the rice starch in exactly the same way as malt does in brewing.

The introduction of the koji process to the western world is chiefly due to the work of Takamine, which started about 1891. His early work was unsuccessful, but later he developed a process for making a fungal diastase that could be operated successfully on a large scale (Takamine, 1914). This involved the growth of *Aspergillus oryzae* not on rice, but on wheat bran, making a preparation called Taka-koji. The process consisted first in moistening and then steaming the bran, thus both sterilizing and gelatinizing the starch. This was cooled and inoculated with spores of *Aspergillus oryzae* and then spread on the floor of a room, or better, on trays with a wire netting bottom so that it could be well aerated. The room was first heated and later, when the reaction had become exothermic, cooled by air to the required temperature. Finally, at the end of the process the product was air-dried so that it could be preserved without contamination. Takamine introduced the technique of acclimatizing the mould to various antiseptics so that these could be added, and sterility thus maintained. Later Takamine carried out the process in rotating drums, at first on a laboratory scale, and then ultimately in an iron drum of 4800 pounds capacity rotating in a horizontal axis once a minute. Large scale trials of the use of Taka-koji instead of malt in distilleries were carried out in the plant of Hiram Walker & Sons in Ontario in 1913. The yields of alcohol were higher than when malt was used, but a disadvantage encountered was a slight off-flavour, which meant that the process could not be used for potable spirit manufacture.

If the Taka-koji is extracted with water and precipitated with alcohol, a much purer enzyme preparation is obtained. Under the name of Taka-diastase this was marketed for many years as a digestive aid.

The transition from Takamine's koji process to the modern use of mould enzymes for saccharifying starch can be traced in a series of papers by Underkofler and others. In 1939 Underkofler and his colleagues at Iowa State College were making highly active amylase preparations by growing *Aspergillus oryzae* on wheat bran in rotating drums, and using this for saccharifying corn mashes prior to yeast fermentation for industrial alcohol production. Later, they were still using bran, but cultivating the mould in aluminium pots of about three litres capacity which were aerated by blowing air through holes in the bottom (Hao, Fulmer & Underkofler, 1943).

At one time two processes for making saccharifying enzymes could be clearly distinguished, but these had very similar origins. The wheat bran process originated from the Japanese koji. In China a rather similar process using different strains of mould had the additional ability of making alcohol from starchy materials. The Chinese first made a dough on which a white mould grew called 'Chinese yeast'. Calmette (1892) studied this and called it *Amyloces rouxii*. Later he isolated various strains of *Mucor* from Chinese yeast and ultimately used a strain of *Rhizopus—Rhizopus delemar*—from the same source. The amylo process was developed by Calmette and Voidin and established in a plant at Seclin near Lille in 1895.

In Calmette's process maize starch or other grain starches were gelatinized under pressure with a small amount of mineral acid present, often after steeping. After sterilizing by boiling in the fermenter and cooling, the mash was inoculated at 40°C and vigorously agitated either mechanically or purely by aeration. This was in fact in many ways a typical modern submerged mould fermentation process. After about 24 h, the broth was cooled to 32°C. Yeast was added, aeration was continued for a further 6 h, and alcohol production proceeded for about a further two days. One of the main difficulties suffered by the amylo process was contamination. At that time not enough was known about the precautions necessary for operating on a large scale under sterile conditions. In one modification of the process, that of Boulard (1931), a very active strain of *Mucor* was used, which it was claimed was resistant to infection, and the mould and the yeast were added simultaneously thereby cutting down fermentation time and enabling the process to be run in open vats. The amylo process was never used in England because the excise laws demanded that the gravity of the wort was determined before alcohol production.

In a later modification of the amylo process Erb & Hildebrandt (1946) grew *Rhizopus delemar* in submerged culture and used the mould in addition to malt for effecting saccharification. *Saccharomyces anamensis* was then added to convert the sugars to alcohol. This work was carried out in Newark, New Jersey, in 1933, but not published for a further 13 years. The process was run in a 2500 gallon pilot plant and ultimately in 16 600 gallon fermenters, and the medium was aerated under 5 lbs pressure. It was claimed to give about 10% more alcohol than when malt was used as the sole saccharifying agent. A rather similar process was described by Le Mense *et al.* (1949). They grew *Aspergillus niger* in submerged liquid culture on a 14 500 gallon scale and used the broth for subsequent saccharification of starch. The next step was naturally the isolation of the enzyme in crude form, and now the majority of enzymes obtained from micro-organisms are made by growing the organism in submerged culture and subsequent isolation of the enzyme. However, the koji process is still used to this day in Japan for the production of certain enzymes.

Gallic acid

Another early mould fermentation process was that used for preparing gallic acid. Gallic acid, which is used in the tanning, printing and other industries, occurs naturally as esters and glycosides in plant tannins, and is prepared either by chemical or enzymatic hydrolysis. In one old method of production gall nuts were piled in heaps and moistened with water. Mould then grew in the heaps and after about a month gallic acid was leached out. It was van Tieghem (1867) who first established the industrial importance of *Aspergillus niger* when he identified this organism as the predominant mould in the gallic acid fermentation and gave it its name. Calmette (1902) worked out a process in which an extract of tannin was fermented by a mould he called *Aspergillus gallomyces*, now believed to be a strain of *Aspergillus niger*. The fungus was grown submerged in an aerated agitated vessel, and is another early example of the submerged fermentation process.

The microbial production of gallic acid from tannins is still being studied and a recent paper (Ikeda *et al.*, 1972) shows that *Aspergillus niger* is still the most satisfactory micro-organism for gallic acid production from Chinese gallotannin, though *Penicillium chrysogenum* is more effective with another type of tannin.

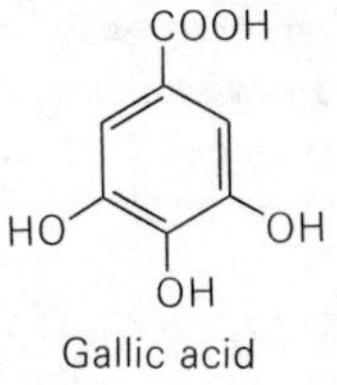

COOH

HO OH

OH

Gallic acid

Fig. 6.1

6.3 Fungal Fermentations in Liquid Culture

Surface culture

CITRIC ACID If a mould is grown in trays, but instead of using a solid medium it is grown on the surface of a liquid medium, one has the conventional surface process, exemplified in particular by the citric acid fermentation. This developed naturally from the growth of moulds in laboratory flasks, a technique going back about 80 years to the early days of Wehmer. Wehmer's work, which started in 1891, first showed that moulds were capable of producing organic molecules such as oxalic, citric and fumaric acids from sugars. Wehmer gave a great impetus to the study of mould biochemistry, and can justly be regarded as the founder of this branch of science. He worked out a process for making citric acid using strains of *Citromyces*—later to be classified as species of *Penicillium*. The manufacture of citric acid by this process was attempted at a factory in Thann, Alsace, which started up in 1893, but was abandoned ten years later. Amongst the difficulties which caused this were degeneration of the organism, contamination, long fermentation time and high costs—penalties for being ahead of one's time! Wehmer believed that citric acid was characteristic of the *Citromyces* and that the natural product of *Aspergillus niger* was oxalic acid. This was disproved by Currie (1917) working at the Bureau of Animal Industry in Washington, who showed that a number of different strains of *Aspergillus niger* produced both citric acid and oxalic acid depending on the experimental conditions. Currie's work was the foundation of the subsequent citric acid fermentation industry, though there is one prior reference to a patent taken out by Zahorski in 1913 for a method for obtaining citric acid from sugar solutions with *Sterigmatocystis nigra*. Zahorski considered that this differed from *Aspergillus niger*, but it was later regarded as the same. Thom & Currie (1916) showed that a number of strains of black *Aspergilli* made citric acid. Currie then carried this far further, and was the first to demonstrate a number of now well known aspects of this fermentation. He worked out a medium—sucrose 125-150 g dm^{-3}, ammonium nitrate 2.0–2.5 g dm^{-3}, potassium dihydrogen phosphate 0.75–1.0 g dm^{-3}, magnesium sulphate heptahydrate 0.2–0.25 g dm^{-3}, with initial acidification to pH 3.4 to 3.5—which is the basis for that used ever since for citric acid production from pure sugar. He was the first to show the importance of using pure reagents for this fermentation (something that many subsequent workers neglected to do and which renders much of their work valueless). He at times used doubly distilled water and recrystallized his reagents, the sugar sometimes as often as five times in order to remove iron and thus to prove its necessity. He showed that the highest yield of citric acid occurred when development of mycelium was restricted and not when it was stimulated. He worked also in shallow pans and he prepared several pounds of calcium citrate. Currie's yields were so good and his process was developed in such detail that it was obviously of great interest to citric acid manufacturers.

Currie later joined Chas. Pfizer & Co. Inc. and thus was subsequently partly responsible for the development of the company's citric acid process. This was first operated on a commercial scale in 1923 and to this day remains one of the best kept industrial secrets. At one time tartaric acid, made from argols and lees, by-products of the wine industry, was the cheapest acid for the soft drinks and similar industries, and citric acid was made in smaller quantities by several of the same companies from precipitated calcium citrate or citrus fruit concentrates. In 1922 Italy produced nearly 90% of the world's supply of calcium citrate and in 1927 the Italian government attempted to safeguard its citric acid industry by putting an embargo on exports of calcium citrate. This was a short sighted policy, because with the advent of the fermentation process, citric acid had become very much cheaper. Soon it was being produced by fermentation in England, Belgium, Czechoslovakia and probably Russia, as well as the United States.

Fernbach, Yuill and Rowntree & Co. Ltd. had taken out patents in 1925 which essentially were based on Currie's work, but which had as a novel feature the prior acidification of the medium to pH 1.8, thereby it is said rendering sterilization unnecessary, a claim that would no longer be accepted as true, though acidification to such a low pH is important for high citric acid yields. This work formed the basis of the first process used by John & E. Sturge Ltd. at their plant in Selby, Yorkshire.

Raistrick examined the metabolic products of the black *Aspergilli* as part of his lifetime's work on the biochemistry of moulds. Over the course of some forty years he and his colleagues published a series of papers, and showed that moulds could produce a vast number of metabolites of many different chemical types. Raistrick, when working under Sir Frederick Hopkins at Cambridge, became interested in mould biochemistry as the result of reading Wehmer's papers, and studied the mechanism of the production of oxalic acid from sugar by *Aspergillus niger* (Raistrick & Clark, 1919). The publication of this work had a great effect, not only on Raistrick's own career, but on the whole subsequent development of mould biochemistry. Subsequently, he was asked to lead a team working in the Ardeer Laboratories of Nobel's Explosive Company to investigate the products of mould fermentations. The company had been impressed by two fermentation processes that looked likely to become of very great importance. One was the acetone-butanol fermentation and the other the production of glycerol by yeast. Many people predicted that Raistrick's investigations would merely give expensive methods for making carbon dioxide from sugar, but an initial series of metabolic studies (Raistrick *et al.*, 1931) showed that moulds produced many products other than carbon dioxide. However, ultimately it appeared that products of practical use to the explosives industry were unlikely to be found, and arrangements were made for Raistrick and his team to transfer in 1929 to the London School of Hygiene and Tropical Medicine. The work on mould biochemistry continued there until 1964.

During this period Raistrick and his co-workers were the first to demonstrate the production of numerous substances by fungi, several of which later achieved industrial importance, for example griseofulvin. Raistrick developed a process for citric acid manufacture, but details of this were never published, though it is referred to in a review by Clutterbuck (1936), who mentions that on a semi-large scale overall yields of 87% were obtained.

Another relatively early manufacturer of citric acid was the Czech company Montan und Industrialwerke of Kasnejov near Pilsen, who began production in 1930 using a strain of *Aspergillus niger* improved by selection from a culture used by Bernhauer, then professor in Prague. The process used here involved the treatment of molasses solutions with ferrocyanide, first tried apparently in an attempt to stop contamination with *Penicillium purpurogenum*, a bugbear to the citric acid producers. The use of ferrocyanide to purify molasses solutions is also covered by a French patent granted to Mezzadroli in 1938. The Czech process was subsequently acquired by the German firm of John A. Benkiser, who were operating it when the British Intelligence Objectives Sub-Committee (B.I.O.S.) visited their Ladenberg factory in 1945. Publication of the B.I.O.S. report gave publicity to this process and many of the plants now operating are believed to use variations of it (see also Chapter 8).

GLUCONIC AND FUMARIC ACIDS The production of mould metabolic products by the submerged fermentation process was very clearly demonstrated in the production of gluconic acid. Bernhauer (1924) showed that certain strains of *Aspergillus niger* would produce gluconate in surface culture if the pH was kept near neutrality, and the operation of this process in submerged culture was reported by Currie, Kane & Finlay (1933) who showed that glucose, maltose or sucrose could be converted to gluconic acid by growing fungi such as *Aspergillus niger* or *Penicillium luteum* in deep culture with aeration and agitation. With *Aspergillus niger* and with glucose as the substrate a yield of 90% of the theoretical was claimed.

Wehmer (1918) was also the first to show that fumaric acid was produced by moulds and he isolated an organism, *Aspergillus fumaricus*, closely allied to *Aspergillus niger*, that would do this. However, by 1928 the mould had changed its habits, had lost completely the power to produce fumaric acid and made citric and gluconic acids instead. Wehmer (1928) discussed possible reasons for this change in behaviour and concluded that the same species may exist in more than one physiological form. This must be one of the earliest authenticated examples of the biochemical degeneration of mould cultures in storage.

Takahashi and his co-workers showed that a number of *Rhizopus* species produced fumaric acid (Takahashi & Sakaguchi, 1927) and the use of *Rhizopus nigricans* to make fumaric acid in submerged culture was patented in 1943 both by Waksman and by Kane, Finlay and Amann working for Merck & Co. Inc. and Chas. Pfizer & Co. Inc. It was later shown that fumaric acid could be made more cheaply by purely chemical means. However, the development of the submerged process for its manufacture was of very considerable importance, and not just of historic interest, since many of the problems that had to be solved in making fumaric acid by submerged culture on a manufacturing scale were the same as those that arose in the production of penicillin. The fermentation process was run near neutrality; it was relatively slow, required aeration and foaming was very difficult to control. All this made it very liable to contamination. By the time that Florey and Heatley travelled to the U.S. with their initial information on the potentialities of penicillin, Pfizer and presumably Merck had acquired considerable expertise in the handling of these problems. It is scarcely surprising that these companies thought of trying to make penicillin in submerged culture, nor that they were among the first to achieve it. Another company which was involved in the early penicillin work in America was Corn Products, whose background expertise was in the acetone-butanol process, but this was an anaerobic bacterial fermentation, which posed quite different problems to those which arose in the penicillin fermentation.

PENICILLIN The story of penicillin, from Fleming's first discovery of it in 1928, through Raistrick's abortive attempts to isolate it (Clutterbuck, Lovell & Raistrick, 1932), to the final triumphant achievements of Chain *et al.* (1940) has been so often told that it will not be repeated in detail here.

Most of the techniques already mentioned were used in one form or another for its manufacture. Spores of *Penicillium notatum* were at one time grown on solid medium in surface culture; Kemball Bishop & Co. Ltd. grew the mould on a liquid medium in surface culture in sterile rooms while the American companies largely used submerged culture. Most of the British companies used surface culture, scaling up the laboratory process by numbers and not by volume, by using bottling plant techniques to fill large numbers of Glaxo flasks, or milk bottles which were subsequently sloped. One story that is not generally known is that when Florey's original papers were read in Holland, then already in German occupation, certain enterprising Dutchmen found a culture of *Penicillium notatum* in the national collection and used it to produce penicillin and to treat patients without German knowledge.

Industries could not suddenly grow in the way the mould fermentation industry did in the years immediately after the 1939–1945 war unless most or all of the technical bottlenecks had been removed. However, it was the discovery of penicillin, soon followed by other antibiotics, that provided the main initial impetus for the rapid expansion of the fermentation industry, but the essential background was ready. During the inter-war years biochemistry had become a major science. Krebs had published his fundamental work on the tricarboxylic cycle in 1937 and this had immediately made clear the relationship of many of the simpler mould metabolic products. The steps in the glycolytic pathways from hexoses to pyruvic acid had also been largely elucidated at about the same time, even if the existence of more than one pathway had obscured the overall picture, which was not to be clarified finally until about ten years later. The discovery of the role of the B-group of vitamins as coenzymes in various fundamental cell systems added to the background knowledge. Study of the mineral requirements of moulds goes back to the work of Raulin (1869) but was followed up by many others. Steinberg in a series of papers (1935, 1936, 1937, 1938) examined the effect of various metals on the growth of *Aspergillus niger*. He appreciated the necessity to treat nutrients extensively in order to attain a high degree of purity.

On the biological side the ubiquitous nature of fungi had been appreciated only surprisingly recently; Waksman had written a paper in 1916 entitled 'Do fungi live and produce mycelium in the soil?' But he had isolated over three hundred species from twenty-five soil samples to prove that they do, and had established that soil samples were among the most obvious places in which to look for new micro-organisms.

Microbial genetics

But the academic work that in retrospect can be seen to have contributed most significantly to the development of the fungal fermentation industry was that on the genetics of micro-organisms. That strains of moulds tend to be unstable had been realized for many years. Over two hundred years ago the Swiss Albrech von Haller referred to them as 'a mutable and treacherous tribe' and, as already indicated, it was realized by 1940 that this mutability applied to their biochemical characteristics as much if not more than it applied to their morphological ones. The use of ultra-violet and other radiations to effect mutations in fungi began in about 1936 and the first serious attempt to deliberately use chemical agents to obtain fungal mutants was described by Thom & Steinberg (1939) though they mention that Schiemann had in fact produced three *Aspergillus niger* mutants by chemical means as far back as 1912. Thom and Steinberg also chose *Aspergillus niger* as the organism they chiefly studied. They wrote 'Of all the moulds which have caught man's fancy as experimental material the black aspergilli—under the name *Aspergillus niger* van Tieghem—have had the greatest share of attention, starting with the basic assumption that they were a single species, hence easy to identify'. Thom and Steinberg also worked with *Aspergillus amstelodami* and *Penicillium caseicolum*, which like *Aspergillus niger* were known to be very stable and not liable to spontaneous mutation. A number of different compounds were used as mutagenic agents. The aspergilli were found to be particularly sensitive to 0.2% sodium nitrite solution, which caused the production of many morphological mutants.

But it was through the work of Beadle and Tatum with *Neurospora crassa* that the real break-through occurred. In their first paper (1941) entitled 'Genetic control of biochemical reactions in *Neurospora*' they stated that 'It is entirely tenable to suppose that these genes are themselves part of the system, control or regulate specific reactions in the system either by acting directly as enzymes or by determining the specificities of enzymes'. Whereas this was not a completely new conception, it had been suggested amongst others by Haldane (1937), it had never been stated so explicitly and certainly never demonstrated so conclusively. Beadle and Tatum also introduced the technique of growing their isolates on a rich and a very restricted medium, introducing the terms complete and minimal media. In their first paper, using X-rays as the mutagenic agent, they identified three auxotrophs (though this term was not then in use), requiring pyridoxine, the thiazole moiety of thiamine, and *p*-aminobenzoic acid for growth. Beadle & Tatum (1945) later carried out much more extensive investigations with *Neurospora crassa*, testing 69 198 single spore isolates after treatment with X-rays and ultra-violet light, and obtained 380 mutants with altered nutritional requirements, mostly for the B-vitamins and for essential amino acids.

The use of mutation techniques for improving the performance of commercially important moulds was extensively used for the first time with *Penicillium notatum* and *Penicillium chrysogenum* in connection with penicillin production. Bonner (1946) found 398 auxotrophs when testing 85 595 isolates of these moulds, with requirements essentially the same as those of the *Neurospora* mutants. But the big improvement in penicillin titres came with the development of the high producing strains X-1612 and Q176 obtained by X-ray and ultraviolet mutation from the strain of *Penicillium chrysogenum* isolated from a mouldy melon bought in Peoria (Raper & Alexander, 1945).

Obtaining better yielding cultures by mutation is the procedure that has probably led to the greatest improvement in costs in the fermentation industry. Originally, purely empirical methods for testing mutants were used, and there are sometimes still no better methods, although increasingly more sophisticated approaches are being employed. For example, if feed-back inhibition is operative, auxotrophs lacking the ability to make the final compound of a series of steps may give enhanced yields of a compound early in the series. Similarly, mutants lacking a particular enzyme may give enhanced yields of the substrate on which the enzyme acts.

Submerged liquid culture

ORGANIC ACID PRODUCTION Development of submerged fermentations on a small scale was greatly helped by the invention of the shake flask technique for the small scale simulation of submerged culture fermentations (Kluyver & Perquin, 1933). This technique has its limitations, particularly if regular sampling is required to follow the course of a fermentation, but it has been extensively used, especially for preliminary testing of strains and mutants, and as the means of growing the first stage inoculum in many industrial fermentations. Perquin (1938) used the shake flask technique to study the production of citric acid by *Aspergillus niger*. He is usually credited to be the first to show that citric acid could be produced in submerged culture, though in fact this had previously been shown by Amelung (1930). Regrettably, Amelung's work has been almost entirely neglected. He worked with a strain of *Aspergillus niger* var. *japonicus* and showed that if a slow stream of air was passed through a 15 cm deep solution of 10% sucrose and salts, the mould grew beneath the surface and produced a considerable amount of citric acid, though not as much as in surface culture. Perquin's work, which was published as a thesis in Dutch just before the beginning of the Second World War, was very much more extensive and gave clearer results. It was consequently not widely read for a further six years and even then the information only spread slowly. Thus Wells & Ward (1939) stated that all reliable evidence indicated the impossibility of using submerged culture techniques for citric acid production and expressed the view that some vital derangement of the enzyme system was responsible.

Subsequent development of the submerged process for citric acid production is chiefly associated with the work of Johnson's group at the University of Wisconsin (Johnson, 1954), with work at the National Research Laboratories, Ottawa, by Martin (1957) and by Clark (1962) both with a number of others and by the series of patents taken out by Miles Laboratories Inc. A number, but not all, of the companies producing citric acid by fermentation now use the submerged culture process, but this remains one of the most difficult fermentations to carry out regularly and successfully.

Most of the commercial antibiotics are produced by *Streptomyces* sp. Since the techniques for large scale production are similar whatever type of micro-organism is used, the vast amount of work carried out in the decade after the war on the development and improvement of these techniques has obviously been applicable to all mould fermentation products. New products have been produced, although the background to the industry has not altered. Despite much talk no filamentous mould product is yet made on an industrial scale by continuous culture, at least to the author's knowledge. Relatively few fairly simple compounds are made by mould fermentation processes. This is scarcely surprising, because in view of the expensive equipment and relatively high operating costs of fermenters compared with chemical plant, and the relative slowness of microbiological processes compared to purely chemical ones, costs tend to be higher. Only if the chemical synthesis involves a number of steps or involves an expensive procedure like the isolation of an optical isomer is microbiological production likely to be cheaper.

The failure of microbiologically produced fumaric acid to compete with that made by chemical synthesis has already been mentioned. Similarly, the mycological production of lactic acid using species of *Mucor* and *Rhizopus* has never been able to compete with the more efficient bacterial fermentation, although the process has a relatively long history and has been thoroughly studied (Ward *et al.*, 1938). Oxalic acid was at one time sold as a by-product of the citric acid fermentation. Kojic acid is made by *Aspergilli* of the *flavus-oryzae* group and itaconic acid by *Aspergillus itaconicus* and *Aspergillus terreus*. Problems in the production of both these acids resemble those of the citric acid fermentation. The production of kojic acid was first detected by Saito (1907) and more thoroughly studied by Yabuta (1912), who later elucidated its chemical structure. It was discovered independently by Raistrick's group in 1923 (Birkinshaw *et al.*, 1931). The development of the itaconic acid fermentation has a rather similar history. It was first shown by Kinoshita (1929) that it was made by *Aspergillus itaconicus* and in 1939 Raistrick and others (Calam, Oxford & Raistrick, 1939) studied its production by *Aspergillus terreus*. A submerged fermentation process was later developed at Peoria (Lockwood & Ward, 1945), and independently by Chas. Pfizer & Co. Inc. (Kane *et al.*, 1945).

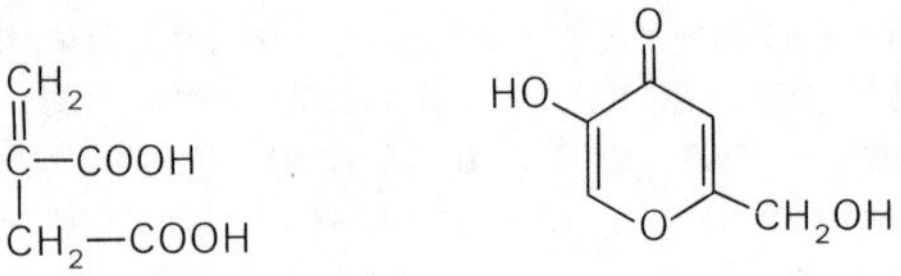

$CH_2{=}C(COOH){-}CH_2{-}COOH$

Itaconic acid

Kojic acid (HO, O, CH_2OH)

Fig. 6.2

No history of the development of industrial fermentations would be complete without mentioning the work carried out at the U.S. Department of Agriculture's Northern Regional Research Laboratory at Peoria, Illinois, and, at what can be regarded as its predecessor, the Colour and Farm Waste Division of the Bureau of Chemistry and Soils at Arlington Farms, Virginia. Amongst mould products worked on at these establishments can be listed citric, itaconic, kojic, gluconic, fumaric and lactic acids, riboflavin, β-carotene, gibberellins, mannitol and various enzymes. The Peoria laboratory was intimately involved in the war-time work on penicillin, and was responsible for the isolation of the parent of all industrial strains of *Penicillium chrysogenum* and for the discovery of the vital effect of corn steep liquor in penicillin production. The rotary drum fermenter was first used by Takamine, but was later developed at Arlington Farms and, although this proved to be of limited importance, it had the effect of focusing attention on submerged culture. A history of the Peoria laboratory has recently been written (Ward, 1970) and it will therefore not be mentioned further here.

ANTIBIOTICS Although the various penicillins as a group are the most widely used of all antibiotics, filamentous moulds in general have not proved to be as satisfactory a source of antibiotics as bacteria, particularly *Streptomyces*. Except for the penicillins the only mould product that has a wide and lasting use in the antibiotic industry is cephalosporin C, closely related to the penicillins, and even it is only used as the starting material for chemical conversion to the much more potent semi-synthetic cephalosporins. Following the original isolation of the strain of *Cephalosporium acremonium* by Brotzu (1948) from the sea near a sewage discharge in Sardinia, work was taken up by Abraham and others at the School of Pathology in Oxford. It was shown that the organism produced three different antibiotics, one of them a penicillin, but that another, while resembling the penicillins, differed in certain essential characteristics such as stability to acids and enzymes. It was finally shown (Abraham & Newton, 1961) to have the structure:

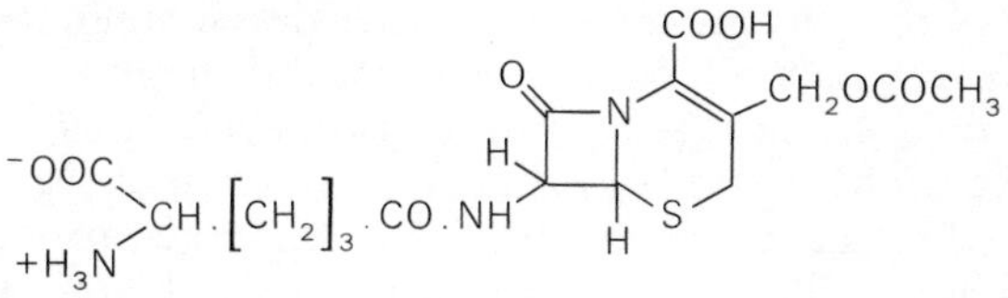

Cephalosporin C

Fig. 6.3

The process for making cephalosporin C was licensed by the National Research and Development Corporation to various pharmaceutical companies, two of which, Glaxo in England and Eli Lilly in the U.S., showed that by replacing the L-aminoadipic acid and acetoxymethyl side chains with other groups therapeutically useful products could be obtained.

Griseofulvin has already been mentioned as one of the products first isolated by Raistrick's group from *Penicillium griseofulvum*—Dierck (Oxford, Raistrick & Simonert, 1939). During the last war an Imperial Chemical Industries team headed by Dr. P. W. Brian was investigating the presence of various moulds in soil from Wareham Heath, Dorset. They isolated an organism, *Penicillium janczewskii* (Brian, Curtis & Hemming, 1946) which made a compound they called 'curling-factor' which caused stunting and excessive branching in germ tubes of *Botrytis allii* and other fungi with chitin walls. They showed this to be chemically identical to griseofulvin (Grove & McGowan, 1947). Later griseofulvin was found to be a metabolic product of a number of species of *Penicillium*. Its production in submerged culture was patented by Glaxo Laboratories Ltd. (Rhodes *et al.*, 1957) and it was first used purely in agriculture for the control of *Botrytis*. This necessitated a study of its mammalian toxicity, and when it was found to be essentially non-toxic its oral use in the control of certain fungal skin diseases was developed.

RIBOFLAVIN The history of the production of the vitamin riboflavin by industrial fermentation goes back to about 1935 when Guilliermond and others were studying a plant parasite, the Ascomycete *Eremothecium ashbyii*. This they found at times to produce yellow crystals, which were identified as vitamin B_2 (riboflavin) by Raffy (1937). Riboflavin can be made relatively easily by chemical synthesis, but it is produced in large amounts by a variety of micro-organisms and production by fermentation is the method of choice. Industrial production in submerged culture with *Eremothecium ashbyii* was developed by several companies about 25 years ago. The morphologically similar organism *Ashbya gossypii* also produces riboflavin in considerable amounts and using this organism production was developed at the Northern Regional Research Laboratories. This process has been stated to be the basis for most commercial riboflavin production at the present time and it has the advantage of using a more stable organism.

β-CAROTENE The production of β-carotene in moulds was first recorded by Schopfer & Jung (1935). They noted that both *Mucor hiemalis* and *Phycomyces blakesleeanus* synthesised β-carotene when grown on media containing asparagine or glycine, and that this occurred to a greater extent with mycelia of the (+) mating type. A later development occurred when Barnett, Lilly & Krause (1956) showed that the amount of carotene produced when the (+) and (−) strains of *Choanephora cucurbitarum* were grown in mixed culture was more than ten times that of either mating type alone. This led to the development of the submerged process by Ciegler (1965) and others at Peoria, first in shake flasks and later on a semi-pilot plant scale using mixed mating types of *Blakeslea trispora*. Mated cultures produce a number of compounds (the so-called β factor) that can stimulate carotene production in unmated cultures (Sutter & Rafelson, 1968). The (+) strain produces the β factor and the (−) strain the extra β-carotene. At one time it was thought that this hormonal action of soluble compounds would not give the same stimulation of carotene production as mixing the mating types, but this was shown not to be the case, stimulation with the equivalent amount of hormones induces as much carotene production in the (−) strain as was obtained with mixed strains (Van den Ende, 1968). The hormones have been shown to be trisporic acids (Caglioti *et al.*, 1966). (−) Trisporic acid B, the most active component, has the formula:

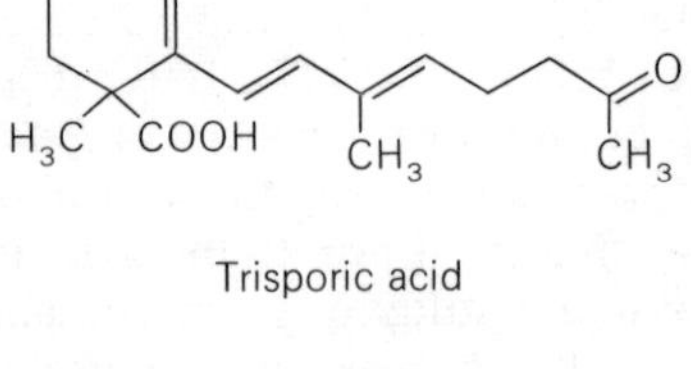

Trisporic acid

Fig. 6.4

The other particularly interesting aspect of this fermentation is the constitution of the medium. Originally it was found that the addition of 0.1% of β-ionone was necessary for maximum yield.

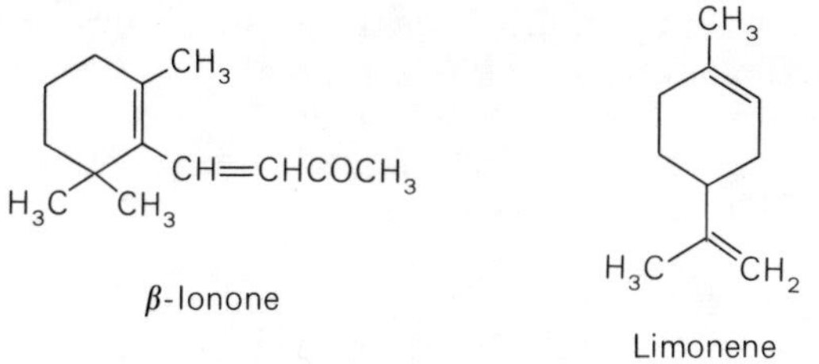

Fig. 6.5

Later it was shown that this could be replaced by cheap citrus by-products (citrus oil, citrus pulp or citrus molasses) giving yields of carotene in the order of 1 g dm^{-3}. Indications are that it is the presence of limonene that is responsible. Spent mycelium of *Blakeslea trispora* from a previous fermentation was found to be an even better source of carotenoid precursors.

In 1963 production costs for the crude dried solids made at a rate of 5000 tons per year were estimated to be $31.35 per kg of β-carotene. But the market for carotene is small; it is primarily used for colouring margarine. What is required is a cheap source of vitamin A, competitive with its synthetic production. However, no efficient *in vitro* method of conversion of β-carotene to vitamin A exists, and the *in vivo* conversion is an inefficient process in many animal species; cows can suffer from vitamin A deficiency on a diet with plenty of carotene.

GIBBERELLINS The story of the gibberellins can be said to have begun when Kurosawa (1926) showed that culture filtrates of *Fusarium moniliforme* stimulated the growth of rice seedlings. It had long been known that infection of rice with this organism, the perfect form of which is *Gibberella fujikuroi*, caused plants to wilt and die, although the first symptom of infection was often stem elongation. One active compound, gibberellin A, was isolated by Japanese workers in 1938. Later, as has been so often the case with fermentation products, it was found that the organism produced several closely related compounds with similar activities. In the early Japanese work *Gibberella fujikuroi* was grown in surface culture and the yield of gibberellins was very low, but later Kitamura *et al.* (1953) achieved a yield of 8 mg dm^{-3} in submerged culture. Gibberellin has now become an article of commerce but its uses have never developed to the degree once predicted.

ERGOT ALKALOIDS The production of the ergot alkaloids has of course a history going back into the middle ages. Outbreaks of ergot poisoning caused by eating food made from infected rye have occurred throughout the centuries, but it was not until 1764 that Munchhausen recognized that the peculiar growth on rye known as ergot was fungal in origin and Tulasne in 1853 showed that ergot was the sclerotial stage of the filamentous fungus he named *Claviceps purpurea.* Attempts to make the various ergot alkaloids in artificial cultures of species of *Claviceps* for a long time were unsuccessful, and ultimately Abe and co-workers in Japan succeeded in isolating a series of clavine alkaloids from one strain (Abe & Yamatodani, 1964). These differ in structure from the lysergic acid derivatives, which are those of medical importance. The major break-through did not come until when Arcamone *et al.* (1960) showed that a strain of *Claviceps paspali* had the ability to produce 10–20 μg dm^{-3} of the required lysergic acid alkaloids when grown in submerged culture. Subsequently, yields have been considerably improved.

MICROBIAL TRANSFORMATIONS The use of fungi in effecting transformations (simple one or two step alterations to relatively complicated molecules as part of a chemical synthesis) stems from the work of Peterson & Murray (1952) who showed that a strain of *Rhizopus arrhizus* would convert progesterone into 11α-hydroxyprogesterone, though the fact that bacteria and yeasts could hydroxylate steroids had been shown many years earlier by Mamoli & Vercellone (1937). Hydroxylation is often the first step in the breakdown of an organic compound by a micro-organism, but it is a reaction that is often difficult, lengthy and expensive to do chemically. Hydroxylation in the 11β-position is an essential step in the manufacture of the anti-rheumatic corticosteroids and this can be done on a manufacturing scale directly with *Curvularia lunata*. Hydroxylation in the 11α-position with *Rhizopus nigricans* goes in better yield, but it is then necessary chemically to epimerize the compound. Micro-organisms have since been found that will hydroxylate in nearly every position in the steroid molecule. But the potential uses of fungi in organic syntheses are not confined to steroids. A further example of a hydroxylation that has been carried out on a manufacturing scale is the conversion of the schistosomicide lucanthone to the more active compound hycanthone which is effected by *Aspergillus sclerotiorum* (Rosi *et al.*, 1967).

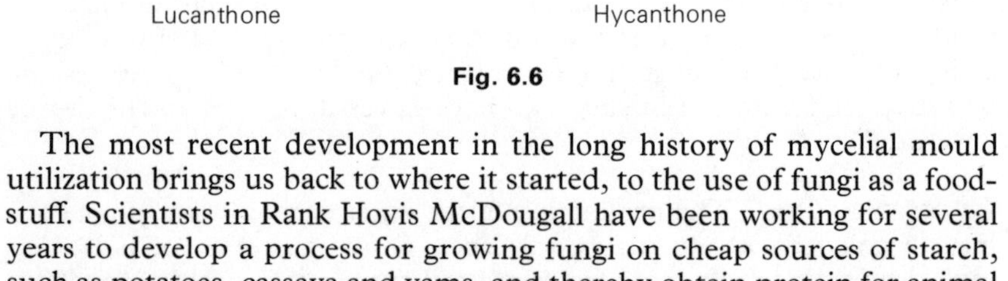

$NH.CH_2.CH_2.N(C_2H_5)_2$ ... CH_3 — Lucanthone → $NH.CH_2.CH_2.N(C_2H_5)_2$ ... CH_2OH — Hycanthone

Fig. 6.6

The most recent development in the long history of mycelial mould utilization brings us back to where it started, to the use of fungi as a foodstuff. Scientists in Rank Hovis McDougall have been working for several years to develop a process for growing fungi on cheap sources of starch, such as potatoes, cassava and yams, and thereby obtain protein for animal feeds. A pilot plant producing about 150 tons a year of a species of *Fusarium* has recently been built and large scale trials of the product will be carried out. One claimed advantage of using fungi rather than bacteria or yeasts is that the fibrous nature of the fungi will make it easier for them to be fabricated into pseudo meat products, for instance steaks. The further development of this project is one that mycologists, as well as food technologists and many others, will be following with great interest over the next few years.

6.4 References

ABE, M. & YAMATODANI, S. (1964). Preparation of alkaloids by saprophytic culture of ergot fungi. *Progress in Industrial Microbiology* **5**, 203–29.

ABRAHAM, E.P. & NEWTON, G.G.F. (1961). The structure of cephalosporin C. *Biochemical Journal* **79**, 377–93.

AMELUNG, H. (1930). Wachstum und Säurebildung von *Aspergillus niger* unter Wasser. *Chemiker-Zeitung* **54**, 118.

ARCAMONE, F., BONINO, C., CHAIN, E.B., FERRETTI, A., PENNELLA, P., TONOLO, A. & VERO, L. (1960). Production of lysergic acid derivatives by a strain of *Claviceps paspali* in submerged culture. *Nature, London* **187**, 238–9.

BARNETT, H.L., LILLY, V.G. & KRAUSE, R.F. (1956). Increased production of carotene by mixed positive and negative cultures of *Choanephora cucurbitarum*. *Science* **123**, 141.

BEADLE, G.W. & TATUM, E.L. (1941). Genetic control of biochemical reactions in *Neurospora*. *Proceedings of the National Academy of Sciences of the United States of America* **27**, 499–506.

BEADLE, G.W. & TATUM, E.L. (1945). *Neurospora*. Methods of producing and detecting mutations concerned with nutritional requirements. *American Journal of Botany* **32**, 678–86.

BERNHAUER, K. (1924). Zum Problem der Säurebildung durch *Aspergillus niger* (Vorläufige Mitteilung). *Biochemische Zeitschrift* **153**, 517–21.

BIRKINSHAW, J.H., CHARLES, J.H.V., LILLY, C.H. & RAISTRICK, H. (1931). Kojic acid. *Philosophical Transactions of the Royal Society of London. Series B* **220**, 127–38.

BONNER, D. (1946). Production of biochemical mutations in *Penicillium*. *American Journal of Botany* **33**, 788–91.

BOULARD, H. (1931). Société d'exploitation des procédés.

BRIAN, P.W., CURTIS, P.J. & HEMMING, H.G. (1946). A substance causing abnormal development of fungal hyphae produced by *Penicillium janczewskii* Zal. (1). Biological assay, production and isolation of 'curling' factor. *Transactions of the British Mycological Society* **29**, 173–8.

BROTZU, G. (1948). Ricerche su di un nuovo antibiotico. *Lavori dell'istituto d'Igiene di Cagliari*.

CAGLIOTI, L., CAINELLI, G., CAMERINO, B., MONDELLI, R., PRIETO, A., QUILICO, A., SALVATORI, T. & SELVA, A. (1966). The structure of trisporic-C acid. *Tetrahedron*, Suppl. No. 7, 175–87.

CALAM, C.T., OXFORD, A.E. & RAISTRICK, H. (1939). Itaconic acid. A metabolic product of a strain of *Aspergillus terreus* Thom. *Biochemical Journal* **33**, 1488–95.

CALMETTE, A. (1892). Annales de l'Institut Pasteur **6**, 605–20.

CALMETTE, A. (1902). Verfahren zur Umwandlung von Tannin in Gallussäure. *German Patent* 129,164.

CHAIN, E., FLOREY, H.W., GARDNER, A.D., HEATLEY, N.G., JENNINGS, M.A., ORR-EWING, J. & SANDERS, A.G. (1940). Penicillin as a chemotherapeutic agent. *Lancet* **2**, 226–8.

CIEGLER, A. (1965). Microbial carotenogenesis. *Advances in Applied Microbiology* **7**, 1–34.

CLARK, D.S. (1962). Submerged citric acid fermentation of sugar beet molasses. *Industrial and Engineering Chemistry Product Research and Development* **1**, 59–62.

CLUTTERBUCK, P.W. (1936). Recent developments in the biochemistry of moulds. *Journal of the Society of Chemical Industry* **55**, 55T–61T.

CLUTTERBUCK, P.W., LOVELL, R. & RAISTRICK, H. (1932). The formation from glucose by members of the *Penicillium chrysogenum* series of a pigment, an alkali-soluble protein and penicillin—the antibacterial substance of Fleming. *Biochemical Journal* **26**, 1907–18.

CURRIE, J.N. (1917). The citric acid fermentation of *Aspergillus niger*. *Journal of Biological Chemistry* **31**, 15–37.

CURRIE, J.N., KANE, J.H. & FINLAY, A. (1933). Process for producing gluconic acid by fungi. *United States Patent* 1,893,819.

ERB, N.M. & HILDEBRANDT, F.M. (1946). Mold as an adjunct to malt in grain fermentation. *Industrial and Engineering Chemistry* **38**, 792–4.

FERNBACH, A., YUILL, J.L. & ROWNTREE & CO. LTD. (1927). Process for the production of citric acid. *British Patents* 266,414 and 266,415.

FLEMING, A. (1929). On the antibacterial action of cultures of a *Penicillium* with special reference to their use in the isolation of *B. influenzae*. *British Journal of Pathology* **10**, 226–36.

GROVE, J.F. & MCGOWAN, J.C. (1947). Identity of griseofulvin and 'curling-factor'. *Nature, London* **160**, 574.

GUILLIERMOND, A., FONTAINE, M. & RAFFY, A. (1935). Sur l'existence dans *l'Eremothecium ashbyii* d'un pigment jaune se rapportant au groupe des flavines.

GUILLIERMOND, A., FONTAINE, M. & RAFFY, A. (*cont.*) *Comptes Rendus Hebdomadaires des Séances de l'Académie des Sciences* **201**, 1077–80.

HALDANE, J.B.S. (1937). The biochemistry of the individual. *Perspectives in Biochemistry*. Ed. J. Needham & D.E. Green. Cambridge University Press.

HAO, L.C., FULMER, E.I. & UNDERKOFLER, L.A. (1943). Fungal amylases as saccharifying agents in the alcoholic fermentation of corn. *Industrial and Engineering Chemistry* **35**, 814–18.

IKEDA, Y., TAKAHASHI, E., YOKOGAWA, K. & YOSHIMURA, Y. (1972). Screening for micro-organisms producing gallic acid from Chinese and Tara tannins. *Journal of Fermentation Technology* **50**, 361–70.

JOHNSON, M.J. (1954). The citric acid fermentation. *Industrial Fermentations*. Ed. L.A. Underkofler & R.J. Hickey. Vol. 1, 420–45. Chemical Publishing Co. Inc.

KANE, J.H., FINLAY, A. & AMMAN, P.F. (1943). Production of fumaric acid. *United States Patent* 2,327,191.

KANE, J.H., FINLAY, A. & AMMAN, P.F. (1945). Production of itaconic acid. *United States Patent* 2,385,283.

KINOSHITA, K. (1929). Formation of itaconic acid and mannitol by a new filamentous fungus. *Journal of the Chemical Society of Japan* **50**, 583–93.

KITAMURA, H., KAWARADA, A., SETA, Y., TAKAHASHI, N., OTSUKI, T. & SUMIKI, Y. (1953). The production of gibberellin in submerged culture. *Journal of the Agricultural Chemical Society of Japan* **27**, 545–9.

KLUYVER, A.J. & PERQUIN, L.H.C. (1933). Methods for the study of the metabolism of molds. *Biochemische Zeitschrift* **266**, 68–81.

KREBS, H.A. & JOHNSON, W.A. (1937). Citric acid in intermediate metabolism in animal tissues. *Enzymologia* **4**, 148–56.

KUROSAWA, E. (1926). Experimental studies on the nature of the substance excreted by 'Bakanae' fungus. *Transactions of the Natural History Society of Formosa* **16**, 213–27.

LE MENSE, E.H., SOHNS, V.E., CORMAN, J., BLOM, R.H., VAN LANEN, J.M. & LANGLYKKE, A.F. (1949). Submerged mold amylase as a saccharifying agent. *Industrial and Engineering Chemistry* **41**, 100–3.

LOCKWOOD, L.B. & WARD, G.E. (1945). Fermentation process for itaconic acid. *Industrial and Engineering Chemistry* **37**, 405–6.

MAMOLI, L. & VERCELLONE, A. (1937). Biochemische Umwandlung von Δ^5-Androstendion in Isoandrostandiol und Δ^4-Testosteron Weiterer Beitrag zur Genese der Keimdrusenhormone. *Berichte der deutschen chemischen Gesellschaft* **70**, 2079–82.

MARTIN, S.M. (1957). Citric acid production by submerged fermentation. *Industrial and Engineering Chemistry* **49**, 1231–4.

MEZZADROLI, G. (1938). Procédé de fabrication d'acide citrique par fermentation d'hydrates de carbone. *French Patent* 833,631.

OXFORD, A.E., RAISTRICK, H. & SIMONART, P. (1939). Griseofulvin, $C_{17}H_{17}O_6Cl$, a metabolic product of *Penicillium griseo-fulvum* Dierckx. *Biochemical Journal* **33**, 240–8.

PERQUIN, L.H.C. (1938). Bijdrage tot de Kennis der oxydative dissimilatie van *Aspergillus niger* van Tieghem. Meinema, Delft.

PETERSON, D.H. & MURRAY, H.C. (1952). Microbiological oxygenation of steroids at carbon 11. *Journal of the American Chemical Society* **74**, 1871–2.

RAFFY, A. (1937). Vitamin properties of the flavin of *Eremothecium ashbyii. Comptes Rendus des Séances de la Société de Biologie et de ses Filiales* **126**, 875–7.

RAISTRICK, H., BIRKINSHAW, J.H., CHARLES, J.H.V., CLUTTERBUCK, P.W., COYNE, F.P., HETHERINGTON, A.C., LILLY, C.H., RINTOUL, M.L., RINTOUL, W., ROBINSON, R., STOYLE, J.A.R., THOM, C. & YOUNG, W. (1931). Studies in the biochemistry of micro-organisms. *Philosophical Transactions of the Royal Society of London, Series B* **220**, 1–367.

RAISTRICK, H. & CLARK, A.B. (1919). On the mechanism of oxalic acid formation by *Aspergillus niger*. *Biochemical Journal* **13**, 329–44.

RAPER, K.B. & ALEXANDER, D.F. (1945). Mycological aspects of penicillin production. *Journal of the Elisha Mitchell Scientific Society* **61**, 74–113.

RAULIN, J. (1869). Études chimique sur la végétation. *Annales des Sciences Naturelles 5me Sér. Botanique* **11**, 93–299.

RHODES, A., CROSSE, R., FERGUSON, T.P. & FLETCHER, D.L. (1957). Griseofulvin. *British Patent* 784,618.

ROSI, D., PERUZZOTTI, G., DENNIS, E.W., BERBERIAN. D.A., FREELE, H., TULLAR, B.F. & ARCHER, S. (1967). Hycanthone, a new active metabolite of lucanthone. *Journal of Medical Chemistry* **10**, 867–76.

SAITO, K. (1907). Über die Säurebildung bei *Aspergillus oryzae* (Vorläufige, Mitteilung). *Botanical Magazine,* Tokyo **21**, 240.

SANDOZ LTD. (1956). Ergotamine, ergotaminine and ergometrine. *British Patent* 755,555.

SCHOPFER, W.H. & JUNG, A. (1935). Recherches sur l'activité vitaminique A du thalle d'une Mucorinée. *Comptes Rendus des Séances de la Société de Biologie et de ses Filiales* **120**, 1033–95.

STEINBERG, R.A. (1935). Nutrient-solution purification for removal of heavy metals in deficiency investigations with *Aspergillus niger*. *Journal of Agricultural Research* **51**, 413–24.

STEINBERG, R.A. (1936). Relation of accessory growth substances to heavy metals, including molybdenum, in the nutrition of *Aspergillus niger*. *Journal of Agricultural Research* **52**, 439–48.

STEINBERG, R.A. (1937). Molybdenum in the utilization of ammonium and nitrate nitrogen by *Aspergillus niger*. *Journal of Agricultural Research* **55**, 891–902.

STEINBERG, R.A. (1938). The essentiality of gallium to growth and reproduction of *Aspergillus niger*. *Journal of Agricultural Research* **57**, 569–74.

SUTTER, R.P. & RAFELSON, M.E. (1968). Separation of β-factor synthesis from stimulated β-carotene synthesis in mated cultures of *Blakeslea trispora*. *Journal of Bacteriology* **95**, 426–32.

TAKAHASHI, T. & SAKAGUCHI, K. (1927). Acids formed by *Rhizopus* species. *Bulletin of the Agricultural Chemical Society of Japan* **3**, 59–62.

TAKAMINE, J. (1914). Enzymes of *Aspergillus oryzae* and the application of its amyloclastic enzyme to the fermentation industry. *Journal of Industrial and Engineering Chemistry* **6**, 824–8.

THOM, C. (1930). *The Penicillia*. Baillière, Tindall & Cox, London.

THOM, C. & CURRIE, J.N. (1916). *Aspergillus niger* group. Oxalic acid production of species of *Aspergillus*. *Journal of Agricultural Research* **7**, 1–15.

THOM, C. & STEINBERG, R.A. (1939). Chemical induction of genetic changes in fungi. *Proceedings of the National Academy of Sciences of the United States of America* **25**, 329–35.

TULASNE, L.R. (1853). *Annales des Science Naturelles, Botanique 3 ser.* **20**, 5.

UNDERKOFLER, L.A., FULMER, E.I. & SCHOENE, L. (1939). Saccharification of starchy grain mashes for the alcoholic fermentation industry. Use of mold amylase. *Industrial and Engineering Chemistry* **31**, 734–8.

VAN DEN ENDE, H. (1968). Relationship between sexuality and carotene synthesis in *Blakeslea trispora*. *Journal of Bacteriology* **96**, 1298–303.

VAN TIEGHEM, P.E.L. (1867). Sur la fermentation gallique. *Comptes Rendus Hebdomadaires des Séances de l'Académie des Sciences* **65**, 1091–4.

WAKSMAN, S.A. (1916). Do fungi live and produce mycelium in the soil? *Science* **44**, 320–2.

WAKSMAN, S.A. (1943). Process for the production of fumaric acid. *United States Patent* 2,326,986.

WARD, G.E. (1970). Some contributions of the U.S. Department of Agriculture to the fermentation industry. *Advances in Applied Microbiology* **13**, 363–82.

WARD, G.E., LOCKWOOD, L.B., TABENKIN, B. & WELLS, P.A. (1938). Rapid fermentation process for dextro-lactic acid. *Industrial and Engineering Chemistry* **30**, 1233–5.

WEHMER, C. (1891). *Botanisches Zentralblatt* **49**, 233–638.

WEHMER, C. (1918). Über Fumarsäure—Garüng des Zuckers. *Berichte der deutschen chemischen Gesellschaft* **51**, 1663–8.

WEHMER, C. (1928). Abnahme des Säuregärungsvermögen und Anderung der Säure bei einem Pilz (Gluconsäure statt Fumarsäure-Gärung). *Biochemische Zeitschrift* **197**, 418–32.

WELLS, P.A. & WARD, G.E. (1939). Fermentation processes. *Industrial and Engineering Chemistry* **31**, 172–7.

WORGAN, J.T. (1968). Culture of the higher fungi. *Progress in Industrial Microbiology* **8**, 73–139.

YABUTA, T. (1912). On kojic acid, a new organic acid formed by *Aspergillus oryzae*. *Journal of the College of Agriculture, Tokyo Imperial University* **5**, 51–8.

ZAHORSKI, B. (1913). Method of producing citric acid. *United States Patent* 1,066,358.

CHAPTER 7

Commercially Important Secondary Metabolites

W. B. TURNER

7.1 Introduction

Micro-organisms share with higher plants the ability to produce secondary metabolites, compounds which serve no obvious function in the cells which produce them. In contrast with the primary metabolites, which are essentially the same for all forms of life, the secondary metabolites are usually species specific. The secondary metabolites of plants and micro-organisms embrace a very wide range of structural types, and their study has played a major role in the development of organic chemistry during the past century. But they are of more than theoretical interest, for many possess biological activity and of these the most important are the antibiotics produced by micro-organisms.

Although it was a filamentous fungus which provided us with penicillin, the first clinically useful antibiotic, in the subsequent search for other antibiotics, the filamentous fungi have proved a less prolific source than have the bacteria, especially the Actinomycetales. Thus by 1967 over 600 antibiotics had been isolated from bacteria and about 150 from fungi (Korzybski, Kowzyk-Gindifer & Kurylowicz, 1967), and the penicillins and the related cephalosporins are the only major antibacterial agents of fungal origin while the Actinomycetales have yielded streptomycin, neomycin, kanamycin, chloramphenicol, the tetracyclines, and the macrolides. On the other hand the development of semi-synthetic penicillins with improved therapeutic properties (see below) has maintained the importance of the penicillins in antibacterial therapy, so that they are still the most widely used group of antibiotics. In 1971 sales of penicillins in the major world markets were £136 m, or 27% of total antibiotic sales (in the U.K. for example, the sales were £10 m and accounted for 56% of total antibiotic sales) and, moreover, the share of the antibiotics market held by the penicillins which has risen over recent years continues to rise. Thus, although in terms of the number of antibacterial agents which they have yielded the filamentous fungi are inferior to the bacteria, the fungal products remain of great importance. Furthermore, in the field of antifungal therapy the fungi have provided us with griseofulvin, the only agent which is orally effective against dermatophytic infections.

Brian (1951) and Broadbent (1968) have surveyed the distribution among fungi of the ability to produce antibiotics, and conclude that the ability is not evenly distributed. In particular, the Moniliales (Fungi Imperfecti) and the Agaricales (Basidiomycetes) produce more antibiotics than do other groups. However, these groups of fungi also produce above-average numbers of secondary metabolites in general (*e.g.* Shibata, Natori & Udagawa, 1964) so that it may be that the ability to produce antibiotics is simply a reflection of the ability to produce secondary metabolites some of which, by chance, happen to possess antibacterial properties. Such conclusions must, of course, always be qualified by the possibility that the known production both of antibiotics and of secondary metabolites may itself be a reflection of the extent to which particular classes of fungus have been examined.

Although, at present, the commercial and practical importance of secondary metabolites is due mainly to the antibiotics, other compounds are produced on a commercial scale. Among these the plant growth regulator gibberellic acid is a fungal product and is discussed in this Chapter. It could be argued that when primary metabolites, such as citric acid, are produced in large excess of physiological quantity they become secondary metabolites; they are, however, a special case and are dealt with in Chapter 8.

7.2 The Biosynthesis of Secondary Metabolites

In spite of the wide range of their chemical structures, the secondary metabolites are formed by only a few basic biosynthetic pathways which are related to the primary metabolic pathways and use the same intermediates (the relationship of primary and secondary metabolism is discussed in Chapters 2 and 3). The biosynthetic pathways are used to varying extents by different biological systems, and in fungi the most important are the polyketide route, the terpenoid route, and processes which utilize the essential amino-acids (Turner, 1971). It is not surprising, therefore, that all the commercially important secondary metabolites of the fungi are produced by one or other of these routes. Even among the fungi the use of the various routes is not evenly distributed; the polyketide route, for example, is particularly associated with the Fungi Imperfecti.

When considering the biosynthesis of a secondary metabolite, there are two questions to be answered. The first concerns the nature of the basic building units, and this is usually readily answered by studying the incorporation of the basic units (*e.g.* acetate, mevalonate, amino-acids), appropriately labelled with isotope, into the metabolite, then using degradative or spectroscopic methods to locate the label within the molecule. The second question concerns the mechanism by which the units are assembled and the steps involved in the biosynthetic process. This is a much more difficult question to answer, because the fact that a supposed intermediate is incorporated into a metabolite does not necessarily mean that it is an obligatory natural intermediate in the biosynthesis of that metabolite. The elusiveness of a clear-cut answer to this second question will become apparent when we discuss the formation of the penicillins; even for compounds as important and as much-studied as these, there are very large gaps in our knowledge of the steps by which the parent amino-acids are assembled into the final products.

Metabolites derived from amino acids

THE PENICILLINS AND CEPHALOSPORINS The penicillins are the most important compounds to have been isolated from fungi. Their discovery marked the beginning of a new era in the treatment of infectious diseases and they have been of enormous commercial importance, an importance which has been maintained by the discovery of 6-aminopenicillanic acid (6-APA) and its conversion to semi-synthetic penicillins with improved therapeutic properties. The literature of the penicillins has been extensively reviewed (*e.g.* Florey *et al.*, 1949; Korzybski *et al.*, 1967; Demain, 1966; Abraham & Newton, 1967), and the history of their development, from Fleming's first observation of the antagonism between a *Penicillium* and a *Staphyllococcus*, to the large-scale production of penicillin G during World War II, makes fascinating reading (Florey *et al.*, 1949; Federal Trade Commission, 1958; Chain, 1971).

The early work on penicillin was carried out with Fleming's strain of *Penicillium notatum.* Subsequently penicillins were isolated from a number of *Penicillia* and *Aspergilli*, and from a *Cephalosporium.* The organisms used for the commercial production of penicillins are mutants of *P. chrysogenum.*

It soon became apparent that 'penicillin' was not a single compound (see Fig. 7.1). The first two penicillins to be obtained pure were F (from surface fermentations in England) and G (from stirred fermentations in America), and penicillins K and X quickly followed. The realization that the limiting factor in penicillin production is the availability of the side-chain precursors led to the addition of carboxylic acids to the fermentation media which gave increased yields of 'natural' penicillins and also to give new penicillins with a variety of side-chains. In this way penicillin V (Fig. 7.1), which is more stable to acid than the early penicillins and therefore suitable for oral dosage, was obtained. The range of side-chains which could be introduced was limited by the toxicity of the precursors to the fungus and by the ability of the fungus to degrade some precursors. The isolation of 6-APA from fermentations on media containing no side-chain precursor removed these limitations and made possible the partial synthesis of new penicillins with improved therapeutic properties, *e.g.* methicillin, with improved penicillinase-resistance, and ampicillin, with activity against Gram negative as well as Gram positive bacteria. Many thousands of semi-synthetic penicillins have been prepared and their structure-activity relationship has been reviewed (Nayler, 1971). 6-APA can also be obtained by enzymatic or chemical hydrolysis of penicillin G and this is the basis of its commercial production (Carrington, 1971).

During a study of the antibiotics produced by a *Cephalosporium* sp., two water-soluble compounds — cephalosporin N and cephalosporin C — were isolated (the same organism also yielded cephalosporin P_1, see p. 133). Cephalosporin N was shown to be a new penicillin and was renamed penicillin N (see Fig. 7.1); it is identical with synnematin B isolated from *Cephalosporium salmosynnematum*, the perfect stage of which is *Emericellopsis.* Penicillin N has also been isolated from *Streptomyces* spp. (Nagarajan *et al.*, 1971) and isopenicillin N (Fig. 7.1), with the L-configuration of the α-aminoadipoyl side-chain, has been isolated from *Penicillium chrysogenum.* Cephalosporin C (Fig. 7.2) has the same side-chain

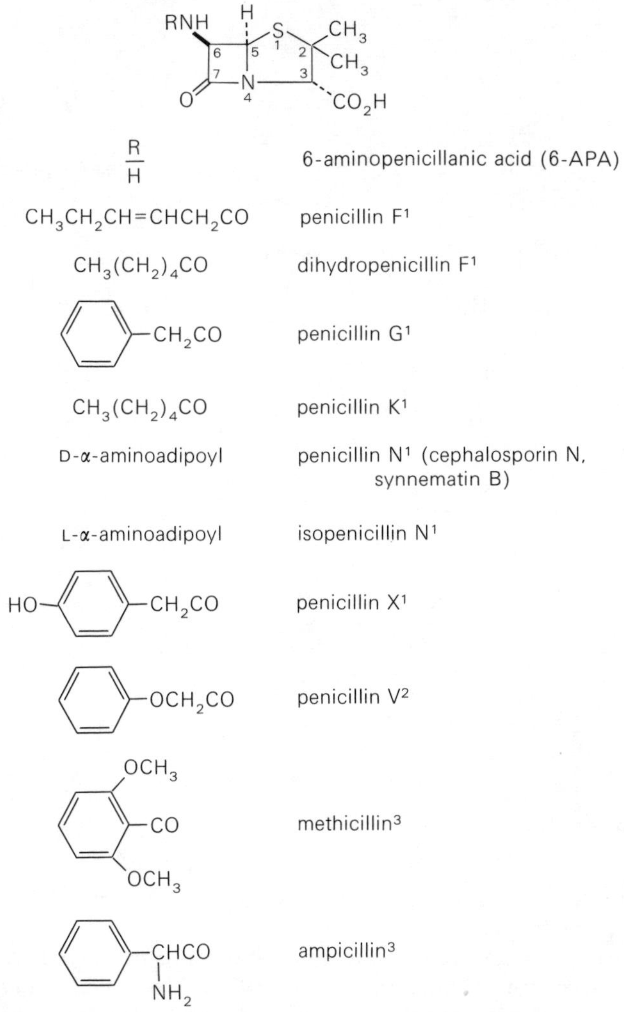

R	
H	6-aminopenicillanic acid (6-APA)
$CH_3CH_2CH{=}CHCH_2CO$	penicillin F[1]
$CH_3(CH_2)_4CO$	dihydropenicillin F[1]
$-CH_2CO$	penicillin G[1]
$CH_3(CH_2)_4CO$	penicillin K[1]
D-α-aminoadipoyl	penicillin N[1] (cephalosporin N, synnematin B)
L-α-aminoadipoyl	isopenicillin N[1]
HO– –CH_2CO	penicillin X[1]
$-OCH_2CO$	penicillin V[2]
OCH_3 –CO OCH_3	methicillin[3]
–CHCO NH_2	ampicillin[3]

Fig. 7.1 The structures of some penicillins. [1] 'Natural' penicillins.
[2] Produced by addition of phenoxyacetic acid to the fermentation.
[3] Produced by acylation of 6-APA.

as penicillin N but a new heterocyclic ring system (the 'cephem' system). Cephalosporin C is a relatively weak antibiotic and, as in the case of the penicillins, semi-synthetic cephalosporins, *e.g.* cephaloridine (Fig. 7.2), with improved antibacterial properties have been produced by acylation of 7-aminocephalosporanic acid (7-ACA), obtained chemically from cephalosporin C (no commercial enzymatic or microbiological process is yet available for the production of 7-ACA). As well as variation of the side-chain, the cephalosporin nucleus has also been modified at the acetoxymethyl group, *e.g.* in cephaloridine and cephalexin (Fig. 7.2). Cephalothin and cephaloridine are not very effective orally; cephalexin, and other desacetoxy compounds, are less active but are orally absorbed.

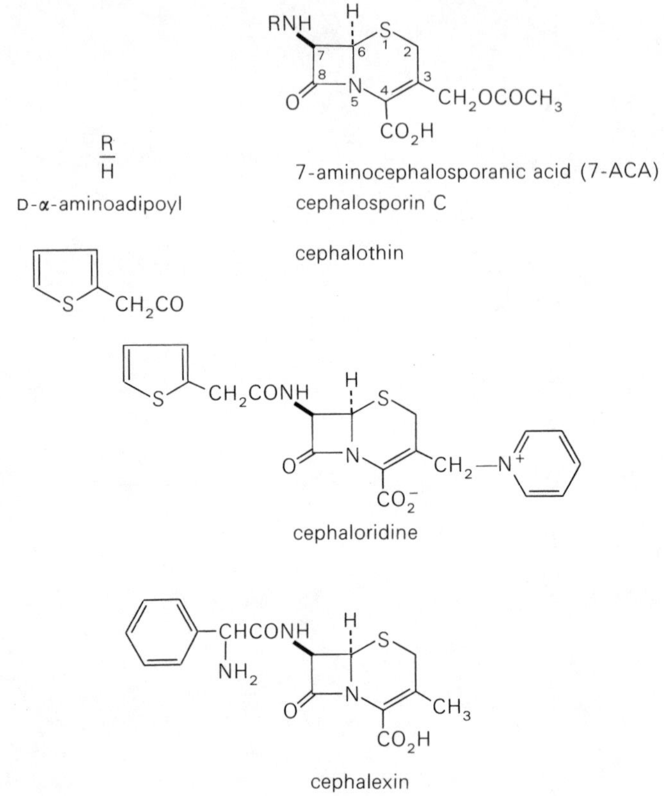

R		
H	7-aminocephalosporanic acid (7-ACA)	
D-α-aminoadipoyl	cephalosporin C	
2-thienyl-CH_2CO	cephalothin	

cephaloridine

cephalexin

Fig. 7.2 The structures of some cephalosporins.

For reviews of the cephalosporins see van Heyningen (1967) and Morin & Jackson (1970).

Several new cephem derivatives have recently been isolated from *Streptomyces* spp. (Nagarajan *et al.*, 1971; Albers-Schönberg, Arison & Smith, 1972) and deacetoxycephalosporin C has been isolated from *Streptomyces* spp. and from several fungi (Higgens *et al.*, 1974).

The biosynthesis of the penicillins and cephalosporins poses two problems — the formation of the heterocyclic ring systems and the introduction of the acyl side-chains. 6-APA is formally derived from cysteine and valine (Fig. 7.3a) and both are incorporated intact into the penicillin molecule. As expected, L-cysteine is a more efficient precursor than D-

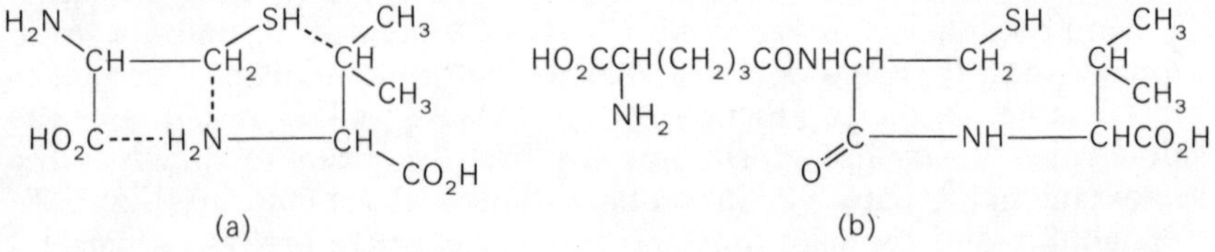

(a) (b)

Fig. 7.3 (a) The formal relationship of cysteine and valine with the penicillin nucleus. (b) The structure of δ-(L-α-aminoadipoyl)-cysteinylvaline.

cysteine, but although C-3 of the penicillin nucleus has the D-configuration, L-valine is incorporated more efficiently than D-valine. The nitrogen atom of valine is retained, though some transamination occurs. The retention of the tritium of [2-^3H]cysteine excludes the intervention of an $\alpha\beta$-unsaturated derivative in penicillin biosynthesis.

The above are the only firmly established facts relating to penicillin biosynthesis. The isolation from *Penicillium chrysogenum* mycelium of the tripeptide δ-(α-aminoadipoyl)cysteinyl-valine (Fig. 7.3b), which is an open-chain analogue of penicillin N or isopenicillin N*, has led to the suggestion that it is a precursor of isopenicillin N which might be transformed, *via* penicillin N and transacylation, into the other penicillins. Support for this hypothesis was provided by the (subsequent) isolation of isopenicillin N from *Penicillium chrysogenum*. δ-(α-Aminoadipoyl)cysteinylvaline is also formed by cell-free extracts of *Penicillium chrysogenum* (Bauer, 1970); the compound has the all-L-configuration, whereas the tripeptide from a *Cephalosporium* sp. (see below) contains D-valine.

The role of 6-APA in penicillin biosynthesis is not clear. According to the above theory, 6-APA is not a precursor of the penicillins but a product of their hydrolysis. On the other hand, *Penicillium chrysogenum* has been shown to contain an enzyme which catalyses the acylation of 6-APA (Brunner, Röhr & Zinner, 1968; Gatenbeck & Brunsberg, 1968; Spencer, 1968) so that 6-APA *can* serve as a precursor of the penicillins (though it may still be formed initially from penicillin N or isopenicillin N).

The biosynthesis of the cephalosporins is clearly related to that of the penicillins and might involve a common intermediate—perhaps δ-(α-aminoadipoyl)cysteinylvaline or a compound with the penicillanic acid ring system (the penicillin ring system can be converted chemically to that of cephalosporin C but the conversion has not been achieved biologically, and there is some evidence that the pathways diverge at an earlier stage). α-Aminoadipic acid, valine, and a mixture of DL- and *meso*-cystine are incorporated into cephalosporin C. Incorporation of valine specifically labelled in one of the methyl groups shows that C-2 of the cephalosporins is derived from the same carbon as the β-methyl group of the penicillins (Neuss *et al.*, 1973; Kluender *et al.*, 1973). As expected acetate is incorporated into the *O*-acetyl and the α-aminoadipoyl groups.

The relationship between penicillin and cephalosporin biosynthesis is emphasized by the isolation (Loder & Abraham, 1971*a*) from mycelium of a cephalosporin C-producing *Cephalosporium* sp. of δ-(L-α-aminoadipoyl)-L-cysteinyl-D-valine along with two higher peptides, P_1 and P_2, which both contain α-aminoadipic acid, cysteine and glycine. P_1 also contains β-hydroxyvaline and P_2 contains valine, and these peptides are probably derived from intermediates on the main biosynthetic pathway. δ-(α-Aminoadipoyl)cysteinylvaline is also formed in a broken-cell preparation of *Cephalosporium* in the presence of δ-(L-α-aminoadipoyl)-L-cysteine and DL-[^{14}C]valine (Loder & Abraham, 1971*b*), but not in the presence of δ-(D-α-aminoadipoyl)-L-cysteine or of DL-α-amino[^{14}C]adipic acid and L-cysteinyl-L-valine or L-cysteinyl-D-valine. These results show that the tripeptide is formed by condensation of valine with α-aminoadipoyl-cysteine and, if the tripeptide is an intermediate in penicillin and cephalo-

* The tripeptide was not obtained in sufficient quantity to permit the determination of its stereochemistry.

sporin biosynthesis, explain the relatively low incorporation of cysteinyl-valine into the penicillins.

Polyketide-derived compounds

The polyketide route to secondary metabolites may be regarded as the biosynthetic route most characteristic of the fungi (especially the Fungi Imperfecti), for not only are more fungal metabolites formed by this route than by any other but also the majority of known polyketide-derived compounds are fungal metabolites. In polyketide biosynthesis a molecule of acetate condenses with three or more molecules of malonate to form a polyketomethylene (polyketide) chain which can undergo a variety of further transformations. Since malonate is formed by carboxylation of acetate, with which it is in equilibrium, in whole cells labelled acetate is usually incorporated into polyketide molecules in a uniform manner — the compounds are thus often referred to as 'poly-acetate' or 'acetate-derived', and the biosynthetic C_2-units as 'acetate-units'. The process is illustrated formally in Fig. 7.4, where the tetraketide chain (A) derived

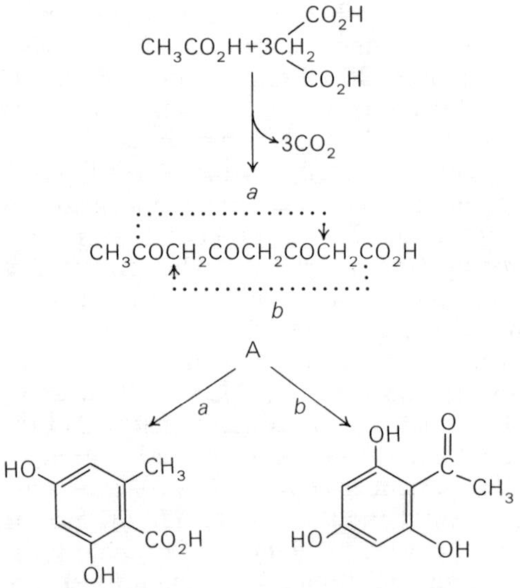

$$CH_3CO_2H + 3CH_2(CO_2H)_2 \xrightarrow{-3CO_2} CH_3COCH_2COCH_2COCH_2CO_2H \text{ (A)}$$

orsellinic acid

acetylphloroglucinol

Fig. 7.4 The formation of orsellinic acid or acetylphloroglucinol by alternative cyclizations of the polyketide intermediate A.

from one molecule of acetate and three molecules of malonate (*i.e.* four 'acetate units') can undergo the alternative intramolecular condensations *a* or *b* to give orsellinic acid or acetylphloroglucinol.

There are two distinct phases in polyketide biosynthesis. The first is the assembly of the polyketide chain and its modification, usually by aromatization or reduction, to give a stable product; this phase involves protein-bound intermediates. In some fermentations this first stable product is the major metabolite, but often it undergoes further transformation(s), the second phase referred to above, to give the major metabolite. This is illustrated by the biosynthesis of griseofulvin, discussed below. A detailed account of polyketide biosynthesis has been presented by Turner (1971).

GRISEOFULVIN Griseofulvin (Fig. 7.5), the only effective antibiotic available for the systemic treatment of fungal infections of skin, hair, and nails, was first isolated from *Penicillium griseofulvum*. Its unusual antifungal properties were not, however, recognized until its re-isolation from *P. janczewski* (≡*P. nigricans*) by Brian, Curtis & Hemming (1946) at the Akers Research Laboratories of I.C.I. Brian *et al.*, (1946) called the compound 'curling factor', a name descriptive of the morphological changes it induces in the hyphae of test fungi). Griseofulvin has subsequently been isolated from about a dozen other species of *Penicillium*, the most important being *P. patulum*, mutant strains of which produce the compound in high yield under stirred aerated conditions and are used for its commercial production. Griseofulvin has also been isolated from *Khuskia (Nigrospora) oryzae* (Furuya, Enokita & Shirasaka, 1967; Giles, Hemming & Lehan, 1970), an organism quite distinct from the *Penicillia*.

The production, biosynthesis, and chemistry of griseofulvin have been extensively reviewed (Brian, 1960; Rhodes, 1963; Grove, 1963*a*, 1964). Here we shall restrict ourselves to a brief account of its biosynthesis, emphasizing its relationship with its co-metabolites (Fig. 7.5).

The structure of griseofulvin suggested that the spiro-ring might arise by oxidative cyclization ('phenol oxidation') of a benzophenone derivative derived by the polyketide route, and the acetate origin of griseofulvin was confirmed by a study of the incorporation of [1-^{14}C]acetate into griseofulvin in *Penicillium griseofulvum*. The *O*-methyl groups are derived from the one-carbon pool, the usual source of biological methyl groups.

In a search for intermediates in griseofulvin biosynthesis the production of griseofulvin by *Penicillium patulum* was inhibited in various ways. This led to the detection of the benzophenones griseophenone A, B, and C and of griseoxanthone C. Further study of the involvement of these compounds in griseofulvin biosynthesis showed that the first aromatic product is griseophenone C which is converted first to griseophenone B and then to griseofulvin. Griseophenones B and C are also converted to griseophenone A, but [^{36}Cl]griseophenone A is not incorporated into griseofulvin. These results led Rhodes *et al.* (1961) to suggest that the final stage in griseofulvin biosynthesis involves the binding of griseophenone B to a multi-enzyme complex which can effect oxidation, reduction, and methylation of protein-bound intermediates. Free griseophenone A is thus a by-product of griseofulvin biosynthesis. Several fungi will transform griseophenone A into (±)-dehydrogriseofulvin, and some give a preponderance of the (+)-isomer (Okuda *et al.*, 1967).

Griseoxanthone C is clearly a by-product of griseofulvin biosynthesis and may be formed by loss of methanol from griseophenone C, a process which proceeds readily *in vitro*. Dihydrogriseofulvin is a co-metabolite of griseofulvin and dehydrogriseofulvin in *Penicillium martinsii* (Kamal *et al.*, 1970).

The introduction of chlorine at a fairly late stage in griseofulvin biosynthesis is paralleled in the biosynthesis of other chlorinated microbial products (Turner, 1971). Dechlorogriseofulvin is a co-metabolite of griseofulvin in surface cultures of *Penicillium griseofulvum* and of some

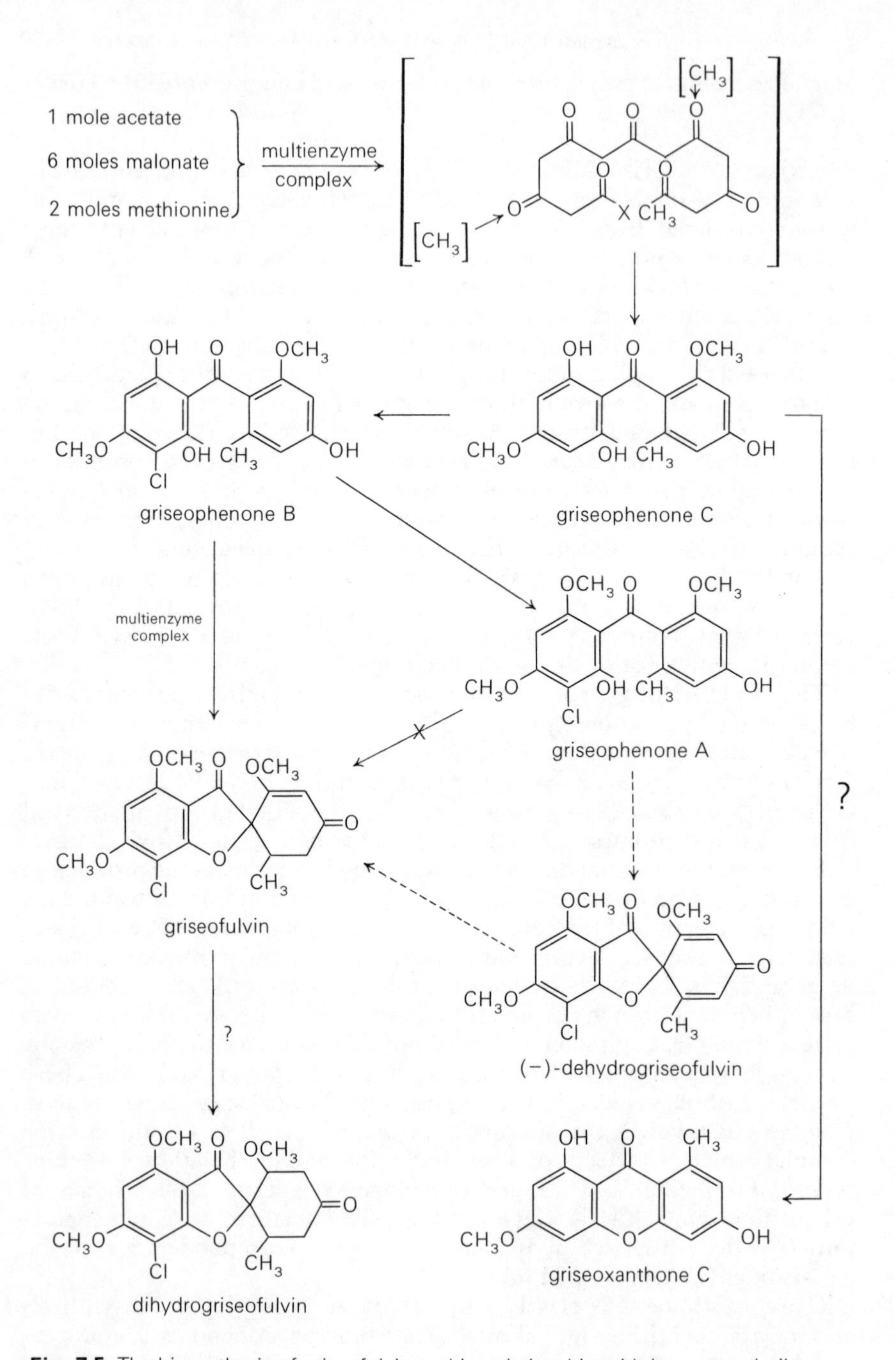

Fig. 7.5 The biosynthesis of griseofulvin and its relationship with its co-metabolites.

strains of *P. janczewskii*; on chloride-free media, or in the presence of inhibitors of chlorination, *P. patulum* accumulates griseophenone C but dechlorogriseofulvin is not produced in any quantity. On media containing bromide in place of chloride, *P. griseofulvum* and *P. janczewskii* produce the bromo-analogue of griseofulvin.

VARIOTIN The antifungal agent variotin (Fig. 7.6) was isolated from *Paecilomyces varioti* var. *antibioticus* and its correct chemical structure was established by Takeuchi, Yonehara & Shoji (1964). Variotin is an *N*-acylpyrrolidone formally derived from a C_{13} branched-chain acid and γ-aminobutyric acid. The derivation of the C_{13}-acid from acetate and methionine (Fig. 7.6) was demonstrated with ^{14}C-labelled precursors and confirmed with ^{14}C-labelled precursors (Tanabe & Seta, 1970). The γ-aminobutyric acid moiety is derived by decarboxylation of glutamic acid (Tanaka, Sashikita & Umezawa, 1962).

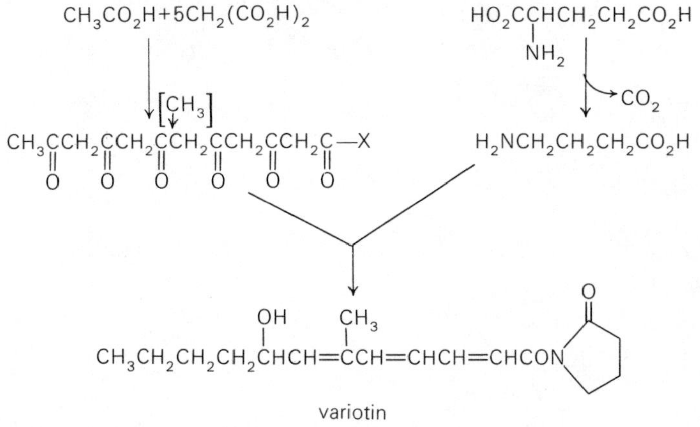

$CH_3CO_2H + 5CH_2(CO_2H)_2$

[CH_3]

$CH_3COCH_2COCH_2COCH_2COCH_2COCH_2CO{-}X$

$HO_2CCH(NH_2)CH_2CH_2CO_2H$

CO_2

$H_2NCH_2CH_2CH_2CO_2H$

$CH_3CH_2CH_2CH_2CH(OH)CH{=}C(CH_3)CH{=}CHCH{=}CHCON$ (pyrrolidone ring, C=O)

variotin

Fig. 7.6 The incorporation of acetate, malonate, methionine and glutamic acid into variotin.

Terpenoid metabolites

Like the polyketides, the terpenes are derived ultimately from acetate, but instead of the linear condensation characteristic of the polyketide route the terpenes are built up from branched C_5-units (the 'isoprene' units) formed from three acetate units. The key intermediate in terpene biosynthesis is mevalonate (Fig. 7.7) and it is this compound, suitably labelled with ^{14}C or ^{3}H, which is most used in the study of terpene biosynthesis. Mevalonate is converted to isopentenyl pyrophosphate and dimethylallyl pyrophosphate, which condense to form geraniol, farnesol, geranylgeraniol, and squalene, which undergo a variety of cyclization reactions to form respectively mono-, sesqui-, di-, and triterpenes.

Apart from the monoterpenes, which seem to be of limited type and distribution, terpenoid compounds are widespread among the fungi, and the gibberellins, which are diterpenes, and fusidic acid, which is a triterpene, are produced on a commercial scale and are discussed below.

THE GIBBERELLINS In the early stages of infection of rice plants by *Gibberella fujikuroi* (the perfect stage of *Fusarium moniliforme*) the plants grow more rapidly than normal. Filtrates from cultures of *G. fujikuroi* were found to produce a similar growth-promoting effect and a crystalline active product, later shown to be a mixture of gibberellins, was isolated. The compounds assumed a wider significance with the discovery that

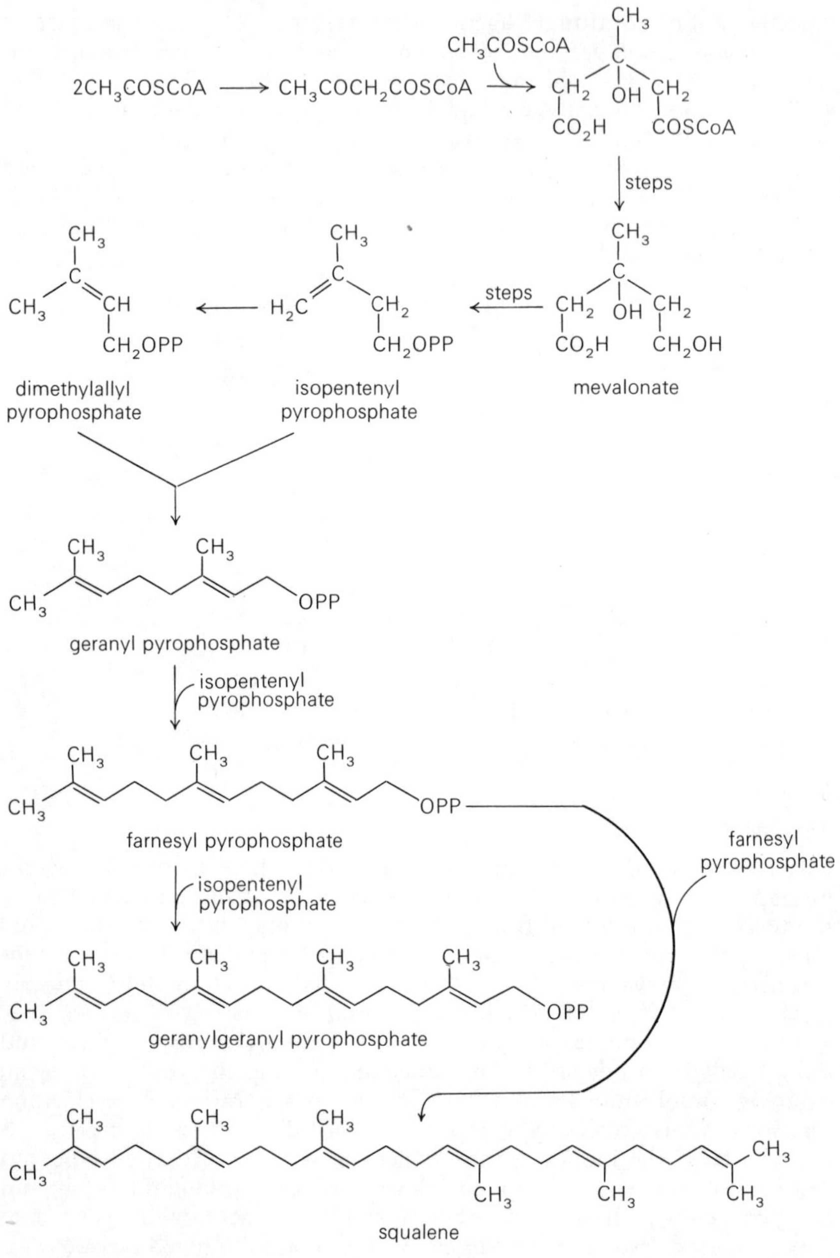

$2CH_3COSCoA \longrightarrow CH_3COCH_2COSCoA$ (+ $CH_3COSCoA$) $\longrightarrow$ CH_3–C(OH)(CH_2CO_2H)($CH_2COSCoA$)

steps

CH_3–C(OH)(CH_2CO_2H)(CH_2CH_2OH)

mevalonate

steps

$H_2C{=}C(CH_3)CH_2CH_2OPP$

isopentenyl pyrophosphate

$(CH_3)_2C{=}CHCH_2OPP$

dimethylallyl pyrophosphate

geranyl pyrophosphate

isopentenyl pyrophosphate

farnesyl pyrophosphate

isopentenyl pyrophosphate

geranylgeranyl pyrophosphate

farnesyl pyrophosphate

squalene

Fig. 7.7 Terpenoid biosynthesis.

gibberellic acid and other gibberellins (many of which have not been detected in *G. fujikuroi*) are endogenous plant hormones. The history, the chemistry, and the biological properties of the gibberellins have been extensively reviewed (Brian, Grove & MacMillan, 1960; Grove, 1961, 1963*b*; Phinney & West, 1960) and a full list of the fungal gibberellins and a detailed account of their biosynthesis has been presented (Turner, 1971). A detailed analysis of the terpenoid metabolites of *G. fujikuroi* has recently been reported (MacMillan & Wels, 1974).

The gibberellins produce a variety of effects on plants depending upon the species, the method of application and the gibberellin used (Turner, 1972). Gibberellic acid has found application in improving the quality and yield of grapes (especially the seedless varieties), oranges and lemons, and for increasing the fruit set of some varieties of pear. It is also widely used in the malting of barley, where it accelerates the process and improves the quality of the malt. These uses, and other minor applications, consume 4–5 tons of gibberellic acid per annum. A mixture of gibberellin A_4 and gibberellin A_7 is also produced on a small commercial scale; the A_4/A_7 mixture has advantages over gibberellic acid for some purposes, particularly in controlling fruit-drop of apples.

The main pathway, as it is presently understood, from geranylgeranyl pyrophosphate to the C_{19} gibberellins *via* (–)-kaurene and the C_{20} gibberellins (*e.g.* gibberellin A_{14}) is outlined in Fig. 7.8. The diterpene origin of gibberellic acid was first confirmed by a study of the incorporation of acetate and mevalonate into the molecule and the intermediacy of (–)-kaurene was subsequently demonstrated. 7β-Hydroxykaurenoic acid and the aldehyde A have been detected in cultures of *Gibberella fujikuroi* by dilution analysis after feeding radioactive (–)-kaurene and are incorporated into gibberellic acid. More recently the intermediacy of gibberellin A_{14} aldehyde has been demonstrated (Hedden & MacMillan, 1974). There is evidence that the ring-contraction stage requires a 6β-leaving group X (Hanson, Hawker & White, 1972).

Since no 4-double bond is involved in the loss of the 4a-methyl group during the formation of the C_{19}-gibberellins from the C_{20}-gibberellins, it has been suggested that an intermediate aldehyde undergoes a Baeyer-Villiger type oxidation to give a formate which is hydrolysed. The introduction of the 2-hydroxyl group proceeds with retention of configuration, the normal process for microbiological hydroxylation. Recent results (Pitel, Vining & Arsenault, 1971) suggest that the main pathway from gibberellin A_4 to gibberellic acid is *via* gibberellin A_7 rather than *via* gibberellin A_1. [See also Bearder, MacMillan & Phinney (1973)].

THE FUSIDANES Fusidic acid, helvolic acid, and cephalosporin P_1 (Fig. 7.9) are antibiotics whose structures are based on the fusidane skeleton. Only fusidic acid is used clinically, being effective orally against Gram-positive infections and used mainly against penicillin-resistant staphylococci. The structure of helvolic acid has undergone several revisions, that given in Fig. 7.9 being the most recent (Iwasaki *et al.*, 1970).

Fusidic acid was first isolated from *Fusidium coccineum* and has since been obtained from *Mucor ramannianus* as 'ramycin', a *Cephalosporium* sp., and most recently from *Isaria kogana* (Hikino *et al.*, 1972). Helvolic acid, the first member of the class to be isolated, was obtained originally from *Aspergillus fumigatus* mut. *helvola*, and has since been isolated from *Cephalosporium caerulens, Emericellopsis terricola,* and *Acrocylindrium oryzae*. Cephalosporin P_1 was isolated from the strain of *Cephalosporium* which yielded cephalosporin C. Thus the fusidane antibiotics have all

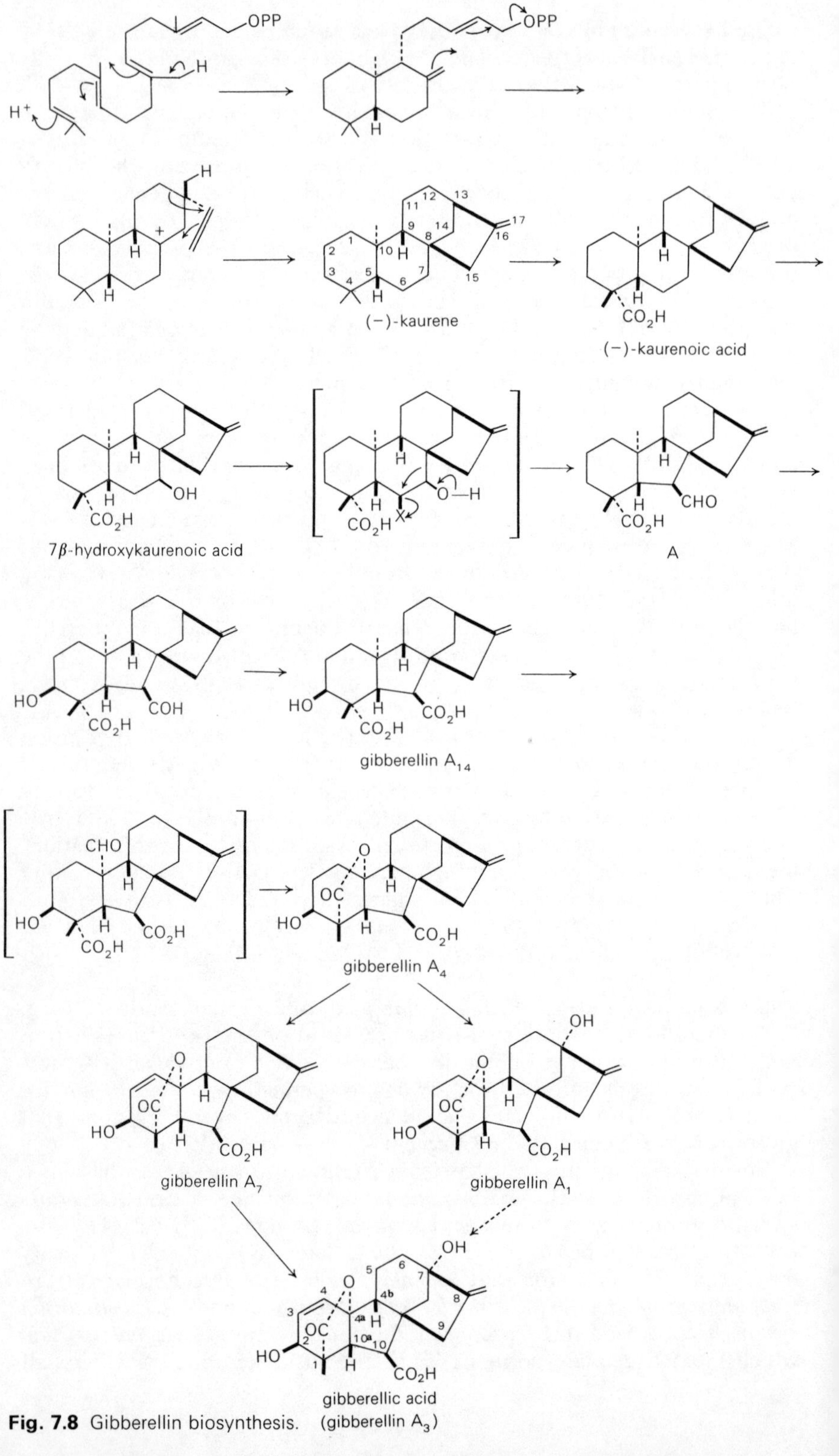

(−)-kaurene

(−)-kaurenoic acid

7β-hydroxykaurenoic acid

A

gibberellin A_{14}

gibberellin A_4

gibberellin A_7

gibberellin A_1

gibberellic acid (gibberellin A_3)

Fig. 7.8 Gibberellin biosynthesis.

been obtained from *Cephalosporium* spp. The isolation of fusidic acid from the Basidiomycete *Isaria kogana* is not surprising in view of the well-known ability of such fungi to produce tetracyclic triterpenes (Turner, 1971).

Viridominic acids A, B and C (Fig. 7.9), compounds which induce chlorosis in higher plants, have recently been isolated from culture filtrates of an unidentified fungus (Kaise *et al.*, 1972*a*, *b*, *c*) along with cephalosporin P_1 which also induces chlorosis. The viridominic acids are oxygenated in rings B and C, in contrast with the antibiotics, which are only oxygenated in one or other of the rings.

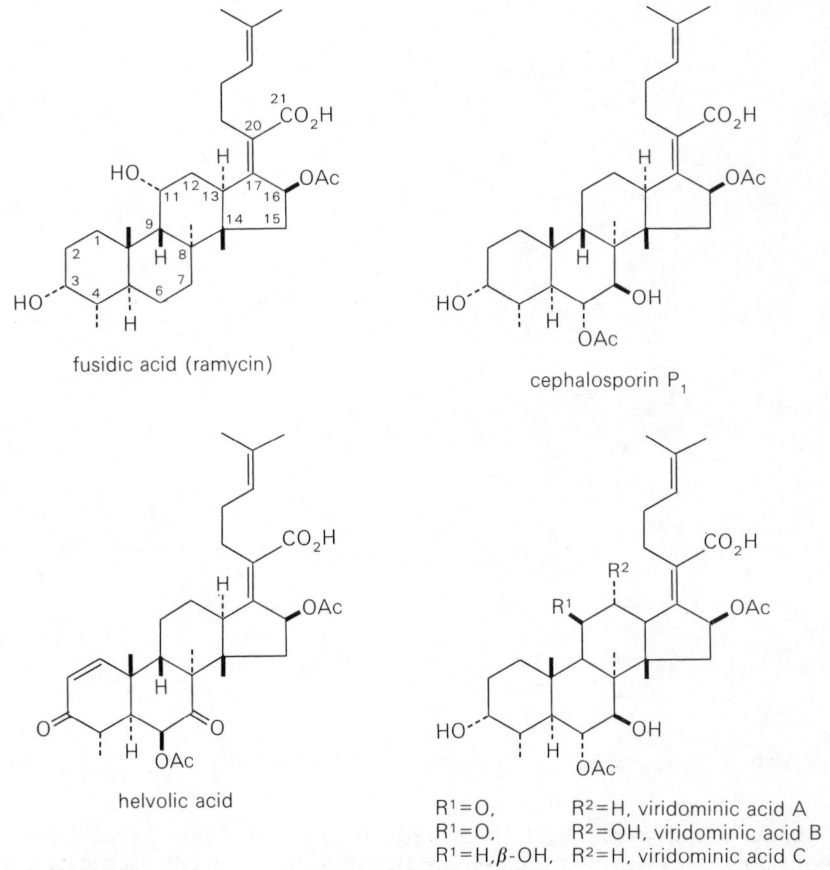

fusidic acid (ramycin)

cephalosporin P_1

helvolic acid

R^1=O, R^2=H, viridominic acid A
R^1=O, R^2=OH, viridominic acid B
R^1=H,β-OH, R^2=H, viridominic acid C

Fig. 7.9 Fusidic acid and related fungal metabolites.

The fusidanes are tetracyclic triterpenes, and it is informative to consider their biosynthesis in relation to that of the other tetracyclic triterpenes and the steroids (Fig. 7.10). The key step in the biosynthesis of the steroids is the acid-catalysed cyclization of squalene oxide to give an intermediate cation (A). Cation (A) then undergoes rearrangement with migration of methyl groups and protons [indicated by arrows in (A)] to give lanosterol which loses three methyl groups to give zymosterol. Further

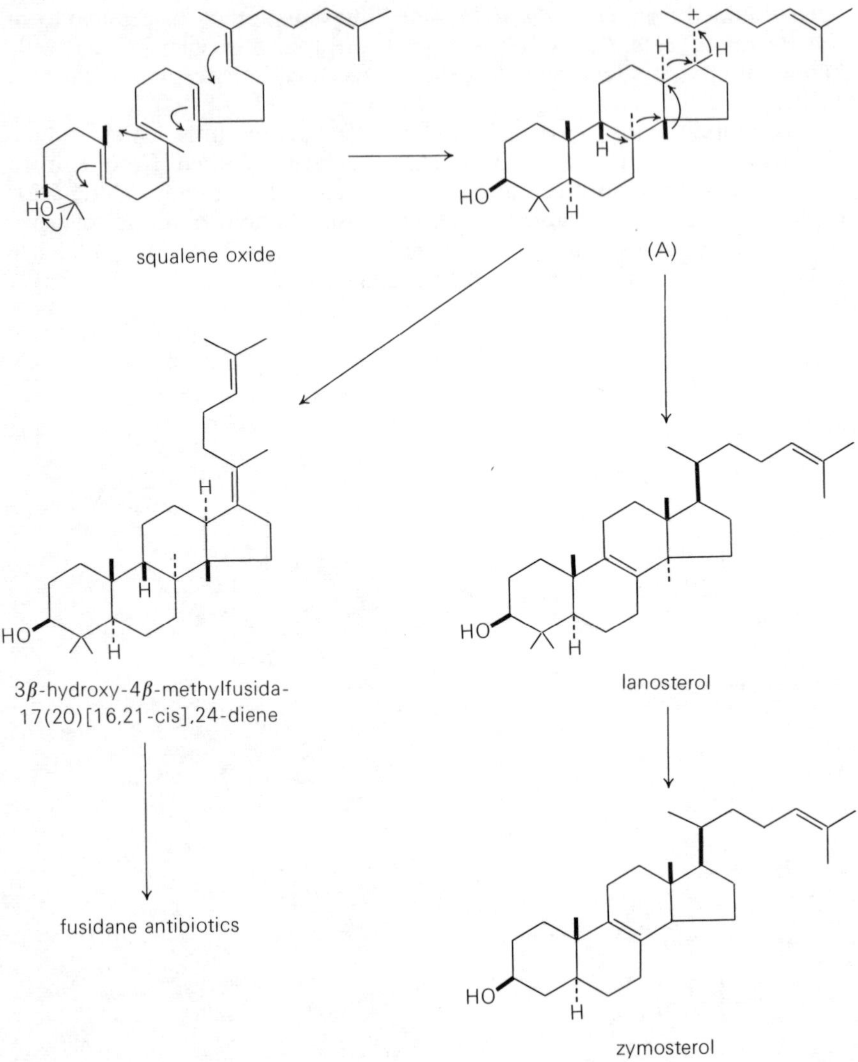

squalene oxide

(A)

3β-hydroxy-4β-methylfusida-17(20)[16,21-cis],24-diene

lanosterol

fusidane antibiotics

zymosterol

Fig. 7.10 The relationship of fusidane and lanosterol biosynthesis.

degradation of zymosterol then leads to the bile-acids, steroid hormones, *etc.* If, instead of undergoing the above rearrangement, the cation (A) simply loses a proton from C-17 it yields 3β-hydroxy-4β-methylfusida-17(20)[16,21-*cis*],24-diene, a co-metabolite of helvolic acid in *Cephalosporium caerulens.* The transformation of this compound into the fusidane antibiotics then merely requires the loss of the 4β-methyl group and various oxidation-reduction reactions, processes for which there is well-established precedent in steroid biosynthesis.

The loss of the 4-methyl group presumably proceeds, as in the conversion of lanosterol to zymosterol, by oxidation of the methyl group to carboxyl and the 3-hydroxy-group to ketone followed by decarboxylation of the resulting β-keto-acid. The formation of an intermediate 3-ketone also makes possible the isomerization of the 3β-hydroxy group present in the precursor to the 3α-configuration of fusidic acid and cephalosporin P_1, and is supported by the fact that tritium from $4R$-[4-^3H]mevalonate is not incorporated into position 3 of fusidic acid (Mulheirn & Caspi, 1971) *i.e.* it is lost in formation of the 3-keto-group. The broad outline of fusidic acid biosynthesis has been confirmed by the specific incorporation of squalene oxide into fusidic acid in *Fusidium coccineum* and many details have been established by experiments with labelled mevalonate (Mulheirn & Caspi, 1971; Caspi *et al.*, 1972; Ebersole *et al.*, 1973), and with [^{13}C]-acetate (Riisom *et al.*, 1974). 3β-Hydroxy-4β-methylfusida-17(2)-[16,21-*cis*],24-diene is biosynthesized from mevalonate by cell-free extracts of *Emericellopsis* spp. (Kawaguchi & Okuda, 1970).

7.3 References

ABRAHAM, E.P. & NEWTON, G.G.F. (1967). Penicillins and cephalosporins. In *Antibiotics,* vol. II, pp. 1–16. Ed. Gottlieb, D. & Shaw, P.D. New York: Springer-Verlag.

ALBERS-SCHÖNBERG, G., ARISON, B.H. & SMITH, J.L. (1972). New β-lactam antibiotics: structure determination of cephamycin A and B. *Tetrahedron Letters,* 2911–14.

BAUER, K. (1970). Zur Biosynthese der Penicilline: Bildung von 5-(2-Aminodipyl)-cysteinyl-valin in Extracten von *Penicillin chrysogenum. Zeitschrift für Naturforschung* **25b**, 1125–9.

BEARDER, J.R., MACMILLAN, J. & PHINNEY, B.O. (1973). Conversion of gibberellin A_1 into gibberellin A_3 by the mutant R-9 of *Gibberella fujikuroi. Phytochemistry* **12**, 2655–9.

BRIAN, P.W. (1951). Antibiotics produced by fungi. *Botanical Reviews* **17**, 357–430.

BRIAN, P.W. (1960). Griseofulvin. *Transactions of the British Mycological Society* **43**, 1–13.

BRIAN, P.W., CURTIS, P.J. & HEMMING, H.G. (1946). A substance causing abnormal development of fungal hyphae produced by *Penicillium janczewskii* Zal. I. Biological assay, production, and isolation of 'curling factor'. *Transactions of the British Mycological Society* **29**, 173–87.

BRIAN, P.W., GROVE, J.F. & MACMILLAN, J. (1960). The gibberellins. *Progress in the Chemistry of Organic Natural Products* **18**, 350–433.

BROADBENT, D. (1968). Antibiotics produced by fungi. *Pest Articles and News Summaries, Section B* **14**, 120–41.

BRUNNER, R., RÖHR, M. & ZINNER, M. (1968). Zur Biosynthese des Penicillins. Untersuchung zur enzymatischen Aktivierung von Phenylessigsaure und Phenoxyessigsaure sowie zur Bildung von Penicillin aus 6-Amino-penicillinansäure und aktivierter Seitenkettensäure durch Mycelhomogenate und zellfreie Extrakte von *Penicillium chrysogenum. Hoppe-Sayler's Zeitschrift für Physiologische Chemie* **349**, 95–103.

CARRINGTON, T.R. (1971). The development of commercial processes for the production of 6-aminopenicillanic acid (6-APA). *Proceedings of the Royal Society, London, Series B* **179**, 321–33.

CASPI, E., EBERSOLE, R.C., GODTFREDSEN, W.O. & VANDEGAL, S. (1972). Mechanism of squalene cyclization; the chiral origin of the C-22 hydrogen atoms of fusidic acid. *Journal of the Chemical Society, Chemical Communications,* 1191–3.

CHAIN, E. (1971). Thirty years of penicillin therapy. *Proceedings of the Royal Society, London, Series B* **179**, 293–319.

DEMAIN, A.L. (1966). Biosynthesis of penicillins and cephalosporins. In *Biosynthesis of antibiotics,* vol. I, pp. 29–94. Edited by Snell, J.F. London and New York: Academic Press.

FEDERAL TRADE COMMISSION (1958). *Economic report on antibiotics manufacture,* pp. 302–54. United States Government Printing Office, Washington.

FLOREY, H.W., CHAIN, E., HEATLEY, N.G., JENNINGS, M.A., SANDERS, A.G., ABRAHAM, E.P. & FLOREY, M.E. (1949). *Antibiotics,* vol. II, pp. 631–72. London: Oxford University Press.

FURUYA, K., ENOKITA, R. & SHIRASAKA, M. (1967). Antibiotics from fungi. II. A new griseofulvin producer, *Nigrospora oryzae. Annual Reports of the Sankyo Research Laboratory* **19**, 91–5.

GATENBECK, S. & BRUNSBURG, V. (1968). Biosynthesis of penicillins. I. Isolation of a 6-aminopenicillanic acid acyltransferase from *Penicillium chrysogenum. Acta chemica Scandinavica* 22, 1059–61.

GILES, D., HEMMING, H.G. & LEHAN, M. (1970). Production of griseofulvin. *British Patent Number* 1,186,507.

GROVE, J.F. (1961). The gibberellins. *Quarterly Reviews* 15, 56–70.

GROVE, J.F. (1963*a*). Griseofulvin. *Quarterly Reviews* 17, 1–19.

GROVE, J.F. (1963*b*). Gibberellins. In *Biochemistry of Industrial Micro-organisms,* pp. 320–40. Ed. Rainbow, C. and Rose, A.H. London and New York: Academic Press.

GROVE, J.F. (1964). Griseofulvin and some analogues. *Progress in the Chemistry of Organic Natural Products* 22, 203–64.

HANSON, J.R., HAWKER, J. & WHITE, A.F. (1972). Studies in terpenoid biosynthesis. Part IX. The sequence of oxidation on ring B in kaurene-gibberellin biosynthesis. *Journal of the Chemical Society. Perkin Transactions I* 1972, 1892–5.

HIKINO, H., ASADA, Y., ARIHARA, S. & TAKEMOTO, T. (1972). Fusidic acid, a steroidal antibiotic from *Isaria kogane. Chemical and Pharmaceutical Bulletin* 20, 1067–9.

IWASAKI, S., IQBAL SAIR, M., IGARASHI, H. & OKUDA, S. (1970). Revised structure of helvolic acid. *Chemical Communications* 1119–20.

KAISE, H., OGAWA, Y., SASSA, T. & MUNAKATA, K. (1972*a*). Studies on the chlorosis-inducing substances produced by a fungus. Part I. Isolation and biological properties of viridominic acids A, B, C and cephalosporin P_1. *Agricultural and Biological Chemistry, Tokyo* 36, 120–4.

KAISE, H., MUNAKATA, K. & SASSA, T. (1972*b*). Structures of viridominic acids A and B, new chlorosis-inducing metabolites of a fungus. *Tetrahedron Letters,* 3789–92.

KAISE, H., MUNAKATA, K. & SASSA, T. (1972*c*). Structure of viridominic acid C, a new steroidal metabolite of a fungus having chlorosis-inducing activity. *Tetrahedron Letters,* 199–202.

KAMAL, A., HUSAIN, S.A., MURTAZA, N., NOORANI, R., QURESHI, I.H. & QURESHI, A.A. (1970). Studies in the biochemistry of micro-organisms. Part IX. Structure of amudane, amudene, and amujane metabolic products of *Penicillium martinsii* Biourge. *Pakistan Journal of Scientific and Industrial Research* 13, 240–3.

KAWAGUCHI, A. & OKUDA, S. (1970). Incorporation of [2-^{14}C]mevalonic acid into the prototype sterol 3β-hydroxyprotosta-17-(20),24-diene with cell-free extracts of *Emericellopsis* sp. *Journal of the Chemical Society. D. Chemical Communications* 1012–13.

KORZYBSKI, T., KOWZYK-GINDIFER, Z. & KURYLOWICZ, W. (1967). *Antibiotics,* vol. II, pp. 1146–93. Oxford: Pergamon Press.

LODER, P.B. & ABRAHAM, E.P. (1971*a*). Isolation and nature of intracellular peptides from a cephalosporin C-producing *Cephalosporium* sp. *Biochemical Journal* 123, 471–6.

LODER, P.B. & ABRAHAM, E.P. (1971*b*). Biosynthesis of peptides containing α-aminoadipic acid and cysteine in extracts of a *Cephalosporium* sp. *Biochemical Journal* 123, 477–82.

MORIN, R.B. & JACKSON, B.G. (1970). Chemistry of cephalosporium antibiotics. *Progress in the Chemistry of Organic Natural Products* 28, 344–403.

MULHEIRN, L.J. & CASPI, E. (1971). Mechanism of squalene cyclization. The biosynthesis of fusidic acid. *The Journal of Biological Chemistry* 246, 2494–501.

NAGARAJAN, R., BOECK, L.D., GORMAN, M., HAMILL, R.L., HIGGENS, C.E., HOEHN, M.N., STARK, W.M. & WHITNEY, J.G. (1971). β-Lactam antibiotics from *Streptomyces. Journal of the American Chemical Society* 93, 2308–10.

NAYLER, J.H.C. (1971). Structure-activity relationships in semi-synthetic penicillins. *Proceedings of the Royal Society, Series B* 179, 357–67.

OKUDA, S., ISAKA, H., IIDA, M., MINEMURA, Y., IIZUKA, H. & TSUDA, K. (1967). Microbial transformation of griseophenone A. *Journal of the Pharmaceutical Society of Japan* 87, 1003–5.

PHINNEY, B.O. & WEST, C.A. (1960). Gibberellins as native plant growth regulators. *Annual Reviews of Plant Physiology* 11, 411–36.

PITEL, D.W., VINING, L.C. & ARSENAULT, G.P. (1971). Biosynthesis of gibberellins in *Gibberella fujikuroi.* The sequence after gibberellin A_4. *Canadian Journal of Biochemistry* 49, 194–200.

RHODES, A. (1963). Griseofulvin: production and biosynthesis. *Progress in Industrial Microbiology* 4, 167–87.

RHODES, A., BOOTHROYD, B., MCGONAGLE, M.P. & SOMERFIELD, G.A. (1961). Biosynthesis of griseofulvin: the methylated benzophenone intermediates. *Biochemical Journal* 81, 28–37.

SHIBATA, S., NATORI, S. & UDAGAWA, S. (1964). *List of Fungal Products,* pp. 147–57. University of Tokyo Press, Tokyo.

SPENCER, B. (1968). The biosynthesis of penicillins: acylation of 6-aminopenicillanic acid. *Biochemical and Biophysical Research Communications* **31**, 170–5.

TAKEUCHI, S., YONEHARA, H. & SHOJI, H. (1964). Crystallized variotin and its revised chemical structure. *The Journal of Antibiotics, Tokyo, Series A* **17**, 267.

TANABE, M. & SETO, H. (1970). Biosynthetic studies with carbon-13; variotin. *Biochimica et Biophysica Acta* **208**, 151–2.

TANAKA, N., SASHIKATA, K. & UMEZAWA, H. (1962). Biogenesis of variotin. III. Incorporation of ^{14}C-glutamic acid into variotin. *Journal of General and Applied Microbiology* **8**, 192–200.

TURNER, J.N. (1972). Practical uses of gibberellins in agriculture and horticulture. *Outlook on Agriculture* **7**, 14–20.

TURNER, W.B. (1971). *Fungal metabolites*, New York and London: Academic Press

VAN HEYNINGEN, E. (1967). Cephalosporins. In *Advances in Drug Research*, vol. 4, pp. 1–70. Ed. Harper, N.J. & Simmonds, A.B. New York and London: Academic Press.

Additional References

EBERSOLE, R.C., GODTFREDSEN, W.O., VANDEGAL, S. & CASPI, E. (1973). Mechanism of oxidative cyclization of squalene. Evidence for cyclization of squalene from either end of the squalene molecule in the *in vivo* biosynthesis of fusidic acid by *Fusidium coccineum*. *Journal of the American Chemical Society* **95**, 8133–40.

HEDDEN, P. & MacMILLAN, J. (1974). Fungal products. Part XII. Gibberellin A_{14}-aldehyde, an intermediate in gibberellin biosynthesis in *Gibberella fujikuroi*. *Journal of the Chemical Society, Perkin Transactions I*, 1970, 587–92.

HIGGENS, C.E., HAMILL, R.L., SANDS, T.H., HOEHN, M.M., DAVIES, N.E., NAGARAJAN, R. & BOECK, L.D. (1974). The occurrence of deacetoxy-cephalosporin C in fungi and Streptomycetes. *The Journal of Antibiotics, Tokyo, Series A* **27**, 298–300.

KLUENDER, H., BRADLEY, C.H., SIH, C.J., FAWCETT, P. & ABRAHAM, E.P. (1973). Synthesis and incorporation of (2*S*, 3*S*)-[4-^{13}C]valine into β-lactam antibiotics. *Journal of the American Chemical Society* **95**, 6149–50.

MacMILLAN, J. & WELS, C.M. (1974). Detailed analysis of metabolites from mevalonic lactone in *Gibberella fujikuroi*. *Phytochemistry* **13**, 1413–7.

NEUSS, N., NASH, C.H., BALDWIN, J.E., LEMKE, P.A. & GRUTZNER, J.B. (1973). Incorporation of (2*RS*, 3*S*)-[4^{13}C]valine into cephalosporin C. *Journal of the American Chemical Society* **95**, 3797–8.

RIISOM, T., JACOBSEN, H.J., RASTRUP-ANDERSEN, N. & LORCK, H. (1974). Assignment of the ^{13}C n.m.r. spectra of fusidic acid derivatives. Biosynthetic incorporation of sodium [1-^{13}C]-acetate into fusidic acid. *Tetrahedron Letters* 1970, 2247–50.

CHAPTER 8

Organic Acid Production

L. B. LOCKWOOD

8.1 Introduction

The commercial production of organic acids by mycelial fungi is, in most instances, limited to members of the genus *Aspergillus*; indeed, for the larger volume fermentations, to one species, *Aspergillus niger* van Tieghem. The production or accumulation of citric acid and D-gluconic acid, the two major metabolites of *A. niger*, is regulated by controlling the level of metallic ions in the culture solution (Shu & Johnson, 1947, 1948*a*; Snell & Schweiger, 1949). A third organic acid, oxalic acid, is of frequent occurrence as a metabolite of *A. niger* and may be formed in large yield. Since oxalic acid can be obtained by less costly processes its production is usually prevented by control of mineral metabolism and pH.

Studies of the mechanism of production of organic acids by the Aspergilli have shown that the major commercial products are either a result of the simple oxidation of monosaccharides, or are components or derivatives of the tricarboxylic acid cycle.

Among the Mucorales citric acid and gluconic acid are not produced in commercially feasible yields. Perhaps this may be explained by the lack of detailed studies on the mineral metabolism of these fungi, and by their lack of ability to grow at low pH. The occurrence of the citric acid cycle enzymes is widespread among them.

8.2 Gluconic Acid, Gluconate Salts, Glucono-δ-lactone, and Glucose Aerodehydrogenase (D-Glucose: Oxygen Oxidoreductase 1,35)

Gluconic acid, gluconate salts, gluconolactone, and the enzyme glucose aerodehydrogenase (glucose oxidase) are produced (Fig. 8.1) by *Aspergillus niger* (Ward, 1967). Since most isolates of *A. niger* produce these products in good yield, together with catalase, an enzyme which is essential to certain commercial uses of glucose aerodehydrogenase, the strains to be used are selected on the basis of their catalase activity.

Gluconic acid was first reported as a product of fungal metabolism by Molliard (1922), although its production had long been known in some of the acetic acid bacteria. A process which uses *Acetobacter suboxydans*, requires about two days to oxidize 10% glucose solution to free gluconic acid. It is used primarily in the manufacture of glucono-δ-lactone.

In the fungal process for the production of gluconic acid (Fig. 8.1), the acid is quickly neutralized on excretion by the fungus, and the fermented solution contains either calcium or sodium gluconate. An early attempt to produce free gluconic acid (Herrick & May, 1928) by a surface culture process using *Penicillium purpurogenum*, gave very low yields of acid, and required a week for completion.

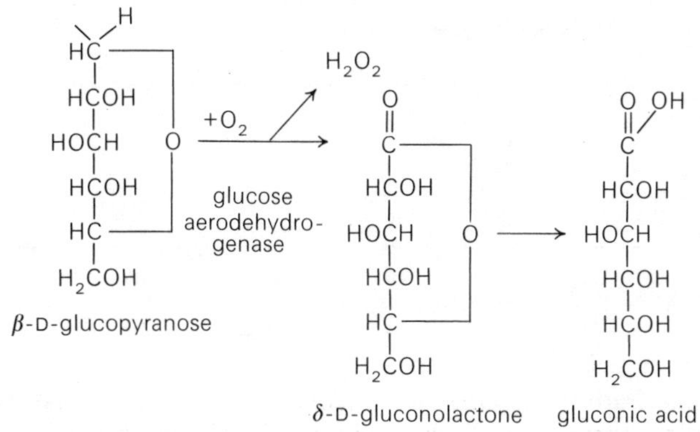

β-D-glucopyranose $\xrightarrow[\text{glucose aerodehydrogenase}]{+O_2 \;\; (H_2O_2)}$ δ-D-gluconolactone $\longrightarrow$ gluconic acid

Fig. 8.1 Reactions of the gluconic acid fermentation.

Modern industrial usage involves the production of calcium or sodium gluconate in the culture solution, together with the enzyme glucose aerodehydrogenase which is recovered from the mycelium as a by-product. Gluconic acid is also the major metabolic product from the growth of the penicillin-producing cultures of *Penicillium chrysogenum* when grown on solutions which contain 5% or more of glucose. Its occurrence is very widespread among the fungi and it is the major metabolite produced from D-glucose by most of the *Aspergillus niger* group, many species of *Penicillium, Gliocadium, Scopulariopsis, Gonatobotrys,* and by the mycelial yeast, *Endomycopsis fibuliger*.

The first process for the production of gluconic acid was a surface culture process (Herrick & May, 1928) in which either *Aspergillus niger* or *Penicillium purpurogenum* was used. More recently, only the submerged fermentation process involving a member of the *A. niger* group has been used. Since this is the surviving process, and is the one which offers the greatest range of operations and variety of products, only this process will be considered here. The raw materials are D-glucose and oxygen. The glucose substrate may be supplied either as a solution of crystalline glucose, or as a syrup obtained from starch or crude starchy materials by the action of the α-amylase of *Bacillus subtilis* followed by the amyloglucosidase of *A. foetidus, A. phoenicis,* or *Endomycopsis fibuliger*. The amyloglucosidase should be free from transglucosidase which, if present, will result in contamination of the gluconate with gentiobiose and isomaltose, with a subsequent reduction in yield of gluconate.

Oxygen is ordinarily supplied by passing air through the fermentation solution (May, Herrick, Moyer & Wells, 1934). The low solubility of oxygen in aqueous media and the speed with which it is removed biologically through its reaction with the glucose aerodehydrogenase system readily makes the oxygen supply a rate limiting factor in the production of gluconate (Table 8.1). Consequently, the conditions of operation are

Table 8.1 Yield of gluconate in relation to air pressure (May *et al.*, 1934)

Pressure (Atmospheres)	Weight yield[a] %
1	42.5
3	80.4
4	82.4
5	81.3
6	86.1

[a] Calculated as gluconic acid. Initial glucose 200 g dm^{-3}, $CaCO_3$ 50 g, aeration rate 40 cm^3 200 cm^{-3} culture, temperature 30°C.

chosen to maintain the maximum concentration of oxygen dissolved in the solution. Fine dispersion of air is achieved by the application of vigorous agitation with a turbomixer or by the action of a cavitator. The efficiency of dissolving oxygen is directly related to the power consumed by the agitator motors. The obvious means of dispersing air in fine bubbles by passage through porous frits has the disadvantage that the mycelium grows into the frit and plugs the pores, causing a considerable pressure drop across the frit. Consequently, frits are now rarely used.

Culture medium composition

The glucose concentration that can be handled depends on whether the calcium or the sodium salt is to be obtained. The solubility of calcium gluconate is only about 4 g cm^{-3} at 30°C. Calcium gluconate fermentation media at harvest are strongly supersaturated when the initial glucose content is only 100 g dm^{-3}. If the initial glucose content exceeds 150 g dm^{-3} the risk of sudden crystallization is considerable, and the contents of the fermentor may set to a solid mass. The solubility of sodium gluconate at 30°C is about 39.6% (Ward, 1967). Consequently, the risk of the material forming a solid mass in the fermentor is avoided, and it is possible to use initial glucose concentrations as high as 28–30%. Other nutrients usually supplied are (Gastrock, Proges, Wells & Moyer, 1938):

Corn steep liquor (commercial concentrate grade)	3.7 g
$MgSO_4 \cdot 7H_2O$	0.17 g
KH_2PO_4	0.2 g
Urea	0.1 g

$(NH_4)_2HPO_4$ 0.4 g
H_2SO_4 to pH 4.5
Tap water to make one litre after inoculation
Antifoam agent as needed.

As the acid is formed, it is neutralized either with $CaCO_3$, which is added in two increments, one at the time of inoculation and the other 6 h later, or with aqueous NaOH solution which is added under continuous automatic pH control.

Essential trace element nutrients are present in the corn steep liquor (the sulphite solution in which grains of maize have been soaked for about 48 h preliminary to the recovery of corn starch, and concentrated to a standard density). If purified solutions of glucose and pure mineral nutrients are used, manganese ion must be included in the solution or glucose aerodehydrogenase will not be formed, and the glucose will be converted to mycelium and a mixture of citric and oxalic acids.

Process operations

Inoculum for gluconate production may be either a sporulated culture grown on bran, beet cossets, a liquid sporulation still culture, or spores germinated in the fermentation solution in a special inoculum vessel (seed tank) similar to a fermenter but of much smaller size. If solid sporulation media are used, the culture is grown in five gallon bottles with air flowing over the surface. The entire sporulated culture mass is made into an aqueous slurry, and transferred aseptically to the fermenter. If germinated spores are used as inoculum, spores are transferred as above to the small seed tank which has been charged with a complete culture medium. Spores are germinated under conditions of vigorous aeration and agitation which give a high efficiency of oxygenation of the culture solution. After about 24 h, the germinated spore culture is transferred to a filled fermenter. The volume of material transferred as inoculum is 3–5% of the volume of the inoculated solution.

The relative advantages of the use of spores directly as inocula and of spores germinated in seed tanks have been the subject of many discussions, but no published conclusions have come to the attention of the writer. Each type of operation has special advantages not found in the other. Direct spore inoculation avoids the cost of installation and operation of the seed tanks. The use of germinated spores reduces the operating cycle for the main fermenter, thus permitting several extra production cycles per year. However, labour costs for the germinated spore inoculum may be almost as great as for the final fermentation. Gluconate produced in the germinated spore inoculum is recoverable in the final fermentation liquor and contributes to the buffering capacity throughout the fermentation period, thus permitting less careful control of pH in the early hours of the fermentation. Thus, the choice of inoculum types depends on the length of cycle of the fermentation process, plant size and capacity, and the availability and cost of labour.

Operational procedures are indicated in the flow sheet (Fig. 8.2), although differences occur depending on whether the calcium or sodium salt is being produced. Culture solutions may be sterilized by either batch or continuous methods. In calcium gluconate production, $CaCO_3$, sterilized as an aqueous slurry, is supplied to neutralize the acid as soon as it is released from the mycelium. During heat sterilization of the slurry, some CO_2 is lost and the pH of the solution may reach 12.5. If a glucose solution is exposed to alkalinity greater than pH 5.5, a reduction in gluconate yield due to the Lobry de Bruyn-van Eckenstein reaction occurs, and the final product may be contaminated with fructose, citrate, and oxalate. Consequently, the $CaCO_3$ slurry is added incrementally so that the pH never becomes alkaline. In sodium gluconate production, the pH is controlled by the addition of NaOH solution by means of an automatic pH controller.

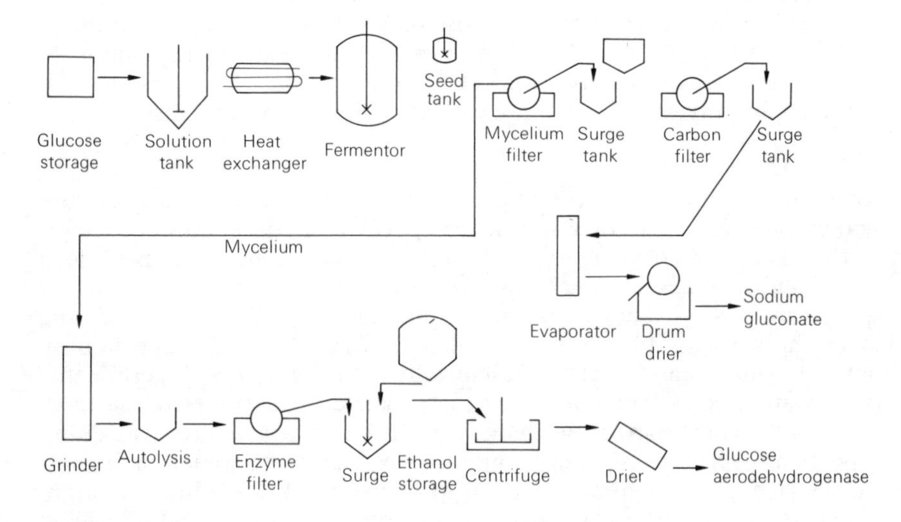

Fig. 8.2 Flow diagram, sodium gluconate and glucose aerodehydrogenase manufacture.

The nutrient solution has considerable buffering capacity, due to the presence of phosphates and corn steep liquor. During the fermentation, the buffering capacity is increased by the presence of gluconate, gluconic acid, and gluconolactones. The initial buffering capacity of the culture medium is sufficient to permit the germination of spores and survival of mycelium if the initial increment of alkaline material is not enough to cause glucose decomposition. Operating conditions for the production of calcium gluconate or sodium gluconate are: temperature 28–30°C, aeration 1–1.5 volumes air/volume solution/min, and vigorous agitation to maintain the maximum oxygenation efficiency. The initial pH is about 6.5, and it is maintained at about 5.5–6.0 by the excess $CaCO_3$ in the production of the calcium salt, or at about 6.8–7.2 by NaOH additions in the production of the sodium salt.

Product recovery

The mycelium is filtered off the solution at the end of either the calcium or sodium gluconate fermentation, and is used for the recovery of glucose aerodehydrogenase. The mycelium is ground sufficiently to disrupt the cells, and the ground mass is permitted to autolyze in phosphate buffer in the range of pH 5.0–9.0. The enzyme occurs in both the active and the inactive bound form within the cell, and autolysis permits greater recovery of the enzyme content. Suspended solids are filtered off from the autolyzate, and the enzyme may be precipitated with cold ethanol; it may be absorbed on diatomaceous earth prior to precipitation, or the enzyme solution may be stabilized by saturating the enzyme with an excess of substrate. Both liquid and dry forms of the enzyme are on the market.

Calcium or sodium gluconate, gluconic acid, and glucono-δ-lactone are recovered from the filtrate obtained when the mycelium is removed. For calcium gluconate production the filtrate is heated with a slight excess of $Ca(OH)_2$, then decolourized with carbon and filtered. On cooling to a temperature below 20°C and seeding with calcium gluconate crystals, the product readily crystallizes. Mother liquors can be evaporated by heat to about 10–15% volume, treated with carbon, filtered, and chilled to obtain a second crop of crystals.

Sodium gluconate is prepared by concentrating the fermented solution filtrate to about 42–45% solids, adjusting the pH to 7.5 with NaOH, and drum drying the material.

Gluconic acid and its delta lactone are recovered from the filtrate obtained when mycelium is removed from the calcium gluconate production solution. The calcium is precipitated by the addition of a stoichiometric quantity of sulphuric acid. Calcium sulphate is filtered off, and the filtrate decolourized with carbon. On removal of carbon, the acid solution is concentrated to 50% acid strength, and sold for use in foods and in metal pickling. The product is a mixture of free gluconic acid and its γ- and δ-lactones. Crystals separating from a gluconic acid solution at a temperature below 30°C and preferably near 0°C, are principally gluconic acid. Between 30°C and 70°C, the material is δ-lactone and above 70°C the γ-lactone predominates (Wang, Alder & Lardy, 1961).

Uses of gluconic acid, gluconates and glucose aerodehydrogenase

Calcium gluconate finds use as a readily available and widely used material for calcium therapy, when calcium deficiency occurs, or when allergy is severe. Sodium gluconate is widely used as a sequestering agent to prevent the precipitation of lime soap scums on cleaned products. Glucose aerodehydrogenase is used for the removal of glucose in the preparation of dried eggs, and for the stabilization of colour and flavour in beer, tinned foods, and soft drinks. It is also the basis for the production of diagnostic aids used in the evaluation of control of diabetic patients. Free gluconic acid is used as a mild acidulant in metal processing, leather tanning, and foods. The δ-lactone is used in baking powders, since it is non-acidic until dissolved in water.

8.3 Citric Acid

In terms of tonnage, citric acid is the major commercial product made by filamentous fungi. World production figures are not available, but it is estimated that it amounts to over 90 000 000 kilos per year. In commercial practice, *Aspergillus niger* or one of its tan coloured mutants is usually used for citric acid production.

Three different commercial processes are being used for citric acid production. They have been held with a great amount of secrecy, and regarding them very little detailed descriptive material has been published. However, examination of the patents issued to various companies gives some suggestions as to the processes used.

The Koji fermentation process

A relatively simple fermentation has been developed in Japan. Solid cooked vegetable residues are spread in winrows on a floor or in trays, and inoculated with a selected strain of *Aspergillus niger* (Hisanaga & Nakamura, 1966; Yamada, 1965). Since the sweet potato residues, or wheat bran used as substrates normally contain enough iron, manganese, cobalt, or nickel to prevent the formation or accumulation of citric acid, strains have been selected which have very little sensitivity to trace metallic ions. Initially, the fibrous residues from the recovery of sweet potato starch were used, but wheat bran has to a considerable extent replaced the sweet potato residue. During incubation, the starch is saccharified by the amylase of *A. niger*, and much of the sugar produced is converted to citric acid. The temperature inside the solid mass should not exceed 28°C (Hisanaga & Nakamura, 1966). A drop in pH to about 1.8–2.0 occurs as citric acid accumulates. After 5–8 days, the mass is extracted with water in percolators, and the citric acid is purified as in other processes. Estimated production by this method is only about 2 500 000 kilos. Hisanaga & Nakamura (1966) found that addition of the filter cake of glutamic acid fermentations to the starchy substrate in an amount to give 3–7% (calculated on pulp dry weight) resulted in improvement in handling and production capability. The acidity was adjusted to pH 4–5, and the moisture content to 70–80% after steam sterilization. On cooling to 60–80°C, α-amylase was added to the mass to liquify the starch. When the temperature reached 30–35°C, the mass was inoculated with *A. niger*, and spread 3–5 cm deep in trays. The most evident advantage is that much more pulp can be used per tray than is possible when glutamic acid fermentation wastes and the amylase treatment are omitted.

The liquid culture shallow pan process

In the older method for citric acid manufacture used in Europe and the Americas, the culture solution is dispersed in shallow pans. It is estimated that more than 12.3 hectares (30 acres) of pans are required in one large citric acid plant. Pans are made of high purity aluminium or stainless steel to avoid problems of corrosion and trace metallic ion contamination. Mallea (1950) has published a fairly complete description of the shallow pan process as used in a plant which he designed and operated in Tucumán, Argentina. Other surface liquid culture plants are believed to be similar, except that they are probably mechanized to a greater extent.

The inoculum is usually spores produced on the surface of a liquid medium designed for spore production, but unsuitable for citric acid production. After about nine days of growth, the fragile sporulating pellicles are transferred to the sterile fermentation solution, and are mechanically dispersed. The inoculated medium is then distributed to fermentation pans containing a medium which favours citric acid production. Alternatively, as a means of inoculation, the spores may be blown over the sterile solutions in the pans.

Beet molasses which has not been subjected to the Steffen's process for sugar recovery is the most commonly used carbohydrate source, but raw sugar, high test syrup, or purified molasses may also be used. It is usually treated with ferricyanide or ferrocyanide to remove traces of iron, and the blue precipitate is filtered off. Such molasses usually contains enough other metallic ions and phosphates to provide adequate nutrient. The medium is acidulated with sulphuric acid to a pH value of about 2.5–4.0 prior to inoculation.

Spores germinate within the first 24 h, and a thin fragile white pellicle of mycelium covers the surface of the solution. Sterile humidified air at about 30°C is blown slowly over the surface of the solution. After five or six days, humidification of the air is discontinued. At this time, the mycelium is a white crinkled or folded mat on the surface of the liquid. Eight to ten days after inoculation, the sugar content of the culture solution will have been reduced from its initial level of about 20% to the range of 1–3%, and the maximum concentration of citric acid reached. Both the fermentation process and evaporation contribute toward the final concentration of citric acid in the culture solution.

If the pH value rises to about 3.5 during the fermentation, considerable oxalic and gluconic acids may be formed, and an inferior yield of citric acid is obtained. The presence of iron also favours the production of oxalic acid, sporulation, and pigment formation. Sporulating mycelia accumulate very little citric acid. Most strains of *Aspergillus niger* produce green or yellow pigments in the mycelium when sporulating. These are often excreted into the culture solution, and are difficult to remove during recovery operations.

The final recovered yield of citric acid is in the neighbourhood of 80–85% of the weight of the initial carbohydrate supplied, recovery efficiency being about 90% of the acid produced in the fermentation.

The submerged fermentation process

The flow sheet (Fig. 8.3) outlines the present submerged process for citric acid manufacture, as developed by Schweiger (1961), Batti (1966, 1967) and Batti & Schweiger (1963). Citric acid is produced by distorted mycelia of *Aspergillus niger* submerged throughout the culture solution in deep tanks. The distorted mycelia are described by Schweiger (1961) as being characterized by short stubby branches which may have swollen tips, and by the presence of chlamydospores. Submerged fermentation is used by Miles Laboratories in America and Israel, and by John E. Sturge, Ltd. in England.

Perquin (1938) made an extensive study of citric acid production by *Aspergillus niger* in shake cultures, and developed the concept that citric acid accumulated in the culture solution as a result of phosphate deficiency. He did not purify his culture solution effectively, and his results may be explained in terms of trace element contamination of his phosphate

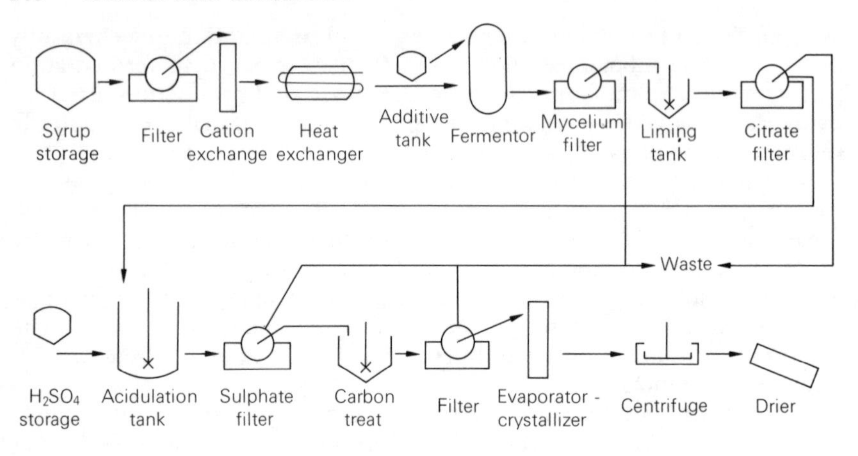

Fig. 8.3 Flow diagram, citric acid manufacture.

source. Shu & Johnson (1947, 1948*b*) showed that in submerged culture, citric acid production occurred only in media deficient in iron and manganese. Nickel and cobalt ions also have a deleterious effect on the fermentation yield. Removal of these ions by the use of ion exchange resins was successfully achieved by Snell & Schweiger (1949) and Woodward, Snell & Nicholls (1949). The use of copper ion as an antagonist to iron was discovered about the same time by Schweiger (1961) (Table 8.2), but the patent was not issued until 1961. The work of Snell, coupled with that of Schweiger, made possible the present successful commercial process. Table 8.2 illustrates the relationships of iron and its copper ion antagonist to yield of citric acid. Clark (1962) has applied precipitation of iron with ferricyanide to the preparation of syrups from cane molasses as raw materials for citric acid manufacture, but the use of cation exchange resins for removal of trace elements appears to be more widely used in the submerged fermentation process.

Raw materials used successfully in the submerged fermentation process include commercial syrups of high glucose content, high test cane syrup (concentrated crude cane juice with about two-thirds of the sucrose inverted), and an enzyme hydrolyzate of grains of maize, grain sorghums, or of crude sweet potato starch (Swarthout, 1966). Such materials are prepared to give sugar concentrations of 15–20 g $100 cm^{-4}$. The syrups are decationized by passage through a sulphonate ion exchange resin in the hydrogen form. The solution after ion exchange is very acidic, and ammonia may be added to adjust the pH to the range of 2.0–4.0. Addition of growth and enzyme inhibitors such as copper sulphate is done at the time of inoculation. The inoculum consists of a sporulated culture of *Aspergillus niger* slurried in water, and aseptically transferred to the sterile medium in the fermenter. Either solid or liquid sporulation cultures can be used, but they must not contain excessive amounts of manganese, iron, cobalt or nickel. The fermentation solution is aerated from inoculation to

Table 8.2 Relationship of ions of copper and iron to the production of citric acid from glucose (Schweiger, 1961)

Fe^{3+}	Cu^{2+}	Yield, citric acid[a]
10	50	77.8
50	50	69.1
100	50	50.7
100	50	14.2
10	100	77.2
50	100	65.4
100	100	53.9
150	100	29.8
10	500	74.0
50	500	65.4
100	500	60.6
150	500	27.6

[a] $\frac{\text{Grammes citric acid produced}}{\text{Grammes glucose supplied}} \times 100$

harvest. The oxygen requirement is very small, but if aeration is stopped for a very brief period, citric acid production stops and is resumed only very slowly unless adjustments in pH, nitrogen supply, and growth inhibitors are made to induce new mycelial growth of suitable metabolic type (Batti & Schweiger, 1961). After growth is established the pH of the solution should never rise above 3.5, or considerable oxalic acid and gluconic acid may contaminate the product, complicate product recovery, and reduce the yield of citric acid. Mechanical agitation of the solution is not required to maintain adequate oxygenation. The duration of the fermentation depends on the initial sugar concentration, subsequent sugar additions, and the amount of growth and enzyme inhibitors added. It may range from 5–14 days, but is uniform for a given set of conditions. Yields as high as 95% w/w have been reported (Lockwood & Batti, 1965), but it is doubtful if this yield is consistently obtained on a commercial scale.

Foam control is necessary to prevent loss of material during the fermentation, and to reduce the hazard of microbial contamination. Sulphonated fatty acid derivatives of sorbitan (Tweens) or lard oil may be used, and must be added as needed throughout the culture period. A role other than that of foam control may be of significance here. Millis, Trumpy & Palmer (1963) found that lipid addition to their cultures resulted in improved yields of citric acid, and Gold & Kieber (1968) obtained a patent on this concept. It is not clear if this is purely a physical effect, or if some of the fatty acid is converted to an intermediate which may enter the citric acid cycle.

A recent development in citric acid production is the use of emulsified hydrocarbons as raw materials. To date, it is believed that this has not been successfully operated on a commercial scale. Several problems remain to be solved. Among these are the low concentration of substrate which must be fed on a continuous basis and the contamination of the product with isocitric acid. In some instances, isocitric acid may represent as much as half the total acidity produced. The organisms usually used are yeasts or yeast like organisms, and are outside the scope of this book.

Product recovery

The mycelium is filtered off from the solution, and is washed on a belt vacuum filter or a rotary drum vacuum filter pre-coated with filter aid, and equipped with an advancing doctor blade (Lockwood & Schweiger, 1967). The filtrate is limed with $Ca(OH)_2$ slurry to the point of complete neutralization of the acid then the calcium citrate is filtered off using a conventional industrial filter, and washed to remove adherent sugar and other impurities (Lockwood & Schweiger, 1967; Mallea, 1950). The washed calcium citrate is suspended in an acidulation tank, and sulphuric acid to about 0.2% excess is added. Calcium sulphate is filtered off, and the dilute citric acid solution is decolourized with carbon. The filtrate is further purified by the use of anion and cation exchange resins, and is then evaporated to 35°–40° Baumé for crystallization. Mother liquors can be recycled back to the initial liming step. When the crystallization temperature exceeds the critical point, 36.6°C, the anhydrous crystals are produced; below this temperature, the monohydrate is formed.

Biochemical mechanisms in the production of citric acid

No clear evidence is available to define the biochemical steps in the conversion of a hexose molecule into an intermediate of the citric acid cycle during the citric acid fermentation. The presence of the Embden-Meyerhof-Parnas glycolytic system is well established in *Aspergillus niger* by many studies (Meyrath, 1967), but these studies have not been made under conditions in which a commercially feasible yield of citric acid is obtained. This is clearly an area which requires further study. Citric acid production appears to result from the normal operation of the citric acid cycle. Citric acid accumulation results from the disruption of that cycle, in which destruction of citric acid is blocked. Low pH, deficiency of metallic ion cofactors for some of the enzymes following citric acid in the cycle, or inhibition of the action of some of these enzymes due to the use of copper ions (Schweiger. 1961), H_2O_2 (Bruchmann, 1961) or various organic compounds (Lockwood & Batti, 1965) all act to hinder the destruction of citric acid. The mechanisms of this blockage and the point or points of inhibition are not known. LaNauze (1966) found that addition of iron enhanced aconitase activity, but this did not result in change in citric acid yield. Bruchmann (1961) holds the opinion that the inhibitory action of hydrogen peroxide on aconitase is a major factor controlling citric acid accumulation. He observed an inverse relationship between catalase activity and citric acid accumulation. He also observed the well known effect of copper and zinc ions on citric acid accumulation, and concluded that the effect of copper ion was aconitase inhibition, and the effect of zinc was the inhibition of some enzyme beyond aconitase in the cycle. Schweiger's observation of an inverse relationship of copper and iron ions indicates that copper inhibition of an enzyme having an iron cofactor is a major element in citric acid accumulation. Doubtless, citric acid accumulation is the result of all these factors, and probably of others not yet defined. The biochemistry of citric acid production has recently been reviewed by Smith, Nowakowska-Waszczuk & Anderson (1974).

8.4 Itaconic Acid

Itaconic acid was first reported as a fungal metabolite by Kinoshita (1931, 1932) who described a new species, *Aspergillus itaconicus*. Callam, Oxford & Raistrick (1939) found that the acid is produced by *A. terreus* and most of the studies since the date of their publication have dealt with *A. terreus* strains. Nubel & Ratajak (1962) found that beet molasses contained some factor which permitted the use of sugar concentrations as great as 20% in *A. terreus* fermentations. Such a factor in beet molasses may also be active in the citric acid fermentation of this material, where other growth and enzyme inhibitors appear to be unnecessary in the surface culture process. Lockwood & Nelson (1946) established the relationship of iron and zinc ions to itaconic acid yield, but did not approach an economically feasible process. Table 8.3 illustrates this response to trace elements. Schweiger &

Table 8.3 The response of *Aspergillus terreus* NRRL 1960 to addition of copper and zinc ions (Batti Schweiger, 1963)

Copper ppm	Zinc ppm	Itaconic Acid Weight[a] Yield (%)
0.0	0.0	16.0
0.0	0.5	43.3
0.0	6.0	50.4
0.5	0.5	55.4
1.0	0.5	51.7
3.0	0.5	52.6
6.0	0.5	55.4

[a] $\frac{\text{Grammes itaconic acid produced}}{\text{Grammes sugar supplied}} \times 100$

Batti realized the close similarity between the citric and itaconic acid fermentations, and applied the combination of cation exchange treatment of culture solutions, and the application of the enzyme inhibitors used in citric acid production. Batti & Schweiger (1961) also added alkaline earth metallic ions to their cultures, and obtained additional benefits in yield.

An additional problem was found in the itaconic acid process which did not occur in the citric acid fermentation. This is the toxicity of itaconic acid to the fermentation process, which limits the maximum concentration of itaconic acid in the broth to about 8%. Fermentation rates are satisfactory until the itaconic acid concentration reaches 7%, beyond which point the reaction proceeds more slowly, and stops completely at about 8%. However, Nubel & Ratajak (1962) showed that partial neutralization of the free acid, to prevent the free acid concentration from exceeding

7–8% makes it possible to continue the fermentation process until the itaconate concentration in the solution reaches 15–20%.

The current commercial process for itaconic acid manufacture is based on the Nubel & Ratajak patent (1962) which requires the use of beet molasses as the substrate for spore germination and the induction of a suitable biochemically active growth form. Spores are germinated in a medium which contains the following:

Beet molasses to give	15 g sugar
$NaSO_4$	1.5 g
$MgSO_4 \cdot 7H_2O$	5.0 g
$CuSO_4 \cdot 5H_2O$	0.02 g
Soybean Oil	0.25 cm^3
Water to 1 dm^3	

The medium is inoculated with spores grown on the surface of a sporulation medium. After inoculation, the spore germination medium is aerated with one-fourth volume of air per minute for eighteen hours to permit spore germination and the initiation of a suitable pattern of growth. The pH drops from about 7.6 to about 5.4. The flow diagram (Fig. 8.4) indicates the process steps in commercial itaconic acid manufacture. The final fermentation medium consists of the following:

Cane molasses to give	150 g sugar
$ZnSO_4$	1.0 g
$MgSO_4 \cdot 7H_2O$	3.0 g
$CuSO_4 \cdot 5H_2O$	0.01 g
Water to 1 dm^3	

Four volumes of final fermentation medium are inoculated with one volume of germinated spore culture. The temperature is maintained at 39–42°C, well outside the most favourable range for growth. The material

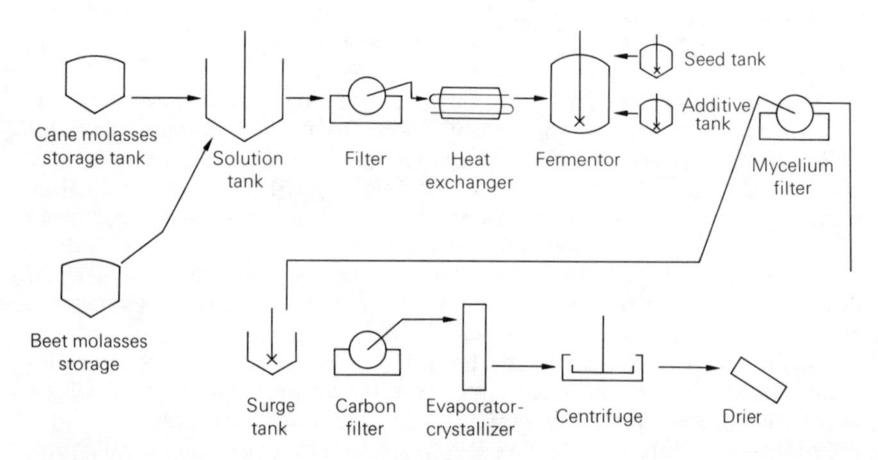

Fig. 8.4 Flow diagram, itaconic acid manufacture.

is aerated at one-quarter to one-half volume of solution per minute. During the first 24 h, the pH drops from 5.1 to about 3.1. It is readjusted to 3.8 with lime or ammonia, and the fermentation is continued for a further 48 h. At that time, the fermentation medium contains about 85 g itaconic acid/l, a weight yield of 56.3%, or about 78.5% of the theoretical yield.

Kobayashi (1967) studied the production of itaconic acid, and presented rate equations based on the use of Waldhof fermenters and steady state fermentations. He used sugar solutions of up to about 20% concentration in single stage fermentations, and obtained yields of 50%. Dilute sugar solutions were used in steady state continuous operations. Three days continuous operation were sufficient to reach a steady state which was continued for 14 days. Cost analysis indicated that continuous operation would be only about 81% as costly as the batch process described by Pfeifer, Vojnovich & Heger (1957) in which the maximum sugar concentration used successfully was 6%.

Recovery

Mycelium is filtered off the finished fermentation solution which is then decolourized with activated carbon, filtered and evaporated with stirring for crystallization. The crude product is sufficiently pure for esterification, but an additional decolourization with activated carbon, and recrystallization are necessary to make a pure white, refined product.

Market

Itaconic acid finds its principal use as a copolymer with acryllic resins. It imparts to the resin the ability to take printer's inks, bonding them much more tightly than do most acryllic resins.

Biochemical mechanisms in the accumulation of itaconic acid

Little attention has been given to the biochemical mechanisms involved in itaconic acid production. It is tempting to assume that the acid is the product of the enzymatic dehydration of cis-aconitic acid of the citric acid cycle, and to assume that, up to that point, the mechanism is identical with that of citric acid production. Unfortunately, little information is available directly bearing on the subject. If itaconic acid arises from the dehydration of cis-aconitic acid, one might expect the inhibition of aconitase by copper ions at the low pH of the fermentation to inhibit the production of itaconic acid by blocking the destruction of the citric acid cycle intermediate. Actually, addition of copper ions at low pH increases the accumulation of itaconic acid (Batti & Schweiger, 1963; Schweiger, 1961). This suggests that the responses to copper ions of the aconitases of *Aspergillus niger* and *A. terreus* are different, and that as in the case of citric acid accumulation, itaconic acid accumulation occurs when enzymes involved in the destruction of the metabolite are inhibited. If the pH of an itaconate solution is raised to about neutral, itaconate becomes an excellent substrate for growth, and is rapidly destroyed. It thus appears that the combination of low pH and copper ions inhibits the destruction of itaconic acid, and that it is a normal intermediary metabolite of *A. terreus*. Itatartaric acid, which is accumulated to a limited extent by some mutants of *A. terreus* NRRL 1960

(Stodola, Friedkin, Moyer & Coghill, 1945) may be the first metabolite in the destruction of itaconic acid by *A. terreus*. The only studies of itaconic acid metabolism that have come to the author's attention have dealt with its destruction by rat or guinea pig liver mitochondria. Here itaconate is converted to pyruvate and acetyl coenzyme A (Wang *et al.*, 1961). This offers the possibility of re-entry into the citric acid cycle, and of recycling of carbon. Itatartrate does not appear to be involved in this reaction, so animal and fungal metabolism appear to differ in this respect. Further research into the mechanisms of destruction of both citric and itaconic acids, as contrasted with their production, is essential to an understanding of these commercial processes. It might also supply guidance to the investigator in the development of processes for the large scale manufacture of other citric acid cycle intermediates and side reaction products. The biochemistry of itaconic acid biosynthesis has been considered recently by Smith *et al.* (1974).

8.5 Miscellaneous Organic Acids

Several organic acids produced by filamentous fungi are marketed in relatively small scale, or have been removed from the market due to competition with chemosynthetic materials. There are also several acids which have never been produced commercially, but which may hold promise for future exploitation.

Fumaric acid, produced by numerous Phycomycetes, and occasionally found as a product of the Aspergilli, was once made commercially with *Rhizopus oryzae*. Oxygenation of the culture solution is essential to fumaric acid accumulation. If the culture is not aerated, ethanol and CO_2 are the principal products of the fermentation. The fermentation of glucose or invert sugar solutions, aerated and agitated, in the presence of suspended $CaCO_3$ results in the crystallization of some calcium fumarate before the fermentation is finished. Advantage is taken of the increased solubility of calcium fumarate with increased temperature. The whole fermented mass is heated until all the calcium fumarate is dissolved, and the suspended solids, mycelium and excess $CaCO_3$, are filtered hot. On cooling of the filtrate, calcium fumarate crystallizes out. The free acid is recovered by acidulation of the calcium fumarate with an excess of sulphuric acid, and removal of calcium sulphate by filtration. On evaporation of the filtrate, fumaric acid crystallizes from the dilute sulphuric acid solution. Microbially produced fumaric acid has now been replaced on the market by fumaric acid made by the catalytic oxidation of benzene. It is doubtful if the fermentation process will be reactivated until benzene becomes much more expensive than it is at present. This is not anticipated for many years.

L(+)-Lactic acid is also produced in submerged fermentation by selected strains of *Rhizopus oryzae*. Special strains are chosen which produce predominantly either lactic or fumaric acid. Procedures and cultures for the fumaric acid and L(+)-lactic acid fermentations are very similar. However, mechanical agitation is not necessary in the L(+)-lactic acid fermentation. If aeration is stopped, the culture produces principally ethanol and CO_2 during anaerobic growth. L(+)-Lactic acid is produced by growing cells. If growth is stopped due to nitrogen exhaustion, chlamydospores are formed abundantly, and lactic acid production stops. Addition of assimilable nitrogen (ammonium or amine, but not nitrate) results in rapid germination of the chlamydospores. Lactic acid production can be detected immediately after chlamydospore germination papilli become visible. Commercially, most fermentation lactic acid is made by bacterial processes which usually give racemic lactic acid. The product of the fungal process is much easier to purify than that of the bacterial process, since the medium for the *Rhizopus* fermentation requires none of the crude nitrogenous materials which are essential to the success of the bacterial processes. These crude nitrogenous materials frequently impart colour to the commercial product of the bacterial fermentations. Weight yields in the bacterial and the *Rhizopus* processes are about equal, 93–95%, and fermentation times are rather similar. In any lactic acid fermentation process, an excess of $CaCO_3$ must always be present. Mild mechanical stirring is required in the bacterial process, but not in the fungal process. Aeration at about one tenth volume of air per minute keeps the $CaCO_3$ suspended in the *Rhizopus* fermentation. The free lactic acid may be recovered from acidified filtrates by extraction with isopropanol, or if only the single optical isomer is present, by direct crystallization from the acidified, decolourized, evaporated culture liquor. L(+)-Lactic acid finds use primarily as a substrate in physiological research, since it is the form which occurs naturally in muscle tissue. The relative ease of metabolism of D(−)- and L(+)-lactic acid is still not established. Lactic acid polymerizes readily in aqueous solution to form the cyclic dimer, lactide, and the linear polymer typified by the dimer, lactoyl lactic acid, but of much greater molecular weight.

L-Malic acid is the principal metabolite of selected cultures of the mushroom, *Schizophyllum commune*. Detailed procedures for the preparation of L-malic acid by this organism have not been described, but it is believed to offer an economical means of producing this naturally occurring isomer. This is also the form which ordinarily is present in the tricarboxylic acid cycle, and is found in apples, strawberries, cherries, and numerous other vegetable materials.

The production of trans-epoxysuccinic acid, a metabolite of *Aspergillus fumigatus* and *Paecilomyces varioti*, has not been studied very extensively. Whether it is a possible member of the tricarboxylic acid cycle or the product of a side reaction, remains to be determined. The patents of Moyer (1953, 1954) indicate that the addition of lower aliphatic alcohols to the culture medium results in improvement in yield in this fermentation, and also in the citric and itaconic acid fermentations. These last two represent a member of the tricarboxylic acid cycle, and a side reaction product of this cycle. Moyer's work suggests similarity of these three fermentations. Trans-epoxysuccinic acid is of interest because, on boiling its aqueous solutions, the acid hydrates to form tartaric acid, a widely used product now available primarily as a byproduct of the wine industry. The demand for tartaric acid may lead into a more detailed future study for process development to produce trans-epoxysuccinic acid.

Three interesting acids, spiculosporic acid, tricholomic acid, and ramigenic acid, are fatty acid substituents of citric or other common organic acid metabolites of fungi. Processes have not been developed to produce these acids, but their potential use in detergents places them in an area of interest for future investigation.

8.6 References

BATTI, M.A. (1966). Process for production of citric acid. *United States Patent* 3,290,227.

BATTI, M.A. (1967). Process for producing citric acid. *United States Patent* 3,335,067.

BATTI, M.A. & SCHWEIGER, L.B. (1961). Process for production of itaconic acid. *Australian Patent* 253,501.

BATTI, M.A. & SCHWEIGER, L.B. (1963). Process for production of itaconic acid. *United States Patent* 3,078,217.

BRUCHMANN, E.E. (1961). Enzymische Untersuchunger über Schimmelpilzgarungen, II. Einige Hemmtoffe der Aconitase und ihre Bedeutung für die Citronensaureanhaufung durch *Aspergillus niger* in Submerskultur. *Biochemische Zeitschrift* **335**, 119–211.

CALAM, C.J., OXFORD, A.C. & RAISTRICK, H. (1939). Studies in the biochemistry of micro-organisms. LXIII. Itaconic acid, a metabolic product of a strain of *Aspergillus terreus* Thom. *Biochemical Journal* **33**, 1488–95.

CLARK, D.S. (1962). Submerged citric acid fermentation of ferricyanide-treated cane molasses. *Biotechnology and Bioengineering* **4**, 17–21.

GASTROCK, E.A., PORGES, N., WELLS, P.A. & MOYER, A.J. (1938). Gluconic acid production on pilot-plant scale, effect of variables on production by submerged mold growths. *Industrial and Engineering Chemistry* **30**, 782–9.

GOLD, W. & KIEBER, R.J. (1968). Process for production of citric acid by fermentation. *United States Patent* 3,372,094.

HERRICK, H.T. & MAY, O.E. (1928). Production of gluconic acid by *Penicillium luteum-purpurogenum* group II, some optimal conditions for acid formation. *Journal of Biological Chemistry* 77, 185–95.

HISANOGA, W. & NAKAMURA, S. (1966). *Japan Patent* 16555/66.

KINOSHITA, K. (1931). Ueber eine neue Aspergillus Art, *A. itaconicus*. *Botanical Magazine (Tokyo)* **45**, 45–61.

KINOSHITA, K. (1932). Über die Produktion von Itaconsaure und Mannit durch einem neuen Schimmelpilz, *Aspergillus itaconicus*. *Acta Phytochimica* 5, 271–87.

KOBAYASHI, T. (1967). Itaconic acid fermentation. *Process Biochemistry* **2**, 61–5.

LANAUZE, J.M. (1966). Aconitase and isocitric dehydrogenase of *Aspergillus niger* in relation to citric acid production. *Journal of General Microbiology* **44**, 73–81.

LOCKWOOD, L.B. & BATTI, M.A. (1965). Biosynthetic process for making citric acid. *United States Patent* 3,189,527.

LOCKWOOD, L.B. & NELSON, G.E.N. (1946). Some factors affecting the production of itaconic acid by *Aspergillus terreus* in agitated cultures. *Archives of Biochemistry and Biophysics* **10**, 365–74.

LOCKWOOD, L.B. & SCHWEIGER, L.B. (1967). Citric and itaconic acid fermentations. In *Microbial Technology*, pp. 183–99. Ed. by Peppler, H.J. London, New York: Reinhold Publishing Company.

MALLEA, O. (1950). La industria de fermentation citrica en La Argentina. *Industria y Quimica* **12**, 264–80.

MAY, O.E., HERRICK, H.T., MOYER, A.J. & WELLS, P.A. (1934). Gluconic acid, production by submerged mold growths under increased air pressure. *Industrial and Engineering Chemistry* **26**, 575–8.

MEYRATH, J. (1967). Citric acid production. *Process Biochemistry* **2**, 25–7, 56.

MILLIS, N.F., TRUMPY, B. & PALMER, B.M. (1963). The effect of lipids on citric acid production by an *Aspergillus niger* mutant. *Journal of General Microbiology* **30**, 365–79.

MOLLIARD, M. (1922). Sur une nouvelle fermentation acide produite par le *Sterigmatoiystis nigra*. *Compte Rendue Academie Science, Paris* **174**, 881–3.

MOYER, A.J. (1953). Effect of alcohols on the mycological production of citric acid in surface and submerged culture, I. Nature of the alcohol effect. *Applied Microbiology* **1**, 1–13.

MOYER, A.J. (1954). Production of organic acids. *United States Patent* 2,674,561.

NUBEL, R.D. & RATAJAK, E.J. (1962). Process for producing itaconic acid. *United States Patent* 3,044,941.

PERQUIN, L.H.D. (1938). Beydrage tot de Kennis der Oxidativen Dissimilatie van *Aspergillus niger* van Tieghem. *Dissertation, Technische Hogeschool, Delft, W.D. Meiniera.*

PFEIFER, V.P., VOJNOVICH, C. & HEGER, E.N. (1957). Itaconic acid by fermentation with *Aspergillus terreus*. *Industrial and Engineering Chemistry* **44**, 2975–80.

SCHWEIGER, L.B. (1961). Production of citric acid by fermentation. *United States Patent* 2,970,084.

SHU, P. & JOHNSON, M.J. (1947). Effect of the composition of the sporulation medium on citric acid production by *Aspergillus niger* in submerged culture. *Journal of Bacteriology* **54**, 161–7.

SHU, P. & JOHNSON, M.J. (1948*a*). The interdependence of medium constituents in citric acid production by submerged fermentation. *Journal of Bacteriology* **56**, 577–85.

SHU, P. & JOHNSON, M.J. (1948*b*). Citric acid production by submerged fermentation with *Aspergillus niger*. *Industrial and Engineering Chemistry* **40**, 1202–5.

SMITH, J.E., NOWAKOWSKA-WASZCZUK, A. & ANDERSON, J.G. (1974). Organic acid production by mycelial fungi. In *Industrial Aspects of Biochemistry*, Vol. 1, pp. 297–317. Ed. by Spencer, B. Amsterdam: Elsevier.

SNELL, R.L. & SCHWEIGER, L.B. (1949). Production of citric acid by fermentation. *United States Patent* 2,492,667.

STODOLA, F.H., FRIEDKIN, M., MOYER, A.J. & COGHILL, R.D. (1949). Itatartaric acid, a metabolic product of an ultraviolet-induced mutant of *Aspergillus terreus*. *Journal of Biological Chemistry* **61**, 739.

SWARTHOUT, E.J. (1966). Citric acid production. *United States Patent* 3,285,831.

WANG, SHU-FUNG, ADLER, J. & LARDY, H.A. (1961). The pathway of itaconate metabolism by liver mitochondria. *Journal of Biological Chemistry* **236**, 26–30.

WARD, G.E. (1967). Production of gluconic acid, glucose oxidase, fructose, and sorbose. In *Microbial Technology*, pp. 200–221. Ed. by Peppler, H.J. London, New York: Reinhold Publishing Company.

WOODWARD, J.C., SNELL, R.L. & NICHOLS, R.S. (1949). Conditioning molasses and the like for production of citric acid by fermentation. *United States Patent* 2,492,673.

YAMADA, K. (1965). Science in Japan. p. 401 in American Association for the Advancement of Science, Washington, D.C.

Additional References

AHMED, S.A., SMITH, J.E. & ANDERSON, J.G. (1972). Mitochondrial activity during citric acid production by *Aspergillus niger*. *Transactions of the British Mycological Society* **59**, 51–60.

WOLD, W.S.M. & SUZUKI, I. (1973). Cyclic AMP and citric acid accumulation by *Aspergillus niger*. *Biochemical and Biophysical Research Communications* **50**, 237–241.

CHAPTER 9

Transformation of Organic Compounds by Fungal Spores

C. VÉZINA and K. SINGH

9.1 Introduction

The fungal spore

Microbial transformation of organic compounds differs markedly from the so-called classical fermentation in which the products, such as antibiotics, enzymes, organic acids and solvents, are the result of complex biosynthetic processes. In microbial transformation of chemical compounds the substrates are added to microbial cultures and usually undergo a simple, specific modification. The first observation that fungal spores can transform organic compounds was made by Gehrig & Knight (1958) who reported that octanoic acid could be converted to 2-heptanone by conidia of *Penicillium roqueforti*. They found that oxygen uptake in ungerminated spores increased directly with methylketone formation (the reaction responsible for development of flavour and aroma in Roquefort cheese), and that methylketone accumulation was directly proportional to the number of spores; germinated conidia and mycelium were inactive. Schleg & Knight (1962) extended this observation to the 11α-hydroxylation of progesterone by conidia of *Aspergillus ochraceus*. Knight (1966) concluded that fungus spores, generally considered as a dormant stage in the life-cycle of fungi, do not only contain enzymes necessary for their metabolic activities, but also possess the enzymic machinery for transformation of substrates apparently unrelated to assimilation and catabolic processes. These findings do not contradict the concept of dormancy which has been defined as any rest period or reversible interruption of the phenotypic development of an organism (Sussman, 1966).

The pioneer work of Knight has led to the 'spore process' which was developed principally at Ayerst Research Laboratories, and has been used in several other laboratories to transform steroids by a variety of reactions, to convert triglycerides and fatty acids into methylketones, to hydrolyze penicillin to 6-aminopenicillanic acid, and soluble starch to glucose. These activities were reviewed by Vézina, Sehgal & Singh (1968) and recently summarized by Marsheck (1971). Fungal spores have also been reported to carry out inversion of sucrose (Nelson, Johnson & Ciegler, 1969), stereo-specific reduction of various pharmacologically-active substances (Sehgal & Vézina, 1969), transformation of glucose to D-mannitol (Nelson, Johnson & Ciegler, 1971), modification of the antimycin A molecule (Singh & Rakhit, 1971), and transformation of certain flavonoids (Ciegler, Lindenfelser & Nelson, 1971). An interesting extension of the spore process has been described by Johnson & Ciegler (1969): using fungal spores entrapped in solid matrices they could operate a fungal spore-continuous column process for inversion of sucrose; this is similar to the 'insolubilization' of purified enzymes on solid supports.

The spores most commonly used in the transformation of organic compounds comprise conidiospores (conidia) of Ascomycetes and Deuteromycetes (Fungi Imperfecti) and sporangiospores of Phycomycetes (especially Mucorales). Conidia of Streptomycetes (mycelial bacteria) are also active, but endospores of bacilli and clostridia are inactive, at least in steroid transformation. During the process of conversion, fungal spores remain in the early pre-germination stage (before swelling and protrusion of the germ tube takes place (Manners, 1966)), and are three to ten times more active than mycelium on a dry weight basis (Vézina *et al.*, 1968). This is not too surprising, since spores can be thought of as finely divided cytoplasm acquired from the mycelium (Hickman, 1965; Mandels, 1965; Raper, 1966). In addition, spores differ from the vegetative mycelium by cessation of cytoplasmic movement, small water content and slow metabolism, and lack of vacuoles (Gregory, 1966). The chemical constituents of spores and mycelium differ quantitatively. In *Neurospora crassa*, conidia are richer in phospholipids than mycelium (Bianchi & Turian, 1967), while in *Penicillium chrysogenum*, the conidia are richer in galactose and amino acids and the hyphae contain more glucosamine (Rizza & Kornfeld, 1969). The amount of copper is higher in spores than in mycelia but zinc and magnesium concentrations are generally higher in mycelia (Dalby & Gray, 1974).

In a series of recent reports Fisher and his collaborators have now shown by chemical and electrophoretic measurements that the surface of *Neurospora crassa* conidia differs from that of the hyphal walls by the presence of phosphate groups in the former (Somers & Fisher, 1967); they also observed that conidia, basidiospores, sporangia and zoospores of various species had characteristic electrophoretic mobilities which they could relate to the presence or absence, as well as proportion of amino, carboxyl and phosphate groups (Fisher & Richmond, 1969). In *Penicillium expansum* conidia grown on a high phosphate medium a polyphosphate layer covers the underlying layer of rodlets (Fisher & Richmond, 1970). The 'rodlet' structure of the conidial surface has been described in detail for several penicillia (Hess, Sassen & Remsen, 1968) and aspergilli (Hess & Stocks, 1969). Fungal spores often contain *n*-alkanes, first reported by Oró, Laseter & Weber (1966) in chlamydospores of rust fungi, and a large variety of lipids (Gunasekaran, Hess & Weber, 1972; Ruiz-Herrera, 1967). Conidia and sporangiospores of several fungi were shown to vary widely in their lipid and hydrocarbon constituents (Fisher & Richmond, 1972; Fisher, Halloway & Richmond, 1972). The lipid, phospholipid, and phosphate content and the charge of conidial and hyphal surfaces could control the passage of very insoluble, large molecular weight substrates, such as steroids, to the site of transformation by specific enzymes into products which are retained or excreted by the cell; therefore, the foregoing observations may become significant in explaining the very complex interactions that take place in microbial transformation.

9.2 Methodology of Transformation

Two methods for the microbial transformation of organic compounds are compared in Fig. 9.1. Moulds, yeasts and bacteria (including streptomycetes) are commonly used. Maintenance of transforming micro-

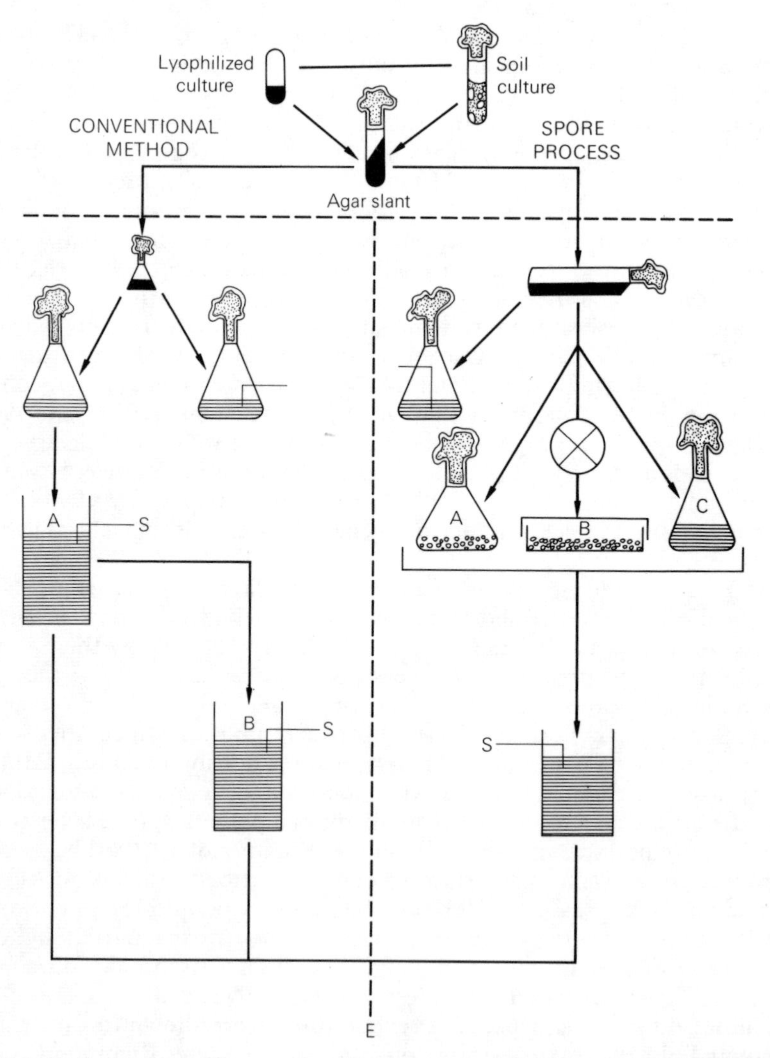

Fig. 9.1 Similarities and differences between two methods for the microbial transformation of organic compounds: the conventional method (left) and the 'spore process' (right).

organisms is afforded by lyophilization, soil and/or slant cultures (above the horizontal broken line). In the conventional method (on the left side of the vertical broken line), vegetative cells or spores are used to inoculate a suitable medium in shake flasks or aerated fermenters (A); substrate (S) is added either at the time of inoculation or when active cells have reached a desired concentration. Incubation is continued until maximum yield of the product is obtained, after which the product is extracted (E). Aseptic conditions are maintained throughout the process. As a modification, the micro-organism is grown, separated from the fermentation broth, re-suspended in buffer (fermenter B), and the substrate is added. In fermenter B, aseptic conditions are not always necessary, and antibiotics can be added as a precaution against contamination. In both cases, transformation is generally complete before the organism sporulates to any extent. Multiple reactions by mixed cultures in a single fermenter is a recent development of the conventional method (Ryu, Lee, Thoma & Brown, 1969). For a detailed description of the conventional method of transformation the interested reader is referred to the reviews by Prescott & Dunn (1959), Stoudt (1960), Kogan (1962), Peterson (1963), Čăpek, Hanč & Tadra (1966), Charney & Herzog (1967), Marsheck (1971), and to the voluminous patent literature since the patent of Murray & Peterson (1952).

The second method, the spore process, is illustrated on the right side of the vertical broken line (Fig. 9.1). In a first step, spores are produced by surface (Roux bottles, Fernbach flasks (A), shallow trays (B)) or submerged (C) culture, and separated from the mycelium by differential centrifugation: they can be stored as a frozen paste until needed. In the second step, spore pastes are suspended in water or buffer to the desired concentration and, if necessary, glucose and antibiotics are added; the substrate (S) is then added, and incubation continued in non-sterile, aerated-agitated vessels until maximum yield of the product is obtained. During transformation, spores remain un-germinated, and no effort is made to maintain aseptic conditions. As a modification, spores can be entrapped in solid matrices and the mixture packed in a suitable column; the substrate (S) is added and the product recovered continuously (Johnson & Ciegler, 1969). In the spore method the medium used for transformation is very simple and generally makes extraction (E) easier than in the conventional method.

Sporulation methods

The capacity of fungi to produce spores (several types may be formed by the same fungus) is a hereditary attribute, and its phenotypic expression, qualitative and quantitative, depends greatly on the environment (Hawker, 1966*a*). All fungi do not respond identically to external factors, and great discrepancy is observed between species of the same genus and even between strains of the same species (Vézina *et al.*, 1968). For transformation, a given spore concentration (1×10^8 to 1×10^9 spores cm^{-3}) is required, and for reasons of economy the optimal sporulation conditions must be used. Quantitative results are seldom reported in the literature, and it is not possible to predict for a given fungus the number of spores that can be produced under certain conditions. The authors have studied quantitatively the relative value of various techniques in mass production of spores.

The following surface and submerged culture methods have given satisfactory yields.

SURFACE SPORULATION ON ARTIFICIAL MEDIA Artificial media dispensed in Roux bottles yield enough spores on a laboratory scale for screening work and for studying factors involved in transformation reactions (Vézina *et al.*, 1968); aseptic conditions are easily maintained throughout the process. Spore yields vary considerably with the media composition. For example, *Aspergillus ochraceus* produces 8×10^9 conidia per bottle on Sabouraud Dextrose Agar, and 1.2×10^{11} on Difco W.L. Nutrient Agar (incubation: 6 days at 28°C and 55–60% relative humidity). Spores are harvested by gently brushing the surface of agar media with a sterile brush in presence of a 0.01% solution of Tween 80 or sodium monolauryl sulphate. Suspensions are then filtered through sterile glass wool to remove mycelial fragments, and to break chains and clumps of spores. Suspensions can be diluted to the desired concentration for transformation, or centrifuged and stored as frozen pastes until needed. Optimal sporulation conditions for *Septomyxa affinis* and *Mucor griseocyanus* have been given by Singh, Sehgal & Vézina (1965, 1967*a*).

SURFACE SPORULATION ON NATURAL MEDIA IN FERNBACH FLASKS Cereal grains, such as barley, cracked corn and hard wheat bran, are non-expensive and suitable for abundant sporulation of many fungi. Critical factors for a given organism are the quantity of water added to the grains before sterilization and the highest possible relative humidity (98% R.H.) of the atmosphere during sporulation. From one 2.8 dm^3 Fernbach flask containing 200 g of moistened 'pot' barley or 100 g of moistened wheat bran inoculated with *Aspergillus ochraceus*, it is usual to harvest, after 6 days of incubation at 28°C and 98% R.H., 6 g of spore paste, 30% dry weight, consisting of 5×10^{11} conidia (Singh, Sehgal & Vézina, 1968). Therefore, one Fernbach flask is equivalent to 5 to 50 Roux bottles. Conidia of *Septomyxa affinis* can be produced by a similar method (Singh *et al.*, 1968). The addition of various supplements, such as mixtures of inorganic salts, honey, glucose, peptone or yeast extract, to cereal grains has no great influence on sporulation of these organisms. Vézina *et al.* (1968) have given a list of Phycomycetes, Ascomycetes, Deuteromycetes, and Basidiomycetes which are capable of sporulating heavily on cereal grains. *Curvularia lunata*, *C. pallescens* and several species of the genus *Corticium*, which are important 11β-hydroxylators of steroids, do not sporulate well on cereal grains. Sansing & Ciegler (1973) have described mass production of conidia of several *Aspergillus* and *Penicillium* species on whole loaves of white bread.

SURFACE SPORULATION ON A LARGE SCALE Methods and equipment which have been described for the mould bran process (Underkofler, 1954; Prescott & Dunn, 1959) can be applied to the scale-up of surface sporulation in drums, pots and trays. Sehgal, Singh & Vézina (1966) and Singh *et al.* (1968) have described the tray method on a semi-industrial scale. Moistened hard wheat bran is sterilized for 1 h at 121°C under slow agitation in a horizontal, squirrel-type, cylindrical screen or in an autoclavable mixer (Fig. 9.2A); the sterile bran is cooled, inoculated with a spore suspension (for example, *Aspergillus ochraceus*) and dispensed in alu-

A B

C

Fig. 9.2 A sporulation unit consisting of an autoclavable mixer (A), truck and trays (B), equipment for harvesting spores (C), mixing tank (left) and solid discharge centrifuge (right).

minium trays. Each tray in Fig. 9.2B is 762 mm × 457.2 mm × 44.5 mm and offers a surface which is twelve times that available in a Fernbach flask; twenty-four trays can be stacked in a truck which occupies 6 square feet of floor space. When maximum sporulation is obtained spores are harvested by transferring the mouldy bran to water or detergent solution contained in a tank equipped with a Turbon mixer (Fig. 9.2C, left).

After vigorous agitation for a few minutes, the slurry of bran and dislodged spores is pumped into a solid discharge centrifuge (Fig. 9.2C, top right), operating at reduced speed, to separate bran and mycelium from the spore suspension; the suspension is filtered through glasswool to remove mycelial fragments and obtain a homogeneous suspension; the suspension can be diluted to the desired concentration and used immediately for transformation, or centrifuged to recover spores as a paste which is kept frozen until needed. Each tray yields 7×10^{12} spores of *A. ochraceus*.

The methods described above are by no means unique; their advantage over many other methods described in the literature lies in their simplicity, high yields, and the homogeneity and purity of spores they afford. With slight modification they could be made suitable to a great variety of fungi including halophilic fungi, such as *Hemispora stellata* (Sala & Burgos, 1972), and sterol-requiring fungi, such as *Phytophthora cactorum* (Elliot, 1972). For producing spores of pathogenic fungi, more aseptic conditions would be required. It must be remembered that human neutrophils are unable to kill ingested *Aspergillus fumigatus* spores (Lehrer & Jan, 1970; Lundborg & Holma, 1972), and that *A. ochraceus* produces ochratoxins (Scott & Hand, 1967; Nesheim, 1967). In our laboratories and pilot plant the sporulation unit is maintained under negative pressure, exhaust hoods are provided above equipment and exhaust air is incinerated before it is returned to the atmosphere; walls, floors and ceilings are sterilized chemically and workers must wear bacteriological masks. Other methods have recently been described for recovery of spores from soil (Bissett & Widden, 1972) and for increased recovery of spores by means of ultrasonic vibrations (Tiner & Golden, 1967).

SPORULATION IN SUBMERGED CULTURE The submerged culture technique affords obvious advantages in scaling up microbiological processes. Submerged sporulation has been studied for *Penicillium notatum* (Foster, McDaniel, Woodruff & Stokes, 1945), *P. roqueforti* (Meyers & Knight, 1958, 1961), *P. griseofulvum* (Morton, 1961) and *Beauveria bassiana* (Samšinákowá, 1966), and was reviewed by Vézina, Singh & Sehgal (1965). In synthetic media, important factors are the nature and concentration of the carbon and nitrogen sources, and the presence of various trace elements (Table 9.1). Certain fungi such as *Aspergillus ochraceus*, sporulate only in complex media enriched with NaCl or other salts. Spore recovery is very simple, because newly produced spores do not germinate in the medium, and mycelium is practically absent. Therefore, broth containing spores can be used either directly or filtered through glasswool, and the recovered spores used for transformation.

Several reports on submerged sporulation have been published since the review by Vézina *et al.* (1965). *Neurospora crassa* conidia were grown in submerged culture, then transferred to solid media and observed for differentiation into conidiophores and conidia (Stine & Clark, 1967); addition of malonate to the growth medium synchronized conidiophore and conidium production; other factors for synchronization were aeration of submerged mycelia, and buffers for treatment of harvested mycelia. Righelato, Trinci, Pirt & Peat (1968) studied submerged sporulation of *Penicillium chrysogenum* in glucose-limited chemostat cultures; maximum

Table 9.1 Sporulation of various filamentous fungi in submerged culture

Species	Medium[a]	Conidia cm^{-3} after 5 days
PHYCOMYCETES		
Conidiobolus sp.	A	6.5×10^7
Cunninghamella blakesleeana Lendner	A	1.7×10^8
Mucor griseo-cyanus Hagem	A	Very few
DEUTEROMYCETES		
Aspergillus fonsecaeus Thom & Raper	B	$<1.5 \times 10^7$
Aspergillus ochraceus Wilhelm	B	2.0×10^8
Coniothyrium sp.	C	3.0×10^7
Curvularia lunata (Walker) Boedjin	B	$<1.0 \times 10^7$
	C	0
Curvularia pallescens Boedjin	C	0
Fusarium conglomerans	B	0
	E	9.0×10^7
	C	3.5×10^8
F. decemcellulare Brick	D	1.6×10^7
F. equiseti (Cda) Saccardo	D	6.4×10^7
F. graminearum Schwabe	D	1.6×10^8
F. merismoides Corda	D	1.2×10^8
	C	2.3×10^8
F. moniliforme Sheldon	D	1.7×10^8
F. oxysporum Schlechtendahl	D	2.5×10^8
Pestalotia funerea Desmazières	C	1.0×10^7
Septomyxa affinis (Sherbakoff) Wollenweber	B	1.8×10^8
	C	5.4×10^8
	E	9.8×10^8
	F	21.0×10^8
Stachylidium theobromae Turconi	B	2.4×10^8
	C	3.0×10^8
	E	3.4×10^8
	G	8.0×10^8
Trichoderma viride Persoon ex Fries	C	5.0×10^7
Trichothecium roseum Link ex Fries	B	2.0×10^7

[a] Medium A: Bacto-malt extract (Difco Laboratories Inc., Detroit, Michigan), 100 g; distilled water to 1 dm^3; pH 4.7. Medium B: corn steep liquor, 5 g; 'blackstrap' molasses, 50 g; 'cerelose', 25 g; NaCl, 25 g; distilled water to 1 dm^3; pH 6.0. Medium C: glucose, 30 g; $NaNO_3$, 3 g; K_2HPO_4, 1 g; $MgSO_4 \cdot 7H_2O$, 0.5 g; KCl, 0.5 g; trace elements: B, 10 μg; Cu, 100 μg; Fe, 200 μg; Mn, 20 μg; Mo, 20 μg; Zn, 2000 μg; distilled water to 1 dm^3; pH 5.5. Medium D: Medium C in which glucose has been replaced by sucrose. Medium E: corn steep liquor, 10 g; sucrose, 30 g; $NaNO_3$, 3 g; K_2HPO_4, 1 g; $MgSO_4 \cdot 7H_2O$, 0.5 g; KCl, 0.5 g; $FeSO_4 \cdot 7H_2O$, 0.01 g; distilled water to 1 dm^3; pH 6.0. Medium F: corn steep liquor, 10 g; edamin, 10 g; Bacto-malt extract, 15 g; distilled water to 1 dm^3; pH 7.0. Medium G: glucose, 10 g; $NaNO_3$, 1 g; KH_2PO_4, 1 g; $MgSO_4 \cdot 7H_2O$, 0.5 g; KCl, 0.5 g; $CaCl_2$, 1 g; trace elements: 5 times the concentration for Medium C; corn steep liquor, 10 g; distilled water to 1 dm^3; pH 4.0. From Vézina *et al.*, 1965; reproduced from *Mycologia 57*, p. 733, with permission of the New York Botanical Garden.

conidia formation (3×10^8 cm^{-3}) was afforded by the optimal glucose feed rate of 0.038 g glucose g^{-1} mycelial dry weight h^{-1}. Galbraith & Smith (1968, 1969*a*) found that submerged sporulation of *Aspergillus niger* occurred in a medium without added nitrogen, in low ammonium nitrogen concentrations, and in a wide range of nitrate nitrogen concentrations.

Ammonium salts were inhibitory to sporulation and their inhibition was overcome by amino acids; glyoxylate and several intermediates of the tricarboxylic acid cycle promoted conidiation in the presence of ammonium ions. Ng, Smith & Anderson (1972) further observed that the pentose phosphate pathway predominates during synchronous conidiophore dedelopment of this fungus. Zeidler & Margalith (1972) could induce synchronous conidiophore and conidia formation in *Penicillium digitatum*; B vitamins and the carbon/nitrogen ratio were found to be important factors. Finally, the morphological changes (Oliver, 1972) and kinetics (Axelrod, 1972) of differentiation into conidiophores and conidiospores in *Aspergillus nidulans* have been reported recently. The biochemistry of asexual reproduction in *Neurospora, Aspergillus* and several other filamentous fungi has been reviewed recently by Smith & Galbraith (1971).

STORAGE OF SPORES FOR TRANSFORMATION Conidia of *Aspergillus ochraceus* can be stored as frozen pastes ($-20°$) for at least 2 years without apparent loss in transforming activity; *Septomyxa affinis* conidia are less stable, but can be kept frozen ($-20°$) for two months. Other spores are more labile and must be used immediately after they are harvested. Macdonald (1972) has reported recently that storage of conidia of *Penicillium chrysogenum* in liquid nitrogen ($-196°$) not only maintains high viability, but also prevents penicillin yield decay by reducing the selection of low producing spontaneous mutants. High stability of steroid 11α-hydroxylating ability is a real advantage. Spores can be produced at a certain place and shipped to another plant where they are used to transform a chemical compound under un-aseptic conditions; thus they can be handled as easily as a biochemical reagent. Another advantage is the constant availability of spores of various fungi needed for various transformation reactions. These microbial reagents can thus be used on short notice to start an experiment or a transformation run.

Transformation methods

The second step of the spore process consists in the transformation of a given substrate into a desired product with spores produced according to one of the methods previously described. To determine precisely the factors involved in transformation or to prevent germination of spores by certain substrates it is desirable or necessary to wash spores one or more times in water or buffers.

TRANSFORMATION IN SHAKE FLASKS Transformation by fungal spores is generally an aerobic process; shake flask culture affords a suitable means to satisfy the aeration requirements of spores and bring the often insoluble substrates, such as steroids, in contact with spores; the technique is convenient for screening work and to study factors of transformation. Fresh, refrigerated, or frozen spore suspensions and pastes are suspended in water or buffer, counted in the haemacytometer, and diluted to the required concentration which lies between 1×10^8 to 1×10^9 spores/cm^{-3} Transformation generally takes place over a wide range of pH values; for steroid transformation pH 5 to 6 is preferred to inhibit growth of bacterial contaminants. For steroid hydroxylations glucose is highly stimulatory and must be added (Vézina, Sehgal & Singh, 1963; Singh *et al.*, 1967*a*).

Since glucose is metabolized to organic acids pH tends to drop during transformation and can be readjusted to the required value by occasional addition of alkali; a suitable buffer concentration may also prevent excessive pH changes without affecting transformation. When long transformation time periods under un-aseptic conditions are anticipated, an antibiotic, such as streptomycin, neomycin, or tetracycline, can be added to reduce contaminants. No nitrogen source is added, because it is never necessary for transformation and would promote spore germination and growth of contaminants. Finally, the substrate is added to spore suspensions. Addition of water-soluble substrates presents no difficulty; water-insoluble substrates, such as steroids, are usually added as solutions in water-miscible solvents (acetone, ethanol, methanol, propyleneglycol, dimethylformamide or dimethylsulphoxide). Steroid substrates can also be added in the form of micronized powders slurried in Tween 80. It is necessary to add water-insoluble substrates at a slow rate while the spore suspension is being stirred vigorously so as to obtain a uniform dispersion of the substrate in very fine particles. Lee *et al.* (1960) have recently studied the influence of mode of steroid substrate addition on conversion of steroids. Shake flasks are then incubated at 25°–30°C on a gyratory (200–300 rev/min) or a reciprocating (100–200 oscillations/min; 1–2 in stroke) shaker. Suitable ratios of volume of suspensions to volume of flasks are: 25/125, 50/250, *etc.*

During transformation spores do not germinate and do not swell to any extent. This absence of germination has been discussed before (Vézina *et al.*, 1968): it was first thought that absence of a nutrient medium, especially of a nitrogen source, was responsible for this observation; however, when spores were suspended in nutrient medium, at the concentration required for transformation, germination was not observed. For a detailed discussion of factors involved in germination, the interested reader is referred to the reviews by Cochrane (1966), Hawker (1966*b*), and Manners (1966). More recent accounts of factors for germination have been given for *Fusarium solani* and *Verticillium albo-atrum* (Lewis & Papavizas (1967)), for *Trichoderma* sp. (Barnes & Parker (1967)), for *Cunninghamella elegans* (Hawker, Thomas & Beckett (1970)), for *Microsporum gypseum* (Leighton & Stock (1970) and Leighton, Stock & Kelln (1970)), for *Sphaerotheca macularis* (Mitchell & McKeen (1970)), for *Cercospora omphakodes* (Judd & Peterson (1972)), and for *Aspergillus flavus* (Pass & Griffin (1972)).

TRANSFORMATION ON SOLID CULTURE MEDIA AND INERT SUPPORTS The shake flask method is very reliable, but only one organism can be studied in a single flask. Vézina, Singh & Sehgal (1969) have described an agar plate method which allows the examination of several micro-organisms growing as macrocolonies on a medium containing a steroid substrate; after suitable incubation a reagent specific for the product under search is sprayed on the surface of the medium and organisms which are capable of transforming the substrate into equilin are detected by the presence of a coloured halo around the colonies. The method can be applied to non-germinating spores, provided a non-nutrient agar is used and a specific reagent is available. In steroid transformations, except for aromatization of a substrate into equilin (Vézina *et al.*, 1969), the latter condition is generally not met. To avoid this difficulty Hafez-Zedan & Plourde (1971) have described an ingenious technique which consists in spotting microlitre portions of steroid solutions (30µg) 0.5 cm apart along the bottom (2 cm from the edge) of a thin layer Silica Gel G Plate; each spot is covered with a spore suspension (2×10^4 spores/spot). Reference steroids and spore suspensions are spotted separately and serve as controls. After incubation, plates are developed in a suitable solvent mixture, dried and sprayed with specific reagents. During development spores remain on the application line and do not germinate. The method is suitable for detecting spores capable of transforming a given substrate and for identifying products obtained with various spores from various substrates.

TRANSFORMATION BY SPORES ENTRAPPED IN SOLID MATRICES Spores bound in a column matrix by ion-exchange resins and polyacrylamide gels afford an ideal stationary phase for enzymic conversion of substrates percolated through the column (Johnson, Nelson, Follstad & Ciegler, 1968*b*); similar to and more economical than column entrapment of 'insolubilized' enzymes (for examples, see Bar-Eli & Katchalski (1960) and Tosa, Mori, Fuse & Chibata (1967)), the use of spores provides a suitable system for continuous column operation. Johnson & Ciegler (1969) reported that optimal conditions for spore retention and flow rate were afforded by a ratio of 5×10^9 to 3×10^{10} spores per 8 grammes of Ecteola-cellulose packed in a column. Spore columns are stable and need only a cursory washing before storage (4°C) or re-use; germination is negligible. The spore column technique is limited to solutions of low viscosity; Johnson & Ciegler (1969) have applied the technique to sucrose inversion by spores of *Aspergillus oryzae*, octanoic acid oxidation by spores of *Penicillium roqueforti*, and starch hydrolysis by spores of *A. wentii*. It can probably be applied to steroid transformation, since Mosbach & Larsson (1970) have been successful in 11β-hydroxylating steroids using entrapped vegetative cells of *Curvularia lunata*.

TRANSFORMATION IN AERATED-AGITATED FERMENTERS An advantage of the spore process is that transformation can be carried out un-aseptically; therefore, expensive fermentors and ancillary equipment to maintain asepsis are not necessary, and can be replaced by simple, inexpensive vessels. Singh *et al.* (1968) have described a modified stainless steel extraction tank which is easily operated (Fig. 9.3); the air inlet serves to sparge air during transformation and inject steam for cleaning the tank at the end of each run. The turbine impeller provides adequate mixing and air incorporation; the upper impeller, which is located slightly below the surface of the suspension, helps to incorporate the foam into the liquid mixture by creating a slight vortex; the blade above the liquid surface serves as a mechanical foam braker. Temperature is kept constant during transformation by circulating water in the coil; agitation is at 250–300 rev/min, and aeration at 52 dm³ of air per minute. During transformation dissolved oxygen is measured and maintained at a minimum of 2%; pH is also recorded and controlled, if necessary, by addition of alkali. Spore suspensions are prepared and substrate and supplements are added as for

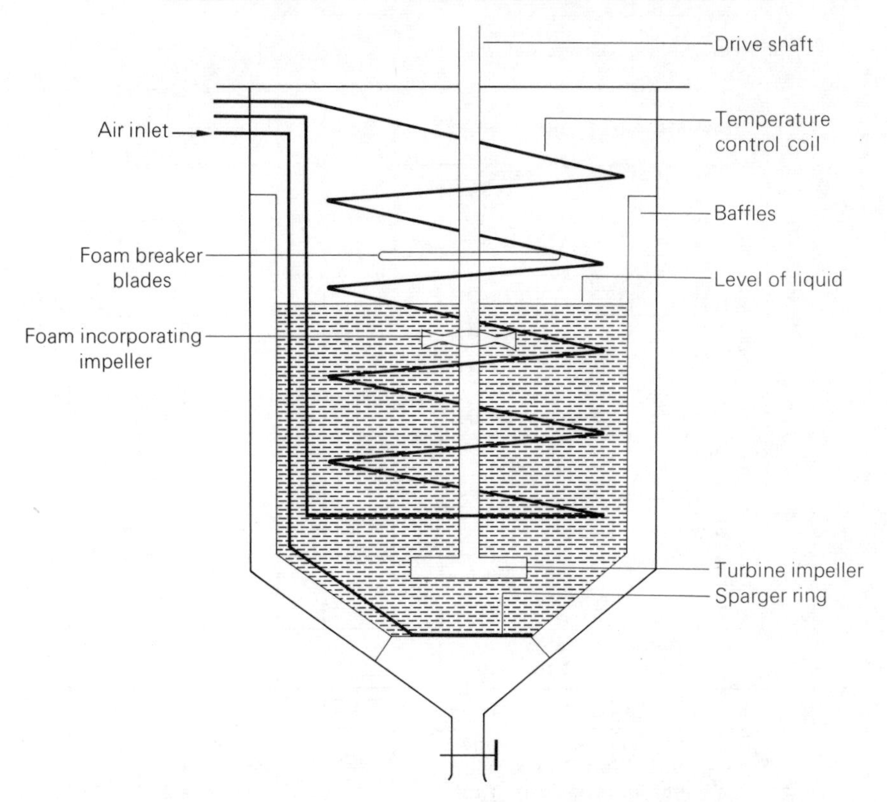

Fig. 9.3 Two-hundred-gallon stainless steel tank for transformation of organic compounds by spores. From Singh *et al.*, 1968; reproduced from *Applied Microbiology* **16**, p. 399, with permission of American Society of Microbiology.

the method in shake flasks; antibiotics can be added, if transformation is expected to last more than 72 hours. During transformation substrate utilization and product accumulation must be followed by appropriate analytical methods to determine optimum time for recovery of the product. The extraction procedure must take into account the nature of the product and its site of accumulation, spores, medium or both. Sometimes, spores can be separated by filtration or centrifugation before extraction and re-used in subsequent runs. Analytical methods, recovery of products and re-use of spores in steroid transformation have been discussed by Vézina *et al.* (1968).

9.3 Organic Compounds Transformed by Spores

Transformation of steroids

Mycelial cultures and spore suspensions of many fungi and streptomycetes have been compared for their transforming activity and were found not to vary qualitatively. On a total nitrogen basis, however, spores seem to be more active than mycelium. When more than one reaction is effected by spores and mycelium the ratio of the products may vary considerably from one form of the fungus to the other. The main reactions observed with

Table 9.2 Some fungal spores capable of transforming steroids

Fungal spores and spores of streptomycetes (references)	I Oxidation																II Reduction							III	IV
	Hydroxylation								1-Dehydrogenation					Side chain cleavage with 17-CO formation	Oxidation of alcohol to ketone		Ketone to alcohol						Reduction of a double bond	Hydrolysis of ester to alcohol	Isomerization of double bond, *e.g.* $\Delta^5 \rightarrow \Delta^4$
	6 β	7 α	11 α	11 β	14 α	15 α	16 α	16 β	Simple Δ^1	+ Side chain cleavage with 17β-OH formation	+ Side chain cleavage with 17-CO formation	+ Side chain cleavage with Ring D expansion and lactone formation	To Ring A aromatization		3β-OH → 3-CO	17β-OH → 17-CO	3-CO → 3α-OH	11-CO → 11β-OH	14-CO → 14α-OH	17-CO → 17β-OH	20-CO → 20α OH	20-CO → 20β OH			
Aspergillus niger (3)			×																						
A. ochraceus (2, 7, 12, 13)	×		×		×									×					×				× (Δ^{16})	×	
A. terreus (3)			×																						
Cylindrocarpon radicicola (3, 4)									×	×	×	×	×			×					×		Epoxide cleavage to 16α-OH and to 16α,17β-di-OH		
Didymella lycopersici (6, 12)			×											×		×						×			
Fusarium javanicum var. *ensiforme* (10)									×	×	×														
F. merismoides (10)									×	×	×														

F. moniliforme (1)						×																			
F. oxysporum var. *lini* (10)									×	×	×														
F. solani (3, 5, 10, 12)									×	×	×	×	×			×					×		Epoxide cleavage to 16α-OH and to 16α,17β-di-OH		
Gliocladium roseum (10)									×	×	×														
Mucor griseo-cyanus (11, 12, 13)		×			×												×								
Penicillium chrysogenum (12)														×											
P. sp. (12, 13)						×																			
Pestalotia funerea (13)								×																	
Septomyxa affinis (8, 9, 10, 12, 13, 14)									×	×	×	×	×	×	×	×		×		×			× (Δ^4)	×	×
Sporotrichum epigaeum (3)																×									
Stachylidium theobromae (12)			×	×	×																				
Streptomyces diastaticus (12)																×									

(*Table continued overleaf*)

Table 9.2 Some fungal spores capable of transforming steroids—*continued*

Fungal spores and spores of streptomycetes (references)	I Oxidation																II Reduction							III	IV
	Hydroxylation								1-Dehydrogenation					Side chain cleavage with 17-CO formation	Oxidation of alcohol to ketone		Ketone to alcohol						Reduction of a double bond	Hydrolysis of ester to alcohol	Isomerization of double bond, *e.g.* $\Delta^5 \rightarrow \Delta^4$
	6 β	7 α	11 α	11 β	14 α	15 α	16 α	16 β	Simple Δ^1	+ Side chain cleavage with 17β-OH formation	+ Side chain cleavage with 17-CO formation	+ Side chain cleavage with Ring D expansion and lactone formation	To Ring A aromatization		3β-OH $\rightarrow$ 3-CO	17β-OH $\rightarrow$ 17-CO	3-CO $\rightarrow$ 3α-OH	11-CO $\rightarrow$ 11β-OH	14-CO $\rightarrow$ 14α-OH	17-CO $\rightarrow$ 17β-OH	20-CO $\rightarrow$ 20α OH	20-CO $\rightarrow$ 20β OH			
S. lavendulae (10, 12)									×	×	×											×			
S. rimosus (12)																×									
S. roseochromogenus (12)							×		×		×											×			
S. viridis (12)							×																		

References

(1) Casas-Campillo & Bautista (1965)
(2) Deghenghi, Boulerice, Rochefort, Sehgal & Marshall (1966)
(3) Hafez-Zedan & Plourde (1971)
(4) Hafez-Zedan, El-Tayeb & Abdel-Aziz (1970)
(5) Plourde, El-Tayeb & Hafez-Zedan (1972)
(6) Sehgal, Singh & Vézina (1963)
(7) Sehgal, Singh & Vézina (1968)
(8) Singh & Rakhit (1967)
(9) Singh, Sehgal & Vézina (1963)
(10) Singh, Sehgal & Vézina (1965)
(11) Singh, Sehgal & Vézina (1967*a*)
(12) Vézina, Sehgal & Singh (1963)
(13) Vézina, Sehgal & Singh (1968)
(14) Vézina, Sehgal, Singh & Kluepfel (1971)

Also see:
Charney & Herzog (1967)
Iizuka & Naito (1967)
Vézina & Rakhit (1974)

spores are listed in Table 9.2. Several reactions are sometimes reported for a given organism: that does not mean that all those reactions are effected at the same time on the same substrate, but it is noticed that C-1-dehydrogenators, such as *Cylindrocarpon radicicola, Fusarium solani* and *Septomyxa affinis,* usually yield a wider variety of products than hydroxylators. Schleg & Knight (1962) were the first to observe 11α-hydroxylation of steroids by spores of *Aspergillus ochraceus*; Vézina *et al.* (1963) further reported 11α-hydroxylation of a large number of substrates with these spores. The 11α-hydroxy derivative is generally the only product, but when progesterone is the substrate some 6β,11α-dihydroxy derivative also accumulates; by increasing the charge of progesterone to 5,10 or 15 g dm^{-3}, only traces of the dihydroxy product are obtained. Acetylated substrates are transformed as efficiently as their corresponding alcohols, but are hydrolyzed before they are hydroxylated. The rate and degree of transformation also vary with the spore concentration and the nature of the substrate. The solubility of the substrate, its crystalline form and particle size are certainly involved (Lee *et al.*, 1970). The mode of addition is especially important: a selected set of substrates consisting of 16α-hydroxy-compound S, 16α,17α-isopropylidinedioxyprogesterone and 6α-fluoro-17α-methylpregn-1-ene-3,20-dione, when added as solutions in organic solvents, give only traces of transformation products; when added as micronized powders they yield 15–50% transformation products. *A. ochraceus* spores can transform almost all the 3-CO-Δ^4 steroids of the androstane and pregnane series tested (Vézina *et al.*, 1968); they sometimes transform steroids that do not possess this conjugation: 3β-hydroxy-pregn-4-en-20-one, 16α,17α-epoxypregnenolone, and 6β-hydroxy-3,5-cyclopregnan-20-one are hydroxylated (Sehgal, Singh & Vézina, 1968).

Mucor griseo-cyanus, a 14α-hydroxylator, also transforms a variety of substrates mainly into their 14α-hydroxy derivatives. However, 17α-alkyl substrates are transformed by spores as well as mycelium into their 7α-hydroxy derivatives: the proximity of a bulky 17α-alkyl substituent to the 14α-position probably hinders hydroxylation at carbon 14 and directs it to position 7 (Singh *et al.*, 1967*a*). All steroid hydroxylators require a minimum concentration of glucose (0.2–0.4%) for optimal conversion: D(+)xylose is as effective as glucose; sucrose and starch can replace glucose partially, whereas ethanol and propylene glycol, often used as carrier solvents, have a slight effect.

Introduction of a double bond between carbon 1 and carbon 2 of steroids is a common reaction effected by fungal spores and mycelium. *Septomyxa affinis* is particularly effective, but it has the tendency, like all other 1-dehydrogenators, to cleave the side chain with the formation of various 17-derivatives, to carry out various oxidation and reduction reactions and sometimes to destroy the steroid nucleus (Table 9.2). Singh, Sehgal & Vézina (1963) reported that side chain cleavage does not take place when the substrate bears a 17-alkyl substituent. It can also be prevented by increasing the temperature to 32°C during transformation. The side-chain cleavage of pregnanes follows a pathway similar to the non-enzymatic Bayer-Villager oxidation of ketones by peracids (Singh & Rakhit, 1967). Spores of *S. affinis* actively metabolize glucose, but do not show any requirement for any carbon source to effect steroid transformation (Singh *et al.*, 1965). *S. affinis* spores also transform 19-nor and 19-substituted steroids into the corresponding aromatic ring A derivatives; the introduction of a double bond into these substrates leads to the formation of an unstable intermediate which undergoes a non-enzymatic rearrangement into an aromatic ring A product (Vézina, Sehgal, Singh & Kluepfel, 1971). Singh, K. (*unpublished results*) also showed that *S. affinis* spores can simultaneously open the epoxy group and cleave the side chain of 16α,17α-oxidoprogesterone to 16α-hydroxyandrost-4-ene-3,17-dione; the product is further transformed to the 1-dehydro derivative which undergoes reduction of the 17-ketone into 17β-hydroxy and even ring D expansion and lactone formation. Using spores of *Fusarium solani* and *Cylindrocarpon radicicola*, Plourde, El-Tayeb & Hafez-Zedan (1972) observed that 16α,17α-oxidoprogesterone was first reduced to the 20α-hydroxy derivative, then 1-dehydrogenated; thereafter, the epoxy group was opened to yield 16α-hydroxy-1-dehydro-testolalactone (ring D expansion) or 16α,17β-dihydroxyandrosta-1,4-dien-3-one.

In summary, spores of fungi and streptomycetes can carry out the six principal steroid transformation reactions presently used in industry: 11α (*Aspergillus ochraceus*), 11β (*Stachylidium theobromae*), 14α (*Mucor griseo-cyanus*) and 16α (*Streptomyces viridis*) hydroxylations, as well as 1-dehydrogenation, including ring A aromatization (*Septomyxa affinis*), and isomerization with formation of 3-CO-Δ^4 conjugation (*S. affinis* and others).

Formation of ketones from fatty acids and triglycerides

The oxidation of fatty acids to methylketones by *Penicillium roqueforti* was first reported by Stärkle (1924). Gehrig & Knight (1958, 1961) observed that the spores, but not the mycelium of *P. roqueforti* and several other fungi of the genera *Aspergillus, Penicillium, Paecilomyces* and *Scopulariopsis* could transform fatty acids to methylketones with one less carbon atom. Knight (1966) postulated that the reaction takes place *via* typical β-oxidation. The optimum pH for transformation by spores lies between 5.5 and 7.0 and optimum temperature is 27°C (Lawrence, 1966); aeration is essential. The ability of washed spores of *P. roqueforti* to oxidize fatty acids decreases rapidly with age of spores, probably because of inactivation of oxidative enzymes. However, 3–4 day-old spores can be stored for more than 36 months at 4°C without losing their transforming ability (Gehrig & Knight, 1963). Lawrence (1966) also found that the rate of oxidation increases with the ratio of spore number to initial acid concentration; the lag phase decreases with increasing spore number and can be eliminated by pre-incubating the spores with low concentration of octanoic acid. He further surmised that oxidative enzymes are activated and not induced during the lag phase, since ketone formation is not inhibited by chloramphenicol, an inhibitor of protein synthesis. Lawrence (1966) further observed that the slow rate of formation of 2-heptanone from octanoic acid by resting spores of *P. roqueforti* was markedly increased by the addition of those amino acids (L-alanine, L-serine and L-proline) and sugars (glucose, galactose, xylose and sucrose) that stimulate germination of spores. Lawrence & Hawke (1968) reported that the mycelium also can transform low concentration of fatty acids to methylketones; fatty acids are more toxic for the mycelium than for the spores of *P. roqueforti*: they found toxicity is dependent upon the concentration and chain length of the fatty acid and the pH value of the medium; C_6 to C_8 acids were less toxic than C_9 to C_{12} fatty acids; and the toxic effect was less pronounced against mycelium which had been previously aerated for an extended period in phosphate buffer. Therefore, the mycelium can transform fatty acids to methylketones, but the conditions are more rigorous than for transformation with spores.

Triglycerides are also transformed into methylketones by spores of *Penicillium roqueforti* (Lawrence, 1967), but only after they have been esterified into fatty acids by the spores. Here again, transformation is stimulated by the same factors that promote germination of the spores.

Transformation of antibiotics

Singh, Sehgal & Vézina (1967*b*) reported that penicillin V, but not penicillin G, was transformed to 6-aminopenicillanic acid (6-APA) by spores of *Fusarium conglomerans* and *F. moniliforme*. About 80% of the substrate were transformed in 48 h under the following conditions: 5×10^8 spores cm^{-3} of 1% phosphate buffer, pH 7.8, and 2 mg cm^{-3} penicillin V. For transformation aeration is necessary, but glucose is not: spores can be used 2 or 3 times without significant loss in activity.

Singh & Rakhit (1971) using mycelium and spores of several aspergilli and streptomycetes could bring about transformation of antimycin A, a piscicidal antibiotic (Lennon & Vézina, 1973). Spores of *Aspergillus ochraceus* (1×10^9 cm^{-3}) suspended in 0.02M phosphate buffer, pH 8, were particularly effective: 0.25 mg of the antibiotic cm^{-3} spore suspension were transformed in 72 hours at 28°C in the presence of glucose. The transformation products result from the opening of the dilactone ring and consist of a mixture of two acids which showed a much reduced activity when tested against *Saccharomyces cerevisiae* and *Candida albicans*.

Transformation of carbohydrates

Johnson, Nelson & Ciegler (1967, 1968*a*) found that conidia of *Aspergillus wentii* could hydrolyze soluble starch to glucose at an initial rate faster than the rate of glucose oxidation. The glucoamylase-like enzyme responsible for the reaction, in which maltose is not involved as an intermediate, reaches its peak activity in three days, then rapidly declines. Maximum yields of 40% were observed for a spore concentration of 5×10^8 cm^{-3} and 5 mg of soluble starch per cm^3. Optimum pH is 4 to 5; as glucose is metabolized pH drops and must be re-adjusted above 3 to prevent germination. Frozen spores are not suitable and starch hydrolysis decreases rapidly with spore age; optimum temperature is 20°C, and aeration is necessary. Addition of iodoacetate is useful to inhibit utilization of the glucose formed and increases product yield.

Nelson *et al.* (1969) compared several species of *Aspergillus* and *Penicillium roqueforti* for their ability to invert sucrose: *A. oryzae* spores were most efficient in carrying out the reaction. Reducing sugar equivalent to 70–80% of an initial 0.5 mol dm^{-3} sucrose solution was produced in 6 h at 35°C and at pH 4 to 5 by 1×10^8 conidia per cm^3. Inversion was insensitive to changes in temperature between 25 and 35°C. Spores could be stored at 5°C for 3 months without loss of activity. Sucrose inversion was used as a model by Johnson & Ciegler (1969) to establish the optimum conditions for the operation of the fungal spore-continuous process previously described.

Nelson *et al.* (1971) have also reported the production of D-mannitol from glucose by conidia of *Aspergillus candidus*. Optimum temperature was 35 to 45°C, pH 6 and glucose concentration of 1 to 2% in phosphate buffer; aeration was necessary. Maximum yields (75%) based on glucose utilized were obtained in 165 h with 5×10^8 conidia per cm^3.

Transformation of flavonoids

The flavonoid naringin was found by Ciegler *et al.* (1971) to be hydrolyzed to naringenin and prunin by several fungi under aerobic conditions at 28°C. The mycelia and spores (1×10^9 cm^{-3}) of *Penicillium charlesii*, *Helminthosporium sativum*, *Cephalothecium roseum*, *Aspergillus flavus* and *Wojnowicia graminis* cleaved the sugar moieties of naringenin and metabolized the sugars obtained through this hydrolysis. When supplied as their aglycone, the flavonoids were not transformed.

9.4 Biochemistry of Transformation by Spores

Factors affecting steroid transformation

Steroid transformations by fungal spores are not strongly influenced by variations in pH, incubation temperature, aeration and agitation within fairly wide limits. Hydroxylation of progesterone is not significantly influenced by pH in the range 4 to 8.5 and temperature between 23 and 34°C (Vézina *et al.*, 1963). However, maximum 1-dehydrogenation of Reichstein's compound S by *Septomyxa affinis* is obtained at pH 6 to 8 and the reaction is significantly influenced by temperature. When the conversion is carried out at 32°C compound S is transformed mainly to 1-dehydro S; only traces of side-chain cleavage products are formed (Singh *et al.*, 1965). This suggests that in *S. affinis* spores the enzyme(s) catalyzing the side-chain degradation is inactivated at this temperature. With spores of *Fusarium solani* optimum 1-dehydrogenation is obtained at pH 6–7 (Hafez-Zedan, 1972). The nature or the proportion of transformation products can sometimes be altered by varying the concentration of steroid substrate. For example, at a progesterone concentration greater than 2 g dm^{-3}, progesterone is almost completely converted into 11α-hydroxyprogesterone; when the charge is less than 0.5 g/l, a certain amount of 6β,11α-dihydroxyprogesterone is formed.

Addition of inorganic or high energy phosphate, metal ions, chelating agents, other co-factors or nitrogen supplements to the transformation medium does not significantly influence steroid transformation by washed fungal spores (Vézina *et al.*, 1963; Singh *et al.*, 1965). This indicates that in spores sufficient reserves of these materials are present. That the spores of fungi contain a ready source of stored energy in the polymetaphosphates and phospholipids has been shown by Shepherd (1956) and Nishi (1961). For hydroxylation of steroids spores of *Aspergillus ochraceus*, *Mucor griseo-cyanus* and other fungi require an exogenous source of carbon (Vézina *et al.*, 1963; Singh *et al.*, 1967*a*). On the other hand, spores of *Septomyxa affinis* actively metabolize glucose, but do not show any requirement for it or any other carbohydrate for catalyzing steroid conversions. For steroid transformation by spores of *Fusarium solani* carbohydrate is not required (Hafez-Zedan, 1972).

It has not been possible to selectively inhibit a particular steroid transformation without affecting the overall metabolic and steroid transformation activity of the spores. None of the antibiotics or metabolic inhibitors tried (Vézina *et al.*, 1963; Singh *et al.*, 1965) can block the formation of 6β,11α-dihydroxyprogesterone by *Aspergillus ochraceus* or inhibit the side-chain cleavage of compound S by *Septomyxa affinis* spores, without inhibiting the overall transformation.

Respiratory activity and enzymes of glucose catabolism

The presence in spores of the respiratory enzymes of the major respiratory pathways has been demonstrated in a number of fungi. The subject has been reviewed by Blumenthal (1965) and Cochrane (1966). Therefore, we will deal mainly with results of studies with spores of *Aspergillus ochraceus* and *Septomyxa affinis*, especially in relation to steroid transformation.

In the ungerminated spore of *Fusarium solani* every known enzyme of the Embden-Meyerhof-Parnas (EMP) sequence has been detected (Cochrane & Cochrane, 1966). Respiration and cellular synthesis by spores of *F. solani* can occur with acetate or other 2-carbon compounds as sole energy sources (Cochrane, Cochrane, Vogel & Coles, 1963*b*); this is presumptive evidence for a functioning glyoxylate system. Presence of isocitratase and malate synthetase was also reported in ungerminated spores.

Caltrider & Gottlieb (1963) found all the components of the EMP-tricarboxylic acid cycle (except pyruvic dehydrogenase and α-ketoglutaric dehydrogenase) in spore extracts of *Penicillium oxalicum*, uredospores of *Puccinia graminis tritici*, *Uromyces phaseoli*, and spores of some other fungi. A terminal electron transport system is also present, since NADH-oxidase, NADH-cytochrome *c* reductase and cytochrome *c* oxidase can be detected. The inhibition of oxygen utilization of spores by fluoride, malonate, azide and cyanide also indicated the presence of an active EMP-tricarboxylic acid (TCA) cycle and electron transport system. Caltrider & Gottlieb (1966) suggested that the EMP pathway is the only oxidative pathway in the early phases of germination, but some of the hexose monophosphate pathway (HMP) and TCA-cycle enzymes are synthesized just prior to germination in teleospores of *Ustilago maydis*. Tokunaga *et al.* (1969) were able to detect TCA-cycle enzymes in spore extracts of *Verticillium alboatrum*. Changes in activities of certain enzymes of the TCA cycle and the glyoxylate cycle during the initiation of conidiation of *Aspergillus niger* have been reported by Galbraith & Smith (1969*b*).

Spores of *Septomyxa affinis* oxidize acetate, glucose, fructose, ethanol and xylose. For 1-dehydrogenation of steroids by spores of *S. affinis* neither glucose nor other exogenous source of carbon is required. The internal reserve of substrate and co-factors in the spore appears to be adequate to sustain the dehydrogenation reaction. However, the reagents that inhibit glucose utilization also inhibit 1-dehydrogenation: for example, antimycin A strongly inhibits glucose and acetate oxidation and steroid transformation by spores of *S. affinis*. Respiration of *S. affinis* spores with various carbohydrates has already been reported (Singh *et al.*, 1965). The 11α-hydroxylation of steroids by spores of *Aspergillus ochraceus*

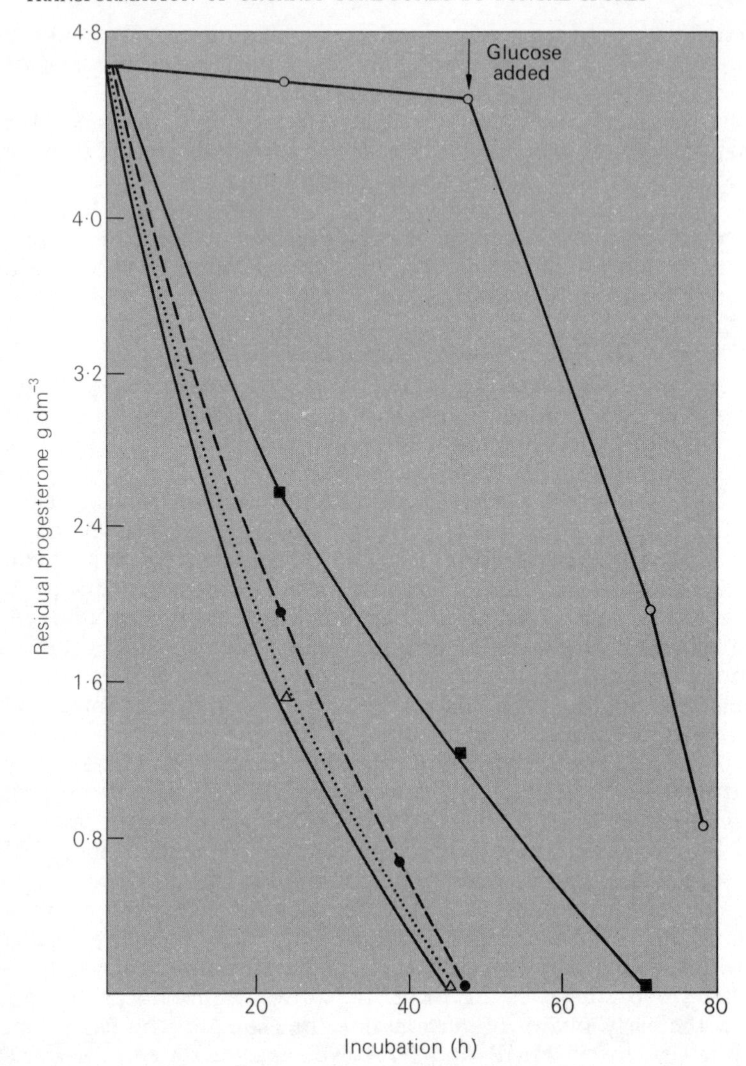

Fig. 9.4 Effect of glucose concentration on hydroxylation of progesterone. *Aspergillus ochraceus* spores, 2×10^8 cm^{-3}; progesterone, 5 g dm^{-3}; pH 5.2; volume, 2.4 dm^{-3}; temperature, 28°C; agitation, 300 rev/min; aeration, 0.6 dm^3 min^{-1}. ○——○, no glucose added at ○ time; ↓ at 43 h, 2.5 g glucose dm^{-3} added. ■——■, glucose, 1.2 g dm^{-3}; ●——●, glucose, 2.5 g dm^{-3}; Δ·······Δ, glucose, 5 g dm^{-3}; Δ——Δ, glucose, 10 g dm^{-3}. From Singh *et al.*, 1968; reproduced from *Applied Microbiology* **16**, p. 397, with permission of the American Society of Microbiology.

is glucose-(or xylose-)requiring (Fig. 9.4). Washed and starved spores of *A. ochraceus* show low endogenous oxygen uptake and effect low progesterone transformation in absence of an exogenous source of energy. Spores transform optimally in the presence of glucose or xylose, which are metabolized rapidly, effecting a considerable enhancement of oxygen uptake. The pentoses are probably metabolized by a pathway suggested for *Penicillium chrysogenum* by Chiang & Knight (1960). A significant activity of the HMP pathway for glucose metabolism is also indicated. The requirement for glucose may be related to the regeneration of NADPH, which is probably essential for the steroid hydroxylation reaction. However, we have no evidence to support this view: in fact, an exogenous NADPH-generating system consisting of NADP, glucose 6-phosphate, and glucose 6-phosphate dehydrogenase has no significant influence on respiration or steroid transformation. Endogenous respiration as well as respiration with glucose or xylose is slightly enhanced by progesterone.

An endogenous respiratory quotient of 0.9 to 1.0 for *Aspergillus ochraceus* and 1.01 for *Septomyxa affinis* indicates a predominant carbohydrate metabolism. Respiratory quotient of 1.1 to 1.2 with glucose is consistent with an oxidative pathway:

$$C_6H_{12}O_6 + O_2 \rightarrow 5(CH_2O) + CO_2 + H_2O$$

Respiration studies with various sugars and phosphorylated intermediates of the glycolytic system generally indicate a relationship between oxygen uptake and steroid hydroxylation. Addition of carbohydrates which are oxidized by spores usually enhances progesterone conversion (Vézina *et al.*, 1968).

Spores of *Aspergillus ochraceus* do not oxidize succinate, fumarate, ketoglutarate, isocitrate, glyoxylate, or pyruvate at pH 4–7. At lower pH (3–4), ketoglutarate is oxidized at a low rate (Fig. 9.5). Spores oxidize acetate rapidly and oxalacetate at a lower rate, but addition of any of the above

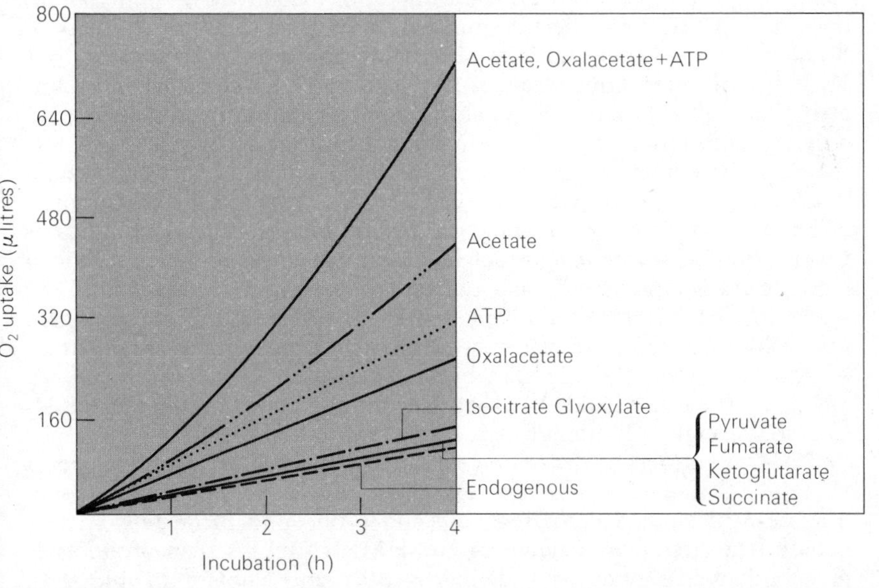

Fig. 9.5 Metabolism of tricarboxylic acid cycle intermediates by spores of *Aspergillus ochraceus.* Spores, 5×10^8 cm^{-3}; substrate (sodium salt), 50 µmoles; 0.05 mol dm^{-3} phosphate, pH 6.0; volume, 2.5 cm^3; temperature, 28°C. From Vézina *et al.*, 1968; reproduced from *Advances in Applied Microbiology* **10**, p. 256, with permission of Academic Press.

intermediates of the TCA cycle has no significant effect on progesterone conversion. Gluconate, glucono-δ-lactone, or 2-ketogluconate are not utilized by spores, indicating that glucose is not metabolized by direct oxidation *via* gluconate and ketogluconate. Similarly, the phosphorylated intermediates tried are not oxidized or are oxidized slowly (Vézina *et al.*, 1968). However, it is known that fungal cells are impermeable to many phosphorylated sugars. Adenosine triphosphate enhances endogenous and glucose respiration of spores but has no effect on progesterone transformation.

Inhibitors are useful tools in metabolic studies, although interpretation of results is difficult, particularly when intact cells, such as spores, are used. Glucose utilization and progesterone hydroxylation by spores of *Aspergillus ochraceus* are insensitive to fluoride. Respiration studies with glucose-1-^{14}C and glucose-6-^{14}C indicate an early loss of $^{14}CO_2$, predominantly from carbon-1 (Vézina *et al.*, 1968). These results and studies with enzymes of cell-free preparations indicate that the spores preferentially utilize the HMP pathway for the oxidation of glucose. In contradistinction to the observations of Cochrane, Cochrane, Collins & Serafin (1963*a*) with spores of *Fusarium solani*, 2,4-dinitrophenol (10^{-3} mol dm^{-3}) does not stimulate oxygen uptake of spores of *A. ochraceus*. Also, steroid conversion is not affected by many antibiotics, such as chloramphenicol, tetracyclines, and streptomycin (Vézina *et al.*, 1963). Oligomycin, which is a strong inhibitor of respiration coupled with phosphorylation (Lardy, Johnson & McMurray, 1958), inhibits progesterone transformation and sugar utilization, but does not inhibit oxygen uptake. Valinomycin (20 µg cm^{-3}), another powerful uncoupler of oxidative phosphorylation, does not significantly affect respiration or progesterone transformation. Sodium azide strongly inhibits respiration and steroid conversion, but KCN inhibition is rather weak; at 10^{-3} mol dm^{-3} KCN has no effect, but at 10^{-2} mol dm^{-3} there is 55% inhibition of progesterone conversion and only 18% inhibition of glucose utilization. Studies with inhibitors suggest the requirement of SH-groups and the involvement of the cytochrome system in spores. However, antimycin A (20 µg cm^{-3}) has a very limited effect on steroid conversion or oxygen uptake with glucose. This indicates that *Aspergillus ochraceus* spores, while having a typical cytochrome-linked electron transport system, does not possess an antimycin-sensitive (AS) site in its electron transport chain (Rieske, 1967), at least as far as glucose utilization and steroid transformation are concerned. However, oxidation of acetate is strongly inhibited by antimycin. In spores of *Septomyxa affinis*, on the other hand, 1-dehydrogenation, glucose and acetate oxidation are strongly inhibited (Singh *et al.*, 1965).

Cell-free extracts of *Aspergillus ochraceus* and *Streptomyxa affinis* spores (Singh, K., *unpublished results*) exhibited relatively higher activities of glucose-6-phosphate dehydrogenase and 6-phosphogluconate dehydrogenase (the first two enzymes of the HMP pathway) than of aldolase, glyceraldehyde-3-phosphate dehydrogenase and enolase (Table 9.3). Some enzymes of TCA cycle were also shown in the cell-free extracts (Table 9.4). An active electron transport system in spores is indicated by the presence of NADH-cytochrome *c* reductase and NADH-oxidase in cell-free extracts (Table 9.5 and 9.6). The NADH-cytochrome *c* reductase

Table 9.3 Enzymes of EMP and HMP pathways in extracts of spores[a]

Enzyme	μmoles substrate transformed/min/mg protein	
	Septomyxa affinis	*Aspergillus ochraceus*
Hexokinase	0.38	0.036
Phosphoglucomutase	0.75	0.19
Phosphoglucoisomerase	1.30	1.10
Aldolase	0.023	0.043
Glyceraldehyde-3-P dehydrogenase	—	0.008
Glucose-6-P dehydrogenase	0.45	0.69
6-P-gluconic dehydrogenase	0.12	0.11
Ribose-5-P isomerase	—	0.2
Enolase	0	0.05
Alcohol dehydrogenase	0	0

— not tested.

[a] Extracts were prepared using a Braun MSK homogenizer.

Table 9.4 TCA-cycle enzymes in cell-free extracts of spores[a]

Enzyme	mμmoles substrate transformed/min/mg protein	
	Aspergillus ochraceus	*Septomyxa affinis*
Citrate synthetase	5.4	12.5
Aconitate hydratase	4.0	—
Isocitrate dehydrogenase	65	57.2
Succinate dehydrogenase	0.3	<0.1
Fumarate hydratase	0.5	221
α-Oxoglutarate dehydrogenase	2.1	—
Malate dehydrogenase	1141	3960
Pyruvate dehydrogenase	0	<0.1

[a] Ten g of spores (wet paste) were mixed with 20 g of glass powder and 20 cm^3 of 0.05 mol dm^{-3} Tris–0.66 mol dm^{-3} sucrose, pH 8.0; extracts were prepared using a Braun MSK homogenizer.

from *A. ochraceus* spores is not significantly affected by antimycin A, whereas the enzyme from spores of *S. affinis*, *F. solani* and *M. griseo-cyanus* (Singh, *unpublished*) was strongly inhibited by antimycin A (Table 9.5). Dowler, Shaw & Gottlieb (1963) reported the presence of antimycin-

Table 9.5 NADH-cytochrome-*c* reductase in cell-free extracts[a]

Organism	mμ/moles cytochrome reduced min/mg protein	
	No antimycin	Antimycin' 6×10^{-6} mol dm^{-3}
Aspergillus ochraceus spores	60	55
Aspergillus ochraceus mycelium	40	8
Penicillium chrysogenum spores	130	13
Mucor griseo-cyanus spores	140	40
Fusarium solani spores	30	0
Septomyxa affinis spores	40	8

[a] Five g of cells were ground with 5 g of sand in a pre-cooled pestle and mortar and extracted with 20 cm^3 0.05 mol dm^{-3} Tris–0.66 mol dm^{-3} sucrose, pH 8.0.

Table 9.6 NADH oxidase in cell-free extracts of spores[a]

Organism	Additions	mμ/moles NADH oxidized/min/mg protein
Aspergillus ochraceus	none	75
	Antimycin 10^{-5} mol dm^{-3}	60
Septomyxa affinis	none	16
	Antimycin 10^{-5} mol dm^{-3}	0

[a] Cell-free extracts were prepared by grinding the cells in a pre-cooled pestle and mortar or by extracting in a French Press.

sensitive NADH-cytochrome *c* reductase in cell-free extracts of ten representative fungi. The enzyme from *A. ochraceus* vegetative growth was also sensitive to antimycin A, pointing out the variations in the nature of the enzyme (or the terminal electron system) between the spore and hyphal stages of the same organism, and spores of different fungi.

Nature of the steroid transforming enzymes in fungal spores

Repeated washing of spores of various fungi and streptomycetes (Vézina *et al.*, 1963) with salt solutions, buffers at various pH values, sucrose, and EDTA did not affect the activity of spores. Moreover, prolonged storage of spore suspensions in buffers did not cause any significant loss of activity or release of exogenous enzymes. Similar observations were made by Hafez-Zedan (1972) with spores of *Fusarium solani*. When spores were washed with aqueous 1 mol dm^{-3} KCl, 1 mol dm^{-3} sucrose, 0.1 mol dm^{-3}

Na_2HPO_4, 0.1 mol dm^{-3} Na_2SO_4, 5m mol dm^{-3} EDTA, citrate-phosphate (pH 3 to 5) and phosphate buffers, no loss in viability or activity to transform steroids was observed. [Invertase activity could be detected in buffer in which spores of *Aspergillus oryzae* were suspended, indicating leakage of invertase activity (Nelson *et al.*, 1969)].

Treatment of intact spores with dilute HCl (pH 1.5) for 1 h did not reduce the 1-dehydrogenation or side-chain degradation activity of *Fusarium solani* spores (Hafez-Zedan, 1972), indicating that the steroid transforming enzyme(s) are located away from the cell wall. Acid treatment presumably destroys or inhibits enzymes located outside the cell membrane, while leaving the internal enzymes fully active. Several investigators have used the acid treatment to locate surface enzymes on fungal spores, e.g. carbohydrases in *Myrothecium verrucaria* and *Aspergillus luchuensis* (Mandels, 1953), diphosphopyridine nucleotidase (Zalokar & Cochrane, 1956), and β-glucosidase (Eberhardt & Beck, 1970).

Mycelial enzymes involved in steroid transformation have been shown to be adaptive in nature (El-Tayeb, 1965; Prairie & Talalay, 1963; Rahim & Sih, 1966; Koepsell, 1962). Perlman, Weinstein & Peterson (1957) demonstrated the adaptive nature of steroid transformations in two streptomycete cultures. The simultaneous addition of a number of antibiotics and steroids prevented steroid conversion. Prairie & Talalay (1963) isolated two steroid-induced enzyme systems responsible for the conversion of testosterone to testololactone from the mycelium of *Penicillium lilacinum*. El-Tayeb (1965) reported the presence of induced 1-dehydrogenation and side-chain degradation enzymes in the cell-free extracts from the mycelium of *Cylindrocarpon radicicola*. Zuidweg (1968) demonstrated 11β- and 14α-hydroxylations of compound S with a cell-free preparation from *Curvularia lunata*. The enzymes were induced by compound S. The 11α-hydroxylation of progesterone by a cell-free preparation obtained from progesterone-induced vegetative growth of *Aspergillus ochraceus* has been demonstrated (Shibahara, Moody & Smith, 1970). Cells that are not induced with progesterone were inactive.

Vézina *et al.* (1963), however, found no evidence that 1-dehydrogenation and side-chain degradation enzymes of *Septomyxa affinis* and 11α-hydroxylase of *Aspergillus ochraceus* are steroid induced. Spores produced on steroid-free media are as active as spores produced in the presence of steroids. None of the antibiotics that inhibit steroid transformation by vegetative cells of streptomycetes (Perlman *et al.*, 1957) had any effect on steroid conversion by spores of *S. affinis* or *A. ochraceus* (Vézina *et al.*, 1968).

Similar results were obtained by Hafez-Zedan (1972) with *Fusarium solani* spores. He showed that the steroid transforming enzymes of the vegetative growth of *F. solani* are inducible. The overall capacity of non-induced mycelium was much lower than that of the progesterone-induced mycelium. The non-induced mycelium also showed a four to five hour lag period in the conversion of steroids, while no lag period was observed with mycelium grown in the presence of progesterone. The transformation of the substrate by non-induced washed mycelium was inhibited by cycloheximide and other antibiotics, whereas conversion of the substrate by progesterone-induced mycelium as well as induced and non-induced spores was unaffected by these antibiotics. Hafez-Zedan (1972) concluded from these observations that the spore enzymes are constitutive, whereas the hyphal enzymes are adaptive.

Singh, K. (*unpublished results*), investigated the effect of chloramphenicol, puromycin, cycloheximide and actinomycin D on progesterone hydroxylation and incorporation of ^{14}C-leucine into hot trichloracetic acid-precipitable fraction by washed spores of *Aspergillus ochraceus* (Table 9.7). Puromycin, actinomycin and chloramphenicol had no effect

Table 9.7 Effect of antibiotics on progesterone hydroxylation and protein biosynthesis by spores of *Aspergillus ochraceus*[a]

Antibiotic	**Concentration** μ**g cm**$^{-3}$	**Inhibition of progesterone hydroxylation**[b] %	**Inhibition of** ^{14}**C-leucine incorporation**[c] %
Puromycin	5	0	—
	20	0	24.2
Actinomycin D	5	0	9.2
	20	0	29.8
Chloramphenicol	20	0	20.0
Cycloheximide	5	0	88.2
	10	7.3	95.6

[a] Washed *A. ochraceus* spores (1×10^8 cm^{-3}) were used for this study.
[b] Progesterone, 1 mg cm^{-3}; spores suspended in 0.05 mol dm^{-3} phosphate, pH 6.0; glucose, 5 mg cm^{-3}; incubation: 18 h at 28°C.
[c] Spores, 1×10^8, were incubated with the antibiotic for 30 min before addition of ^{14}C-leucine; incorporation of label in 1 h into the hot trichloracetic acid-precipitable material was determined.

on progesterone hydroxylation. Cycloheximide, at levels which caused greater than 90% inhibition of protein synthesis, had a very limited effect on progesterone hydroxylation, indicating that for progesterone hydroxylation by spores of *A. ochraceus de novo* synthesis of protein is not necessary.

Here, it is of some interest to mention that induction of some enzymes in dormant conidia has been observed (Sinohara, 1970*a*, *b*, 1972, 1973). In dormant conidia of *Aspergillus oryzae* α-amylase, invertase and glucose dehydrogenase were induced by their respective inducers.

Protein synthesizing activity of fungal spores

During the last few years, studies carried out in the laboratories of Van Etten (Brambl & Van Etten, 1970; Van Etten, 1969; Merlo, Roker & Van Etten, 1972), Staples (Staples, Yaniv, Ramakrishnan & Lipetz, 1971), Horikoshi (Horikoshi & Ikeda, 1969), and Lovett (1968) have demonstrated that many of the macromolecular components, enzymes and informational molecules of the protein-synthesizing apparatus are present in ungerminated spores of fungi.

Van Etten & Brambl (1968) from their studies with extracts of *Botryodiplodia theobromae* conidiospores (prepared with a mechanical homogenizer) concluded that ungerminated spores of *B. theobromae* contain active aminoacyl-tRNA synthetases and transfer enzymes although the activities are low when compared to those of germinated spores. Van Etten, Koski & El-Olemy (1969) found that tRNA fractions isolated from germinated and ungerminated spores of *B. theobromae* and *Rhizopus stolonifer* had acceptor activity for all twenty amino acids commonly found in proteins when tested with enzyme fraction from germinated spores. Accordingly, it is unlikely that the absence of tRNA for a particular amino acid limits protein synthesis in fungal spores. More recent studies of Merlo *et al.* (1972) on the kinetics of aminoacyl-tRNA formation indicate that active aminoacyl-tRNA synthetases and tRNA's for at least eighteen amino acids are present in the ungerminated spores of *R. stolonifer*. In general, spore tRNA preparations from *R. stolonifer* have greater acceptor activity for most amino acids than the corresponding preparations from germinated spores. Van Etten, Roker & Davies (1972) also demonstrated that the preformed messenger ribonucleic acid in *B. theobromae* spores directs the synthesis of several relatively stable polypeptides.

Staples, App, McCarthy & Gerosa (1966) and Staples & Bedigian (1967) investigated the *in vitro* protein synthesis during uredospore germination of *Ustilago phaseoli*. No differences in the biological activities of the ribosomes and protein synthesizing enzymes prepared from germinated and ungerminated spores were observed when polyU was employed as mRNA.

Horikoshi & Ikeda (1968) demonstrated that conidia of *Aspergillus oryzae* contain complete protein synthesizing activity, although this activity is about 25% that of the vegetative cells. From the kinetic studies of phenylalanine incorporation, Horikoshi & Ikeda (1969) concluded that the polyU-directed protein synthesizing machineries in conidia and vegetative cells are essentially identical.

9.5 Concluding Remarks

In the last two decades, fungal spores, generally considered as a dormant stage in the life cycle of fungi, have been shown to contain enzymes necessary for a variety of metabolic activities. More recent studies in the laboratories of Staples, Van Etten, Horikoshi, Lovett and others have demonstrated that enzymes and informational molecules of the protein synthesizing apparatus are present in nongerminating spores of fungi. Before the observations of Knight and his collaborators on transformation of fatty acids and steroids, spores had been used by the microbiologist primarily for the maintenance of cultures and for inoculation purposes, and the possibility of using them for transformation of organic compounds was not envisaged. The extensive work done at Ayerst Laboratories since 1960 on transformation of steroids and some antibiotics, the work of Lawrence (1966) on transformation of triglycerides and the reports of Ciegler and his collaborators on transformation of carbohydrates and flavonoids have demonstrated the widespread and high activity of fungal spores in transforming compounds including many which are apparently unrelated to their metabolic processes. There is no reason to believe that the spores of certain fungi and streptomycetes will not transform other types of compounds, such as hydrocarbons, alkaloids, *etc.*

Transformation by spores is a two-step process. Spores are first produced, harvested and washed. The harvested spores are then used to transform the substrate. This characteristic offers some advantages: spores are generally stable, can be stored and transported as a concentrated biochemical catalyst which is readily available when required for transformation. Spores yield homogenous suspensions which are easy to standardize and from which suitable aliquots can be taken. They give reproducible conversions of substrate. The yield of desired product is usually high and undesirable by-products are not generally formed. The transformation medium is simple (a buffer with or without glucose), and the spores do not germinate and can sometimes be re-used. Due to the absence of nitrogenous and other nutrients in the medium, foaming is minimal; extraction of products is more efficient and offers fewer problems than those encountered in the conventional process where complex media are used. The transformation process can easily be scaled up by using suitably modified, relatively simple plant equipment, since aseptic conditions are not essential for transformation. These characteristics make the 'spore process' a process of choice when the substrate and the product are expensive and large-scale fermentation equipment for aseptic operation is not available.

The use of columns of spores entrapped in solid supports for transformation of carbohydrates (Johnson & Ciegler, 1969) has opened up new possibilities. The demonstrated stability and capacity for repeated use of many fungal spores makes the spore-column continuous transformation of compounds a potentially attractive proposition.

The spore process has certain distinct disadvantages. The process is limited to the organisms that sporulate abundantly. The production of spores necessitates sterilization of the medium before inoculation and proper humidity in the incubation rooms (if trays or flasks are used) or drums. Harvesting of spores obtained from grains or other solid medium has to be done with care to prevent dissemination in the atmosphere; workers must be protected from inhaling the dry spores. These problems can be by-passed if the spores are obtained in submerged fermentation, but, to do so, one would need the same type of equipment as is required for the conventional fermentation process. Transformation by spores has been scaled up to 200-gal vessels, and problems in further scale up are not anticipated. However, large transformation runs, 10 000 gal or more, may require suitable automated procedures for production of spores to avoid excessive labour costs.

In spite of these limitations, spores offer interesting possibilities as catalysts in a variety of biochemical studies in the laboratory and for industrial application.

9.6 References

AXELROD, D.E. (1972). Kinetics of differentiation of conidiophores and conidia by colonies of *Aspergillus nidulans*. *Journal of General Microbiology* 73, 181–4.

BAR-ELI, A. & KATCHALSKI, E. (1960). A water-insoluble trypsin derivative and its use as a trypsin column. *Nature, London* 188, 856–7.

BARNES, M. & PARKER, M.S. (1967). Use of the Coulter counter to measure osmotic effects on the swelling of mould spores during germination. *Journal of General

Microbiology **49**, 287–92.
BIANCHI, D.E. & TURIAN, G. (1967). Lipid content of conidia of *Neurospora crassa*. *Nature, London* **214**, 1344–5.
BISSETT, J. & WIDDEN, P. (1972). An automatic, multichamber soil-washing apparatus for removing fungal spores from soil. *Canadian Journal of Microbiology* **18**, 1399–1404.
BLUMENTHAL, H.J. (1965). Carbohydrate metabolism. I. Glycolysis. In *The Fungi*, vol. I. pp. 229–68. Ed. Ainsworth, G.C. and Sussman, A.F. New York and London: Academic Press.
BRAMBL, R.M. & VAN ETTEN, J.L. (1970). Protein synthesis during fungal spore germination. V. Evidence that the ungerminated conidiospores of *Botryodiplodia theobromae* contain messenger ribonucleic acid. *Archives of Biochemistry and Biophysics* **137**, 442–52.
CALTRIDER, P.G. & GOTTLIEB, D. (1963). Respiratory activity and enzymes for glucose catabolism in fungus spores. *Phytopathology* **53**, 1021–30.
CALTRIDER, P.G. & GOTTLIEB, D. (1966). Effect of sugars on germination and metabolism of teliospores of *Ustilago maydis*. *Phytopathology* **56**, 479–84.
ČAPEK, A., HANČ, O. & TADRA, M. (1966). *Microbial transformations of steroids*. Prague: Academia, and The Hague: Dr. W. Junk, Publisher.
CASAS-CAMPILLO, C. & BAUTISTA, M. (1965). Microbiological aspects in the hydroxylation of estrogens by *Fusarium moniliforme*. *Applied Microbiology* **13**, 977–84.
CHARNEY, W. & HERZOG, H.L. (1967). *Microbial transformations of steroids—A handbook*. New York: Academic Press.
CHIANG, C. & KNIGHT, S.G. (1960). A new pathway of pentose metabolism. *Biochemical and Biophysical Research Communications* **3**, 554–9.
CIEGLER, A., LINDENFELSER, L.A. & NELSON, G.E.N. (1971). Microbial transformation of flavonoids. *Applied Microbiology* **22**, 974–9.
COCHRANE, V.W. (1966). Respiration and spore germination. In *The Fungus Spore*, pp. 201–13. Ed. Madelin, M.F. London: Butterworths.
COCHRANE, V.W. & COCHRANE, J.C. (1966). Spore germination and carbon metabolism in *Fusarium solani*. V. Changes in an aerobic metabolism and related enzyme activities during development. *Plant Physiology* **41**, 810–14.
COCHRANE, V.W., COCHRANE, J.C., COLLINS, C.B. & SERAFIN, F.G. (1963*a*). Spore germination and carbon metabolism in *Fusarium solani*. II. Endogenous respiration in relation to germination. *American Journal of Botany* **50**, 806–14.
COCHRANE, V.W., COCHRANE, J.C., VOGEL, J.M. & COLES, R.S. JR. (1963*b*). Spore germination and carbon metabolism in *Fusarium solani*. *Journal of Bacteriology* **86**, 312–19.
DEGHENGHI, R., BOULERICE, M., ROCHEFORT, J.G., SEHGAL, S.N. & MARSHALL, D.J. (1966). Antiinflammatory Δ^4-pregnenolone derivatives. *Journal of Medicinal Chemistry* **9**, 513–16.
DOWLER, W.M., SHAW, P.D. & GOTTLIEB, D. (1963). Terminal oxidation in cell-free extracts of fungi. *Journal of Bacteriology* **86**, 9–17.
EBERHARDT, B.M. & BECK, R.S. (1970). Localization of the β-glucosidases in *Neurospora crassa*. *Journal of Bacteriology* **101**, 408–17.
ELLIOTT, C.G. (1972). Sterols and the production of oospores by *Phytophthora cactorum*. *Journal of General Microbiology* **72**, 321–7.
EL-TAYEB, O.M. (1965). *Studies on steroid side-chain degradation by microorganisms*. Ph.D. Thesis, University of Wisconsin, Wisconsin, U.S.A.
FISHER, D.J. & RICHMOND, D.V. (1969). The electrokinetic properties of some fungal spores. *Journal of General Microbiology* **57**, 51–60.
FISHER, D.J. & RICHMOND, D.V. (1970). The electrophoretic properties and some surface components of *Penicillium* conidia. *Journal of General Microbiology* **64**, 205–14.
FISHER, D.J. & RICHMOND, D.V. (1972). Fatty acids and hydrocarbons on the surface and in the walls of fungal spores. *Proceedings of the Society for General Microbiology* (Imperial College, London, April 1972), xi.
FISHER, D.J., HOLLOWAY, P.J. & RICHMOND, D.V. (1972). Fatty acid and hydrocarbon constituents of the surface and wall lipids of some fungal spores. *Journal of General Microbiology* **72**, 71–8.
FOSTER, J.W., MCDANIEL, L.E., WOODRUFF, H.B. & STOKES, J.L. (1945). Microbiological aspects of penicillin. V. Conidiospore formation in submerged cultures of *Penicillium notatum*. *Journal of Bacteriology* **50**, 365–8.
GALBRAITH, J.C. & SMITH, J.E. (1968). Induction of asexual sporulation in *Aspergillus niger* in submerged liquid culture. *Proceedings of the Society for General Microbiology* (Imperial College, London, April 1968), x.

GALBRAITH, J.C. & SMITH, J.E. (1969*a*). Sporulation of *Aspergillus niger* in submerged culture. *Journal of General Microbiology* **59**, 31–45.

GEHRIG, R.F. & KNIGHT, S.G. (1958). Formation of ketones from fatty acids by spores of *Penicillium roqueforti*. *Nature, London* **182**, 1237.

GEHRIG, R.F. & KNIGHT, S.G. (1961). Formation of 2-heptanone from caprylic acid by spores of various filamentous fungi. *Nature, London* **192**, 1185.

GEHRIG, R.F. & KNIGHT, S.G. (1963). Fatty acid oxidation by spores of *Penicillium roqueforti*. *Applied Microbiology* **11**, 166–70.

GREGORY, P.H. (1966). The fungus spore: what it is and what it does. In *The Fungus Spore*, pp. 1–13. Ed. Madelin, M.F. London: Butterworths.

GUNASEKARAN, M., HESS, W.M. & WEBER, D.J. (1972). The fatty acid composition of conidia of *Aspergillus niger* v. Tiegh. *Canadian Journal of Microbiology* **18**, 1575–6.

HAFEZ-ZEDAN, H. (1972). *A contribution to the study of 1-dehydrogenation and side-chain degradation of steroids by microbial spores*. Ph.D. Thesis, University of Montreal, Montreal, Canada.

HAFEZ-ZEDAN, H. & PLOURDE, R. (1971). 'Spore plate method' for transformation of steroids by fungal spores entrapped in Silica Gel G. *Applied Microbiology* **21**, 815–19.

HAFEZ-ZEDAN, H., EL-TAYEB, O.M. & ABDEL-AZIZ, M. (1970). Transformation of progesterone with the vegetative growth and non-germinating spores of *Cylindrocarpon radicicola*. *Second Conference of Microbiology* (Cairo, Egypt), pp. 71–2.

HAWKER, L.E. (1966*a*). Environmental influences on reproduction. In *The Fungi*, vol. II, pp. 435–69. Ed. Ainsworth, G.C. & Sussman, A.S. New York and London: Academic Press.

HAWKER, L.E. (1966*b*). Germination: morphological and anatomical changes. In *The Fungus Spore*, pp. 151–61. Ed. Madelin, M.F. London: Butterworths.

HAWKER, L.E., THOMAS, B. & BECKETT, A. (1970). An electron microscope study of structure and germination of conidia of *Cunninghamella elegans* Lendner. *Journal of General Microbiology* **60**, 181–9.

HESS, W.M. & STOCKS, D.L. (1969). Surface characteristics of *Aspergillus* conidia. *Mycologia* **61**, 560–71.

HESS, W.M., SASSEN, M.M.A. & REMSEN, C.C. (1968). Surface characteristics of *Penicillium* conidia. *Mycologia* **60**, 290–303.

HICKMAN, C.J. (1965). Fungal structure and organization. In *The Fungi*, vol. I, pp. 21–45. Ed. Ainsworth, G.C. & Sussman, A.S. New York and London: Academic Press.

HORIKOSHI, K. & IKEDA, Y. (1968). Studies on the conidia of *Aspergillus oryzae*. VII. Development of protein synthesizing activity during germination. *Biochimica et Biophysica Acta* **166**, 505–11.

HORIKOSHI, K. & IKEDA, Y. (1969). Studies on the conidia of *Aspergillus oryzae*. IX. Protein synthesizing activity of dormant conidia. *Biochimica et Biophysica Acta* **190**, 187–92.

IIZUKA, H. & NAITO, A. (1967). *Microbial transformation of steroids*. University of Tokyo Press, Tokyo.

JOHNSON, D.E. & CIEGLER, A. (1969). Substrate conversion by fungal spores entrapped in solid matrices. *Archives of Biochemistry and Biophysics* **130**, 384–8.

JOHNSON, D.E., NELSON, G.E.N. & CIEGLER, A. (1967). Metabolic transformation by fungal spores. *Bacteriological Proceedings* (67th Annual ASM Meeting, New York), p. 34 G69.

JOHNSON, D.E., NELSON, G.E.N. & CIEGLER, A. (1968*a*). Starch hydrolysis by conidia of *Aspergillus wentii*. *Applied Microbiology* **16**, 1678–83.

JOHNSON, D.E., NELSON, G.E.N., FOLLSTAD, M.N. & CIEGLER, A. (1968*b*). Enzymatic reactions by fungal spores entrapped on ion-exchange columns. *Bacteriological Proceedings* (68th Annual ASM Meeting, Detroit, Mich.), pp. 7–8 A39.

JUDD, R.W. & PETERSON, J.L. (1972). Temperature and humidity requirements for the germination of *Cercospora omphakodes* spores. *Mycologia* **64**, 1253–7.

KNIGHT, S.G. (1966). Transformation: a unique enzymatic activity of mold spores and mycelium. *Annals of the New York Academy of Sciences* **139**, 8–15.

KOEPSELL, H.J. (1962). 1-Dehydrogenation of steroids by *Septomyxa affinis*. *Biotechnology and Bioengineering* **4**, 57–63.

KOGAN, L.M. (1962). Microbiological reactions of steroids. *Russian Chemical Reviews* (English translation) **31**, 294–308.

LARDY, H.A., JOHNSON, D. & MCMURRAY, W.C. (1958). Antibiotics as tools for metabolic studies. I. A survey of toxic antibiotics in respiratory, phosphorylative and glycolytic systems. *Archives of Biochemistry and Biophysics* **78**, 587–97.

LAWRENCE, R.C. (1966). The oxidation of fatty acids by spores of *Penicillium roqueforti*. *Journal of General Microbiology* **44**, 393–405.

LAWRENCE, R.C. (1967). The metabolism of triglycerides by spores of *Penicillium roqueforti*. *Journal of General Microbiology* **46**, 65–76.

LAWRENCE, R.C. & HAWKE, J.C. (1968). The oxidation of fatty acids by mycelium of *Penicillium roqueforti*. *Journal of General Microbiology* **51**, 289–302.

LEE, B.K., BROWN, W.E., RYU, D.Y., JACOBSON, H. & THOMA, R.W. (1970). Influence of mode of steroid substrate addition on conversion of steroid and growth characteristics in a mixed culture fermentation. *Journal of General Microbiology* **61**, 97–105.

LEHRER, R.I. & JAN, R.G. (1970). Interaction of *Aspergillus fumigatus* spores with human leukocytes and serum. *Infection and Immunity* **1**, 345–50.

LEIGHTON, T.J. & STOCK, J.J. (1970). Biochemical changes during fungal sporulation and spore germination. I. Phenyl methyl sulfonyl fluoride inhibition of macroconidial germination in *Microsporum gypseum*. *Journal of Bacteriology* **101**, 931–40.

LEIGHTON, T.J., STOCK, J.J. & KELLN, R.A. (1970). Macroconidial germination in *Microsporum gypseum*. *Journal of Bacteriology* **103**, 439–46.

LENNON, R.E. & VÉZINA, C. (1973). Antimycin A, a piscicidal antibiotic. *Advances in Applied Microbiology* **16**, 55–96.

LEWIS, J.A. & PAPAVIZAS, G.C. (1967). Effects of tannins on spore germination and growth of *Fusarium solani* f. *phaseoli* and *Verticillium albo-atrum*. *Canadian Journal of Microbiology* **13**, 1655–61.

LOVETT, J.S. (1968). Reactivation of ribonucleic acid and protein synthesis during germination of *Blastocladiella* zoospores and the role of the ribosomal nuclear cap. *Journal of Bacteriology* **96**, 962–9.

LUNDBORG, M. & HOLMA, B. (1972). *In vitro* phagocytosis of fungal spores by rabbit lung macrophages. *Sabouraudia* **10**, 152–156.

MACDONALD, K.D. (1972). Storage of conidia of *Penicillium chrysogenum* in liquid nitrogen. *Applied Microbiology* **23**, 990–3.

MANDELS, G.R. (1953). Localization of carbohydrases at the surface of fungal spores by acid treatment. *Experimental Cell Research* **5**, 48–55.

MANDELS, G.R. (1965). Kinetics of fungal growth. In *The Fungi*, vol. I, pp. 599–612. Ed. Ainsworth, G.C. & Sussman, A.S. New York and London: Academic Press.

MANNERS, J.G. (1966). Assessment of germination. In *The Fungus Spore*, pp. 165–73. Ed. Madelin, M.F. London: Butterworths.

MARSHECK. W.J. (1971). Current trends in the microbiological transformation of steroids. In *Progress in Industrial Microbiology* **10**, 49–103.

MERLO, D.J., ROKER, H. & VAN ETTEN, J.L. (1972). Protein synthesis during fungal spore germination. VI. Analysis of transfer ribonucleic acid from germinated and ungerminated spores of *Rhizopus stolonifer*. *Canadian Journal of Microbiology* **18**, 949–56.

MEYERS, E. & KNIGHT, S.G. (1958). Studies on the nutrition of *Penicillium roqueforti*. *Applied Microbiology* **6**, 174–8.

MEYERS, E. & KNIGHT, S.G. (1961). Studies on the intracellular amino acids of *Penicillium roqueforti*. *Mycologia* **53**, 115–22.

MITCHELL, N.L. & MCKEEN, W.E. (1970). Light and electron microscope studies on the conidium and germ tube of *Sphaerotheca macularis*. *Canadian Journal of Microbiology* **16**, 273–80.

MORTON, A.G. (1961). The induction of sporulation in mould fungi. *Proceedings of the Royal Society* **153**, 548–69.

MOSBACH, K. & LARSSON, P. (1970). Preparation and application of polymer-entrapped enzymes and microorganisms in microbial transformation processes with special references to steroids 11β-hydroxylation and Δ^1-dehydrogenation. *Biotechnology and Bioengineering* **12**, 19–27.

MURRAY, H.C. & PETERSON, D.H. (1952). Oxygenation of steroids by mucorales fungi. U.S. Patent 2,602,769.

NELSON, G.E.N., JOHNSON, D.E. & CIEGLER, A. (1969). Inversion of sucrose by fungal spores. *Developments in Industrial Microbiology* **10**, 284–9.

NELSON, G.E.N., JOHNSON, D.E. & CIEGLER, A. (1971). Production of D-mannitol by conidia of *Aspergillus candidus*. *Applied Microbiology* **22**, 484–5.

NESHEIM, S. (1967). Note on ochratoxins. *Journal of the Association of Official Agricultural Chemists* **50**, 370–1.

NG, W.S., SMITH, J.E. & ANDERSON, J.G. (1972). Changes in carbon catabolic pathways during synchronous development of conidiophores of *Aspergillus niger*. *Journal of General Microbiology* **71**, 495–504.

NISHI, A. (1961). Role of polyphosphate and phospholipid in germinating spores of *Aspergillus niger*. *Journal of Bacteriology* **81**, 10–19.

OLIVER, P.T.P. (1972). Conidiophore and spore development in *Aspergillus nidulans*. *Journal of General Microbiology* **73**, 45–54.

ORÓ, J., LASETER, J.L. & WEBER, D. (1966). Alkanes in fungal spores. *Science* **154**, 399–400.

PASS, T. & GRIFFIN, G.J. (1972). Exogenous carbon and nitrogen requirements for conidial germination by *Aspergillus flavus*. *Canadian Journal of Microbiology* **18**, 1453–61.

PERLMAN, D., WEINSTEIN, M.J. & PETERSON, G.E. (1957). Effect of antibiotics on oxidation of progesterone by two streptomycetes. *Canadian Journal of Microbiology* **3**, 841–6.

PETERSON, D.H. (1963). Microbial transformations of steroids and their application to the preparation of hormones and derivatives. In *Biochemistry of Industrial Micro-organisms*, pp. 537–606. Ed. Rainbow, C. & Rose, A.H. London and New York: Academic Press.

PLOURDE, R., EL-TAYEB, O.M. & HAFEZ-ZEDAN, H. (1972). Reduction of the 20-carbonyl group of C-21 steroids by spores of *Fusarium solani* and other microorganisms. *Applied Microbiology* **23**, 601–12.

PRAIRIE, R.L. & TALALAY, P. (1963). Enzymatic formation of testolactone. *Biochemistry* **2**, 203–8.

PRESCOTT, S.C. & DUNN, C.G. (1959). *Industrial microbiology*. New York: McGraw-Hill Book Company.

RAHIM, M.A. & SIH, C.J. (1966). Mechanism of steroid oxidation by microorganisms. XI. Enzymatic cleavage of the pregnane side chain. *Journal of Biological Chemistry* **241**, 3615–23.

RAPER, J.R. (1966). Life cycles, basic patterns of sexuality, and sexual mechanisms. In *The Fungi*, vol. II, pp. 473–511. Ed. Ainsworth, G.C. & Sussman, A.S. New York and London: Academic Press.

RIESKE, J.S. (1967). Antimycin A. In *Antibiotics*, vol. I, pp. 542–84. Ed. Gottlieb, D. & Shaw, P.D. Springer-Verlag, New York.

RIGHELATO, R.C., TRINCI, A.P.J., PIRT, S.J. & PEAT, A. (1968). The influence of maintenance energy and growth rate on the metabolic activity, morphology and conidiation of *Penicillium chrysogenum*. *Journal of General Microbiology* **50**, 399–412.

RIZZA, V. & KORNFELD, J.M. (1969). Components of conidial and hyphal walls of *Penicillium chrysogenum*. *Journal of General Microbiology* **58**, 307–15.

RUIZ-HERRERA, J. (1967). Chemical components of the cell wall of *Aspergillus* species. *Archives of Biochemistry and Biophysics* **122**, 118–25.

RYU, R.Y., LEE, B.K., THOMA, R.W. & BROWN, W.E. (1969). Transformation of steroids by mixed cultures. *Biotechnology and Bioengineering* **11**, 1255–70.

SALA, F.J. & BURGOS, J. (1972). Simple method for mass production and collection of conidia from *Hemispora stellata*. *Applied Microbiology* **24**, 504–5.

SAMŠINÁKOVÁ, A. (1966). Growth and sporulation of submerged cultures of the fungus *Beauveria bassiana* in various media. *Journal of Invertebrate Pathology* **8**, 395–400.

SCHLEG, M.C. & KNIGHT, S.G. (1962). Hydroxylation of progesterone by conidia from *Aspergillus ochraceus*. *Mycologia* **54**, 317–9.

SCOTT, P.M. & HAND, T.B. (1967). Method for the detection and estimation of ochratoxin A in some cereal products. *Journal of the Association of Official Agricultural Chemists* **50**, 366–70.

SEHGAL, S.N. & VÉZINA, C. (1969). Stereospecific reduction of organic compounds with microorganisms. *Bacteriological Proceedings* (69th Annual ASM Meeting, Miami Beach, Fla.), p. 3 A5.

SEHGAL, S.N., SINGH, K. & VÉZINA, C. (1963). Transformation of Reichstein's compound S with *Didymella lycopersici*. *Steroids* **2**, 93–7.

SEHGAL, S.N., SINGH, K. & VÉZINA, C. (1966). Process for producing spores. *U.S. Patent 3,294,647*.

SEHGAL, S.N., SINGH, K. & VÉZINA, C. (1968). 11α-Hydroxylation of steroids by spores of *Aspergillus ochraceus*. *Canadian Journal of Microbiology* **14**, 529–32.

SHEPHERD, C.J. (1956). Changes occurring in the composition of *Aspergillus nidulans* conidia during germination. *Proceedings of the Society for General Microbiology* (University of Exeter, Exeter, September 1956), i.

SHIBAHARA, M., MOODY, J.A. & SMITH, L.L. (1970). Microbial hydroxylations. V. 11α-Hydroxylation of progesterone by cell-free preparations of *Aspergillus ochraceus*. *Biochimica et Biophysica Acta* **202**, 172–9.

SINGH, K. & RAKHIT, S. (1967). Mechanism of side-chain degradation of C-21 steroids by spores of *Septomyxa affinis*. *Biochimica et Biophysica Acta* **144**, 139–44.

SINGH, K. & RAKHIT, S. (1971). Transformation of antibiotics. I. Microbial transformation of antimycin A. *The Journal of Antibiotics* **24**, 704–5.

SINGH, K., SEHGAL, S.N. & VÉZINA, C. (1963). C-1-dehydrogenation of steroids by spores of *Septomyxa affinis*. *Steroids* **2**, 513–20.

SINGH, K., SEHGAL, S.N. & VÉZINA, C. (1965). Transformation of Reichstein's compound S and oxidation of carbohydrates by spores of *Septomyxa affinis*. *Canadian Journal of Microbiology* **11**, 351–64.

SINGH, K., SEHGAL, S.N. & VÉZINA, C. (1967*a*). Transformation of steroids by *Mucor griseo-cyanus*. *Canadian Journal of Microbiology* **13**, 1271–81.

SINGH, K., SEHGAL, S.N. & VÉZINA, C. (1967*b*). Conversion of penicillin V to 6-aminopenicillanic acid by the use of spores. *U.S. Patent 3,305,453.*

SINGH, K., SEHGAL, S.N. & VÉZINA, C. (1968). Large-scale transformation of steroids by fungal spores. *Applied Microbiology* **16**, 393–400.

SINOHARA, H. (1970*a*). Induction of enzymes in dormant spores of *Aspergillus oryzae*. *Journal of Bacteriology* **101**, 1070–2.

SMITH, J.E. & GALBRAITH, J.C. (1971). Biochemical and physiological aspects of differentiation in the fungi. *Advances in Microbial Physiology* **5**, 45–134.

SOMERS, E. & FISHER, D.J. (1967). Effect of dodine acetate on the electrophoretic mobility of *Neurospora crassa* conidia. *Journal of General Microbiology* **48**, 147–154.

STAPLES, R.C. & BEDIGIAN, D. (1967). Preparation of an amino acid incorporation system from uredospores of the bean rust fungus. *Contributions Boyce Thompson Institute* **23**, 345–7.

STAPLES, R.C., APP, A.A., MCCARTHY, W.J. & GEROSA, M.M. (1966). Some properties of the ribosomes from uredospores of the bean rust fungus. *Contributions Boyce Thompson Institute* **23**, 159–64.

STAPLES, R.C., YANIV, Z., RAMAKRISHNAN, L. & LIPETZ, J. (1971). Properties of ribosomes from germinating uredospores. In *Morphological and Biochemical Events in Plant-Parasite Interaction,* pp. 59–90. Ed. Akai, S. & Ouchi, S. The Phytopathological Society, Tokyo, Japan.

STÄRKLE, M. (1924). Die Methylketone im oxydativen Abbau einiger Triglyceride (bzw. Fettsäuren) durch Schimmelpilze unter Berüchsichtigung der besonderen Ranzidität des Kokosettes. *Biochemische Zeitschrift* **151**, 371–412.

STINE, G.J. & CLARK, A.M. (1967). Synchronous production of conidiophores and conidia of *Neurospora crassa*. *Canadian Journal of Microbiology* **13**, 447–53.

STOUDT, T.H. (1960). The microbiological transformation of steroids. *Advances in Applied Microbiology* **2**, 183–222.

SUSSMAN, A.S. (1966). Dormancy and spore germination. In *The Fungi*, vol. II, pp. 733–64. Ed. Ainsworth, G.C. & Sussman, A.S. New York and London: Academic Press.

TINER, J.D. & GOLDEN, A. (1967). Process for harvest of spores from aerial microbial structures. *U.S. Patent 3,300,390.*

TOKUNAGA, J., MALCA, I., SIMS, J.J., ERWIN, D.C. & KELN, N.T. (1969). Respiratory enzymes in the spores of *Verticillium albo-atrum*. *Phytopathology* **59**, 1829–32.

TOSA, T., MORI, T., FUSE, N. & CHIBATA, I. (1967). Studies on continuous enzyme reactions. IV. Preparation of DEAE-Sephadex-Amino-acylase column and continuous optical resolution of acyl-DL-amino acids. *Biotechnology and Bioengineering* **9**, 603–15.

UNDERKOFLER, L.A. (1954). Fungal amylolytic enzymes. In *Industrial Fermentations*, vol. II, pp. 97–121. Ed. Underkofler, L.A. & Hickey, R.J. New York: Chemical Publishing Company.

VAN ETTEN, J.L. (1969). Protein synthesis during fungal spore germination. *Phytopathology* **59**, 1060–4.

VAN ETTEN, J.L. & BRAMBL, R.M. (1968). Protein synthesis during fungal spore germination. II. Aminoacyl-soluble ribonucleic acid synthetase activities during germination of *Botryodiplodia theobromae* spores. *Journal of Bacteriology* **96**, 1042–8.

VAN ETTEN, J.L., KOSKI, R.K. & EL-OLEMY, M.M. (1969). Protein synthesis during fungal spore germination. IV. Transfer ribonucleic acid from germinated and ungerminated spores. *Journal of Bacteriology* **100**, 1182–6.

VAN ETTEN, J.L., ROKER, H.R. & DAVIES, E. (1972). Protein synthesis during fungal spore germination: differential protein synthesis during germination of *Botryodiplodia theobromae* spores. *Journal of Bacteriology* **112**, 1029–31.

VÉZINA, C. & RAKHIT, S. (1974). Microbial transformation of steroids. In *CRC Handbook of Microbiology*, vol. 3, pp. 117–441. Ed. Laskin, A.I. & Lechevalier, H. The Chemical Rubber Company, Cleveland.

VÉZINA, C., SEHGAL, S.N. & SINGH, K. (1963). Transformation of steroids by spores of microorganisms. I. Hydroxylation of progesterone by conidia of *Aspergillus ochraceus*. *Applied Microbiology* **11**, 50–7.

VÉZINA, C., SINGH, K. & SEHGAL, S.N. (1965). Sporulation of filamentous fungi in submerged culture. *Mycologia* **57**, 722–36.

VÉZINA, C., SEHGAL, S.N. & SINGH, K. (1968). Transformation of organic compounds by fungal spores. *Advances in Applied Microbiology* **10**, 221–68.

VÉZINA, C., SINGH, K. & SEHGAL, S.N. (1969). Agar plate method for detecting microorganisms which produce equilin and other estrogens from various steroids. *Applied Microbiology* **18**, 270–1.

VÉZINA, C., SEHGAL, S.N., SINGH, K. & KLUEPFEL, D. (1971). Microbial aromatization of steroids. *Progress in Industrial Microbiology* **10**, 1–47.

ZALOKAR, M. & COCHRANE, V.W. (1956). Diphosphopyridine nucleotidase in the life cycle of *Neurospora crassa*. *American Journal of Botany* **43**, 107–10.

ZEIDLER, G. & MARGALITH, P. (1972). Synchronized sporulation in *Penicillium digitatum* (Sacc.). *Canadian Journal of Microbiology* **18**, 1685–90.

ZUIDWEG, M.H.J. (1968). Hydroxylation of Reichstein's compound S with cell-free preparations from *Curvularia lunata*. *Biochimica et Biophysica Acta* **152**, 144–158.

Additional References

BACON, C.W., SWEENEY, J.G., ROBBINS, J.D. & BURDICK, D. (1973). Production of penicillic acid and ochratoxin A on poultry feed by *Aspergillus ochraceus*: temperature and moisture requirements. *Applied Microbiology* **26**, 155–60.

DALBY, D.K. & GRAY, W.D. (1974). Inorganic ion content of fungal mycelia and spores. *Canadian Journal of Microbiology* **20**, 935–6.

GALBRAITH, J.C. & SMITH, J.E. (1969*b*). Changes in activity of certain enzymes of the tricarboxylic acid cycle and the glyoxylate cycle during the initiation of conidiation of *Aspergillus niger*. *Canadian Journal of Microbiology* **15**, 1207–12.

HAFEZ-ZEDAN, H. & PLOURDE, R. (1973). Steroid l-dehydrogenation and side-chain degradation enzymes in the life cycle of *Fusarium solani*. *Biochimica et Biophysica Acta* **326**, 103–15.

MARTINELLI, S.D. (1973). Conidiation of *Aspergillus nidulans* in submerged liquid culture. *Proceedings of the Society for General Microbiology* (Imperial College, London, April 1973), vi.

MARTINELLI, S.D. (1974). Glucose inhibition of *Aspergillus nidulans* in submerged culture. *Proceedings of the Society for General Microbiology* (Imperial College, London, April 1974), 72.

PLOURDE, R. & HAFEZ-ZEDAN, H. (1973). Distribution of steroid l-dehydrogenation and side-chain degradation enzymes in the spores of *Fusarium solani*: causes of metabolic lag and carbohydrate independence. *Applied Microbiology* **25**, 650–8.

SANSING, G.A. & CIEGLER, A. (1973). Mass propagation of conidia from several *Aspergillus* and *Penicillium* species. *Applied Microbiology* **26**, 830–1.

SAXENA, R.K. & SINHA, U. (1973). Conidiation of *Aspergillus nidulans* in submerged liquid culture. *Journal of General and Applied Microbiology* **19**, 141–6.

SINGH, K., SEHGAL, S.N. & VÉZINA, C. (1969). Hydrolysis of phenoxymethyl penicillin into 6-aminopenicillanic acid with spores of Fusaria. *Applied Microbiology* **17**, 643–644.

SINOHARA, H. (1970*b*). Induction of enzymes in dormant spores of *Aspergillus oryzae*. II. Effects of some aliphatic and aromatic alcohols. *Biochimica et Biophysica Acta* **224**, 541–52.

SINOHARA, H. (1972). Induction of enzymes in dormant spores of *Aspergillus oryzae*. III. Effects of kanamycin and streptomycin. *Biochimica et Biophysica Acta* **281**, 425–33.

SINOHARA, H. (1973). Induction of enzymes in dormant spores of *Aspergillus oryzae*. IV. Effects of ethidium, aminoacridines, and related compounds. *Biochimica et Biophysica Acta* **299**, 662–8.

ZEIDLER, G. & MARGALITH, P. (1973). Modification of the sporulation cycle in *Penicillium digitatum* (Sacc.). *Canadian Journal of Microbiology* **19**, 481–3.

CHAPTER 10

Industrial Enzyme Production

J. A. BLAIN

10.1 Introduction

Processes such as brewing and breadmaking have for millennia involved the unrecognized use of enzymes so that traditional practices and technologies involving enzymic conversions were well established before any coherent body of knowledge on their rational application could exist. It was natural that early observations on enzymic action centred around yeast fermentation and the diastatic processes for conversion of starch to sugar, although the nature of digestive juices also evoked interest.

When in 1837 Berzelius wrote on the nature of catalysis he cited processes which we now know to be enzymic and crude precipitates of diastase from malt and pepsin from gastric juice had already been studied.

While the use of yeast and malt formed much of the foundation for enzyme technology in the West, there were in Asia corresponding processes for the modification of food sources which hinged on certain enzymic conversions associated with fungal growth. These led to a major step in the deliberate use of microbial enzymic material when Takamine towards the end of the 19th century used the amylase preparation 'Takadiastase'. This was obtained from *Aspergillus oryzae* grown on wheat bran.

It was natural that further development of enzyme additives was in general to provide enhancement of traditional processes rather than to open up novel possibilities. Even now it remains the case that production of crude enzymes in bulk is devoted largely to those which hydrolyse the glycosidic links of carbohydrates such as starch and pectins and to the proteases which have an analogous hydrolytic action on the peptide links of proteins.

On the other hand there has been a steady proliferation of uses and suggested uses for more highly purified enzyme preparations in industrial processing, clinical medicine and laboratory practice. The range of pure enzymes which are commercially available is still expanding rapidly. The use of these highly refined preparations is in general limited by price which, because of the number of processes involved in their isolation, is of the order of hundreds of pounds per gramme. In contrast the crude enzymes, which may have less than 1% active material would be costed rather in pounds per kilogramme. In many operations such as clarifying wines and juices, chill-proofing beer and improving bread doughs the use of crude enzymes is likely to add very little to the cost of the product.

Most of the crude enzymes used at present are those secreted by cells to act upon their substrates in an external environment as do the digestive enzymes of animals. The enzymes of similar function in micro-organisms—those which split large molecules into an assimilable form—are generally secreted into the fermentation media. The amylases of seeds have a similar role in breaking down starch reserves on germination. Traditional sources were animal pancreas for proteases, amylases and lipases and malted barley for amylases. The fermentation media from the culture of micro-organisms then become major sources for proteases, amylases and to a lesser extent cellulases, lipases and other enzymes. Long established sources of vegetable proteases have been figs (ficin) and papaya (papain) the latter being used very extensively.

It will be seen then that, of the six major classes into which enzymes are divided according to the proposals of the Enzyme Commission of the International Union of Biochemistry in 1965, those in industrial use belong mostly to group III, the Hydrolases. Such enzymes are capable of acting without complex co-factors, are readily obtained from micro-organisms without rupturing the cell-walls and are water-soluble.

Microbial sources

Utilization of microbiological source material for enzyme production has increased steadily for the following reasons:

1. There is, in general, high specific activity per unit dry weight of product.
2. Seasonal fluctuations of raw material and possible shortages do not apply.
3. There is a wide spectrum of characteristics such as optimal pH and temperature resistance available for selection.
4. There are possibilities for greatly optimizing enzyme yield through strain selection, mutation, induction and selection of growth conditions.

The corresponding animal and plant enzymes come from relatively few sources and some are from by-products, limited in supply.

Consequently, while pancreatic enzymes and papain, for example, are still widely used it is to be anticipated that future developments in enzyme technology will rely largely on microbial sources. Even in the malting process of brewing where the amylases of germinated barley which hydrolyse starch are from an inexpensive source, around which existing brewing technology has developed, there are now competitive processes involving microbial enzymes.

Types of micro-organism selected

The rationale of selection between bacterial and fungal sources is not always obvious since a number of ill-defined factors may be involved. These range from the economics of cultivation and separation to specific characteristics desired in the enzymes and also whether other enzymes incidentally present are harmful or otherwise. Thus, in a number of commercial enzyme preparations, both amylases and proteases will be found. Selection of a particular source may depend upon the desired balance of such enzymes in relation to a particular process.

For some food manufacturing procedures, it is important to have protease low in amylase, as in biscuit manufacture. However, the presence of protease may be desirable in amylase used for the production of brewing worts while again when amylases are used in textile treatment inclusion of proteases can be useful if gelatin removal from starching material is also necessary.

According to source material, enzymes will differ greatly in their stability to temperature and to extremes of hydrogen ion concentration. Thus *Bacillus subtilis* proteases which are relatively heat-stable and active under alkaline conditions have been most suitable as soap-powder additives (Wieg, 1969). On the other hand fungal amylases, because of their greater sensitivity to heat, have been found more useful in the baking industry, as will be discussed subsequently.

A comprehensive list of microbial extracellular enzymes which have been prepared has been published (Davies, 1963) and probably the most extensive account of industrial food enzymes from all sources is that by Reed (1966).

Discussion here will be limited to known or proposed uses of enzymes from moulds except where reference to other types is necessary for the establishment of a rational context.

10.2 Production of Fungal Enzymes

Cultivation techniques

Since the production and sale of industrial enzymes is highly competitive, manufacturers are obviously reluctant to release details of either production or processing and it is not easy to estimate the extent to which different methods of cultivation are at present being used.

Fungal enzymes have been produced increasingly over the last twenty years by deep culture methods which permit greater control of environmental factors such as temperature and hydrogen ion concentration than do the earlier methods of cultivation of moulds in which growth takes place on the surfaces of bran or other solids. However surface culture has survived in major use and may often be advantageous in yielding higher intitial enzyme concentration. It is also the case that some moulds which grow well in the laboratory in shake cultures are very difficult to grow in fermenters due to wall growth. Among enzymes produced from surface cultures are the amyloglucosidase of *Rhizopus* sp. and the acid-resistant proteases of *Aspergillus niger* (Arima 1964; Underkofler, 1969).

In general, surface cultures are carried out on sterilized moistened bran with added nutrients. This material is spread on trays up to a depth of 4 cm, total quantities being of the order of thousands of kilogrammes. Series of trays are enclosed in a larger vessel in which the moulds are incubated after inoculation. Temperature is controlled by the circulation of suitably humidified air and in most cases subsequent extraction with water produces enzyme in solution. In deep culture methods for enzyme preparation, quantities of 1000–30 000 gal or more may be involved (Sakaguchi, Vernura & Kinoshita, 1971). Various nitrogen sources such as soya fractions and casein have been described and Arima (1964) gives a fairly detailed account of nutrient compositions for the production of various mould enzymes. Incubations are generally shorter than those used for laboratory cultures often being only from one to two days although possibly up to 5 days (Beckhorn, Labee & Underkofler, 1965). Usually the enzyme is extracted from the fermentation liquor after separation of mycelia by the use of plate and frame presses or by rotary vacuum filters.

Extraction of enzymes from fermentation media

In contrast to laboratory preparations where enzyme purity is normally a major objective, the industrial methods seek to obtain a product which satisfies requirements with the minimum of processing. Each precipitation stage is not only costly but being more protracted than corresponding small-scale operations will expose enzymes to greater denaturation.

To concentrate activity into small bulk is not necessarily desirable since subsequent mixing of the enzyme into the material to be processed could be less easy. The type of product sought is likely to have between 0.1–10% of the active enzyme and the simplest preparations may consist of the filtered fermentation broth with additives and with some measure of concentration. Such concentration is unlikely to involve more volume reduction than to one fifth or one sixth that of the original broth because the total soluble solids would be inconveniently high. Very crude extracts of this type can also be spray-dried.

Most frequently enzyme fractions are obtained from the filtered broth (or in the case of solid surface cultures from aqueous extracts of the bran after incubation) by the precipitation techniques commonly used for protein isolation. Such methods are based on the tendency for proteins to be precipitated by water miscible solvents or strong salt solutions, particularly at hydrogen ion concentrations close to their iso-electric points. Despite a number of theoretical studies which have been made on such precipitations there are few generalizations which apply to a large range of proteins. Procedures in use have mostly been adapted from empirical studies.

The salts most commonly used for industrial enzyme precipitations have been ammonium sulphate and sodium sulphate. The latter is less soluble and may require temperatures above ambient for adequate solubility since its use may demand concentrations of over 40%. Ammonium sulphate can be used at higher concentrations.

Over 90% of amyloglucosidase can be extracted from fungal mycelia grown on a wheat bran surface culture using two or three volumes of water and then precipitated down by ammonium sulphate at 70% saturation (Arima, 1964).

Solvent precipitation has been carried out with methanol, ethanol, isopropanol, acetone and methyl ethyl ketone adding quantities of the order of one to three volumes according to the nature of the enzyme to be precipitated. Such precipitates tend to be sticky and difficult to handle so that carriers such as starch, kieselguhr or lactose are added to correct this.

After precipitation the filter cake may be washed free of salt with dilute ethanol and then tray-dried and ground or extracted with dilute salt solution to give a concentrated liquor and this spray-dried to a powder.

Various stabilizers are used to minimize enzyme loss during processing and these include sodium chloride, propylene glycol, calcium salts (for heat stability) and sometimes sulphites if the enzyme is prone to oxidation.

Processing developments

Techniques which have proved useful in enzyme separation and use at the laboratory or pilot-plant scale and which are likely to influence industrial-scale production and use of enzymes are those associated with membrane technology, affinity chromatography and insolubilized enzymes.

In recent years considerable advances have been made in the use of membranes which by passage of water and small molecules under pressure can be used to concentrate enzymes and other proteins without corresponding increase of salt concentration. (Chian & Selldorff, 1969; Thomson, 1971). Membranes may be selected to pass smaller protein molecules and retain larger ones. Current advances in these ultrafiltration techniques would lead to anticipation of their wider use in large scale processes. Similarly, the removal of water only by techniques involving 'reverse osmosis' is also very relevant to enzyme concentration on a large scale.

The extensive work which has been carried out over the last few years on insolubilized enzymes (Kay, 1968; Barker & Epton, 1970; Katchalski, Silman & Goldman, 1971) could conceivably lead, in the future, to a greater demand for enzymes of high purity at the expense of crude enzymes. Insolubilized or immobilized enzymes are formed by physical entrapment, adsorption or covalent binding of enzymes to physical supports.

Such binding usually diminishes specific activity somewhat while rendering the enzyme more stable to denaturation by temperature and on storage. The same sample of enzyme may be used repeatedly since it is separable at the end of a process and thus it may become feasible to use, for industrial conversions, enzymes which would otherwise be uneconomic. Such immobilized enzymes could be deployed in columns or other types of reactor (Lilly & Dunnill, 1972).

While the various patents registered for specific uses of insolubilized enzymes include large-scale processing such as the chillproofing of beer, it is possible that the relatively high cost of preparation may at present limit their wider use.

In affinity chromatography solutions are passed through a column containing compounds which specifically abstract certain enzymes. Such compounds are often analogues of the enzyme's true substrate. Recovery of enzymes so isolated is possible and purification of over a hundredfold may be achieved. It has been possible to purify staphylococcal nuclease directly from culture medium to a state of homogeneity in one step (Cuatrecasas & Afinsen, 1971). It may be hoped that in some cases such methods can be scaled up for industrial use and the possibilities inherent in separation of enzymes on a basis of their specificities are impressive.

10.3 Major Industrial Uses of Fungal Enzymes

Producers of industrial enzymes often limit description of products to terms such as 'fungal' or 'bacterial' without specifying the nature of the organism. It is however well known that *Aspergillus oryzae* and *A. niger* have been the most widely exploited sources of fungal enzymes, although species of *Mucor, Rhizopus, Penicillium* and others have been cited (Table 10.1). A list of American enzyme preparations and their manufacturers, which in some cases give the general type of organism used, has been compiled (de Becze, 1970).

Table 10.1 Some fungal sources of industrial enzymes

Fungus	Type of Enzyme
Aspergillus oryzae	Amylase, pectinase, protease, adenylic acid deaminase
A. niger	Amylase, amyloglucosidase, pectic enzymes, protease, glucose oxidase, naringinase
Rhizopus sp.	Amyloglucosidase, lipase, rennet protease, pectic enzymes
Penicillium chrysogenum	Glucose oxidase
P. citrinum	5'phosphodiesterase
Mucor sp.	Rennet protease
Trichoderma viride	Cellulase
A. foetidus	Amyloglucosidase

It will be useful here to group the enzymes in major use according to whether their action is principally on starches and sugars, on pectic substances or on proteins and subsequently to itemize those of minor or more tentative applications.

Enzymes acting on starches and sugars

α-AMYLASE α-Amylase is an endo-enzyme which hydrolyses the internal 1,4-glycosidic links of both amylase and amylopectin constituents of starch, but not the α1.6 linkage of amylopectin, with degradation of the starch to dextrins and oligosaccharides. The main bacterial source is *Bacillus subtilis* and the fungal enzyme is obtainable from *Aspergillus niger* and *A. oryzae*, the latter being frequently grown on a semi-solid medium at temperatures close to 30°C and yielding suitable enzyme levels in 24–48 h (Beckhorn, 1967). Contaminating protease can be removed when undesirable by various methods such as heating at high pH values where the protease is less stable (Miller & Johnson, 1954) or in general by differential solvent precipitation methods.

AMYLOGLUCOSIDASE OR GLUCO-AMYLASE This enzyme hydrolyses α-1,4-glucan links in polysaccharides removing glucose units from the non-reducing ends of the chains and also hydrolyses the 1.6-glycosidic links of amylopectin. It thus converts starch and dextrins to glucose, action rates increasing to a certain extent with dextrin chain-length, but the amyloglucosidases of *Aspergillus niger* and *A. oryzae* will hydrolyse maltose to glucose.

Rhizopus niveus is reported as a source of this enzyme (Arima, 1964) as well as *Aspergillus oryzae* and *A. niger*, for which deep culture methods have been used successfully for production (Aschengreen, 1969). A detailed description of the development of a commercial process using an *A. foetidus* mutant as an amyloglucosidase source has been published (Underkofler, 1969).

β-AMYLASE The enzyme which hydrolyses alternate α-1,4-glucan links in starch, so as to remove maltose units from the non-reducing ends of chains, is not of consequence as a fungal enzyme but is listed since its function in alcohol production may be to some extent replaceable by amyloglucosidase.

APPLICATIONS OF α-AMYLASE AND AMYLOGLUCOSIDASE These enzymes are utilized in production of sugars or oligosaccharides from starches in various industrial contexts.

Bread-making In breadmaking an important source of sugar for yeast fermentation is from the enzymic hydrolysis of damaged starch granules in flour, the undamaged being unsusceptible to attack in the doughing processes. The α-amylase from added malt and also from the wheat flour will convert the starch to dextrins and thus promote the action of β-amylase in forming maltose as a yeast nutrient. Thus the aeration of bread will depend on adequate α-amylase and β-amylase activity. Wheat flour α-amylase varies according to climatic conditions at harvest since a wet harvest promotes some germination of the wheat seeds and results in a considerable increase of α-amylase activity.

Addition of α-amylase is most likely to be necessary in countries where harvesting conditions do not promote germination and this enzyme becomes a limiting factor in fermentation (Pomeranz, 1968). The enzyme may be added as malted (germinated) grain but fungal enzyme preparations from *Aspergillus oryzae* are also used in Western countries while Russian sources refer to the use of enzyme preparations from *A. niger* and *A. awamori* (Reed, 1966).

The effects of α-amylase are not only on sugar formation but if excessive will impair loaf quality by overdextrinization. Fungal α-amylases are, in this respect, superior to bacterial amylases which, because of their greater heat-stability, are not inactivated rapidly in the earlier stages of baking and continue to act at temperatures which gelatinize starch in the loaf. The gelatinized starch is very susceptible to enzyme attack and too much dextrin is produced (Miller, Johnson & Palmer, 1953; Amos, 1955). The fungal enzyme is effective in promoting sugar production for the fermentation stages but is substantially inactivated in the oven before starch gelatinizing temperatures are reached.

Production of beer and spirits. In the traditional processes the sugar which yeast converts to alcohol is liberated from cereal starch by the action of α-amylase and β-amylase from malted barley, various minor enzymic conversions also occurring. In brewing barley malt has normally been used in quantities sufficient to ensure that hydrolysis of starch to sugar is not a limiting factor in fermentation. Addition of fungal enzymes has not therefore been extensively practised. Processes using fungal enzymes have however been patented (Dennis & Quittenton, 1962) and interest in substitution of fungal enzymes for malt continues (Wieg, Hollo & Varga, 1969; Harrison & Rowsell, 1970; Stentebjerg-Olesen, 1971). Processes include the use of the α-amylase of *Bacillus subtilis* (*liquefaciens*), with fungal amyloglucosidase. In this process 5 l of the enzyme preparation converts 1000 kg of starch to fermentable sugars in three days. The amyloglucosidase, which has a temperature optimum of 60–65°C and is active at pH values as low as 3, is obtained from a strain of *Aspergillus niger* (Aschengreen, 1969).

In some countries, such as Germany, legal definition of beer may restrict substitution of enzymes for malt. There would appear to be less difficulty in spirit production.

In spirit production, bran cultures of *Aspergillus awamori* and submerged cultures of *A. niger* have been used as sources of saccharifying enzymes to act on sorghum while in Germany and Russia fungal enzymes have been used to convert starch from potatoes, wheat and other sources. In the United States not only grain neutral spirits but also rye and whisky have been prepared (Reed, 1966).

Glucose syrup production. Glucose syrups are the products of starch hydrolysis and consist of concentrated solutions of glucose, maltose and higher oligosaccharides in varying proportions. The degree of hydrolysis is usually described in Dextrose Equivalents (DE) which is the reducing sugar content calculated as glucose and expressed as a percentage of the total solids. Such syrups have been prepared in the United States largely from corn starch, in Europe largely from potato starch and in Japan from both potato and sweet potato starch. Their uses in the food industry may require not only particular DE values but for specific purposes the ratio of glucose to maltose is also of importance (Palmer, 1970).

The acid process for hydrolysing starch was used from the early 19th century and gave a product having about 60% of glucose, maltose and oligosaccharides in roughly equal parts the rest being dextrins (the DE for this being a little over 40). Higher levels of hydrolysis were possible but tended to lead to bitter tastes. Over twenty years ago the use of fungal enzymes subsequent to acid hydrolysis raised glucose syrups to a DE of over 60 and it was subsequently found that combinations of α-amylase, malt β-amylase and amyloglucosidase would enable ratios of glucose to maltose to be varied (Langlois, 1959). Newer methods of using only enzymes for hydrolysis favour use of bacterial α-amylase which is stable at high temperatures followed by fungal amyloglucosidase. In one such process starch, in a slurry containing 27–33%, is heated for over half an hour at gelatinizing temperature and at pH 5.5–7 with bacterial α-amylase. After cooling to 60°C and adjusting the pH to 3.5–5, incubation with fungal amyloglucosidase for 72–96 h at 60°C takes place giving DE values over 97% (de Becze, 1970).

Transglycosylation. The hydrolytic enzymes acting on carbohydrates could be regarded as transferring enzymes in which the glycosyl group is transferred from a donor molecule such as a disaccharide to water acting as an acceptor molecule. At high concentrations of sugars enzymes normally acting as hydrolases may act as transglycosylases and produce oli- or oligosaccharides such as iso-maltose and iso-maltotriose.

This type of conversion is undesirable in glucose production from starch. Transglucosylase activity which can in some instances be attributed to specific enzymes is encountered in amyloglucosidase preparations from *Aspergillus* and *Rhizopus* species and it has been considered important to select strains so as to minimize this type of activity. Various methods of processing crude amyloglucosidase so as to diminish transglycolase activity have also been proposed (Kooi, Harjes & Gilkison, 1962; Hurst & Turner, 1962). A very extensive account of transglycosidase reactions by fungal and other microbial enzymes has been published (Spencer & Gorin, 1968).

Other fungal carbohydrases

β-GLUCANASE The use of a preparation of this enzyme from a fungal source to improve the filtration characteristics of beer has been described (Enkelund, 1972). This improvement is associated with hydrolysis of the β-glucans which are chains of glucose residues linked by β-1.4 and β-1.3 glycosidic links and therefore not susceptible to hydrolysis by α-1.4 glucosidases such as amylase or amyloglucosidase.

LACTASE OR β-GALACTOSIDASE This enzyme hydrolyses the β-1.4 glycosidic link of the disaccharide lactose (milk sugar) producing glucose and galactose on hydrolysis, thus giving a sweeter and more soluble product. *Aspergillus niger* and *A. oryzae* have been used as fungal sources and the fungal lactase has, at pH 5, a lower pH optimum and also higher activity at elevated temperatures than the corresponding enzymes from yeast and bacterial sources (Pomeranz, Robinson & Shellenberger, 1963).

INVERTASE Since sucrose is both an α-glucoside and a β-fructoside it can be hydrolysed to glucose and fructose by both α-glucosidases and β-fructosidases. Either or both enzymes may be found in fungal sources. While invertases may be obtained from *Aspergillus oryzae* and *A. niger* and other fungi, yeast would appear to be the usual source for major processes involving sucrose conversion in the confectionery industry, the object in general being a product which is sweeter and more soluble than sucrose.

The systematic names and Enzyme Commission numbers for the enzymes discussed in this section are given in Table 10.2.

Table 10.2 Systematic nomenclature and Enzyme Commission numbers for industrial carbohydrases

Enzyme	Enzyme Commission number	Systematic nomenclature
α-Amylase	EC.3.2.1.1	α-1,4-Glucan 4-glucanohydrolase
β-Amylase	EC.3.2.1.2	α-1,4-Glucan maltohydrolase
Amyloglucosidase	EC.3.2.1.3	α-1,4-Glucan glucohydrolase
Invertase (β-fructofuranosidase)	EC.3.2.1.26	β-D-Fructofuranoside fructohydrolase
Lactase (β-galactosidase)	EC.3.2.1.23	β-D Galactoside galactohydrolase
β-Glucanase	EC.3.2.1.6	β-1,3(4) Glucan glucanohydrolase
Transglycosylases	EC.2.4.1 enzymes, hexosyltransferases—but hydrolytic carbohydrases may show this activity.	

Pectinolytic enzymes

The pectic substances consist mainly of D-galacturonic acid residues linked by α-1,4 glycosidic links in linear chains. The carboxyl group may be esterified with methanol. According to accepted usage (American Chemical Society, 1944) the unmethylated polygalacturonic acid polymer is *pectic acid*, the substantially methylated colloid is *pectinic acid* and the insoluble pectic material as it occurs in the plant tissues is *protopectin*. Pectins are water-soluble pectinic acids of varying methyl ester content which may form gels with sugar and acid. In considering the action of hydrolytic enzymes on pectic substances one must bear in mind that mono-saccharide residues other than galacturonic acid are also present in the molecule (McCready & Gee, 1960; Aspinall, 1970).

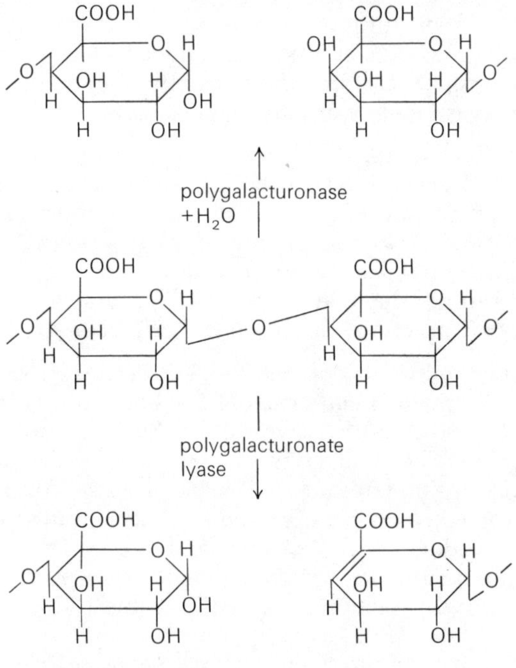

Fig. 10.1 Depolymerizing pectinolytic enzymes.

Since the pectic substances occur notably in the middle lamellae of plant cells they have a structural function in binding and supporting the cells. The nature and stability of this protopectin material has been of great interest in food technology (Joslyn, 1962).

Because of the complexity of the substrates development of precise knowledge of pectinolytic enzymes was slow. These enzymes occur widely in plants and micro-organisms and their nature and distribution is the subject of a recent comprehensive review (Rombouts & Pilnik, 1972). The industrial uses of pectic enzymes have also been discussed recently (Fogarty & Ward, 1972).

It was believed that the action of pectin-degrading enzymes was entirely hydrolytic until it was shown that transeliminative splitting of glycosidic links could occur by the action of a lyase (Albersheim, Newkom & Deul, 1960). Comparison of a lyase with the corresponding hydrolase is seen in Fig. 10.1.

The pectinolytic enzymes which split or hydrolyse the α-1,4 glycosidic bonds of the pectic substances can be classified on the basis of whether they are *exo* or *endo*, whether they are hydrolytic or trans-eliminative, and whether their specificity is towards methylated or unmethylated galacturonic acid polymers. The eight types are shown in Tables 10.3 and 10.4.

Table 10.3 Depolymerizing pectinolytic enzymes hydrolysing α-1,4-glycosidic links

Enzyme	Mode of action
1. Endo-polygalacturonase (Endo-PG)	Random hydrolysis of pectic acids
2. Exo-polygalacturonase (Exo-PG)	Hydrolysis at terminal linkage of pectic acids
3. Endo-polymethylgalacturonase (Endo-PMG)	Random hydrolysis of pectins.
4. Exo-polymethylgalacturonase (Exo-PMG)	Hydrolysis at terminal linkage of pectin chain

Table 10.4 Depolymerizing lyases splitting α-1,4-glycosidic links by trans-elimination producing C_4–C_5 unsaturation (as in Fig. 1)

Enzyme	Mode of action
1. Endo-polygalacturonate lyase (Endo-PGL)	Random splitting of pectic acids
2. Exo-polygalacturonate lyase (Exo-PGL)	Splitting at terminal linkage of pectic acids
3. Endo-polymethylgalacturonate lyase (Endo-PMGL)	Random splitting of pectins
4. Exo-polymethylgalacturonate lyase (Exo-PMGL)	Splitting at terminal linkage of pectin chain

In addition to these are the pectinesterases (EC.3.1.1.11) which hydrolyse off the methoxyl groups from the pectins.

Plants, fungi and bacteria possess different groups of these enzymes. The fungi possess most notably the endo-polygalacturonases which are of much less frequent occurrence in the bacteria. Many bacteria possess endo pectate lyases which are relatively infrequent in fungi although reported in *Fusarium* sp. (Rombouts & Pilnik, 1972). The authors doubt the existence of polymethylgalacturonases and suggest that those described in the earlier literature were probably pectin lyases. *Endo* pectin lyases which have been studied in various fungi including *Aspergillus* sp. (Ishii & Yokotsuka, 1971; Edstrom & Phaff, 1964) are not in general found in bacteria.

Exo-polygalacturonases occur also in a number of fungi including *Aspergillus niger* (Mill, 1966).

INDUSTRIAL USE Addition of pectolytic enzymes in industrial processes involve mostly conversions in fruit pulp and fruit juice. The fact that the fungal enzymes have a low pH optimum have made them particularly suitable for this purpose.

The organisms which have been used include various species of *Botrytis, Mucor, Penicillium* and the *Aspergillus*. Preparations vary in their proportions of the different pectolytic enzymes. It would appear that there is a more extensive breakdown of fruit pectins in commercial processing when the pectolytic enzymes are used together (Nyiri, 1969).

The extraction of berry fruit juice from pulp is aided and improved by destruction of the natural pectic substances present. Pectolytic enzymes can be added for this purpose normally at temperatures about 45°C and can, in a period of 2–4 h, produce breakdown of the natural pectic substances to D-galacturonic acid, digalacturonic acid and trigalacturonic acid. The pulp can then be pressed to give a readily filterable juice (Robbins, 1968). The various enzymes of this group will contribute to the degradation.

The uses of pectolytic enzymes for clarification of fruit juices and wines involve a number of complex factors which have been discussed by Reed (1966).

While commercial pectolytic preparations have been obtained mostly from fungi grown on solid media, Nyiri (1968, 1969) has described production of a commercial enzyme preparation from *Aspergillus ochraceus* grown in deep culture. The medium contained 2.0% wheat bran, 2.0% ammonium sulphate, 0.25% potassium dihydrogen phosphate, 0.25% yeast extract and 0.5% apple-pectin. (The addition of pectin as an inducer is common to most processes described.) Incubation was at pH 4 and 80 kg of crude pectolytic enzyme could be isolated from 10 000 dm^3 of filtered fermentation liquid.

Cellulases

Cellulases form a group of enzymes hydrolysing β-1,4 glycosidic links in cellulose or its derivatives. In general, the cell-free extracts of enzymes produce slow degradation of the substrate in contrast with intact microorganisms which may cause fairly rapid destruction. The advantages that would arise from the use of potent cellulase preparations to convert the enormous amounts of available cellulose to glucose are obvious. However, the situation is not analogous to the breakdown of starch by amylases since the β linkage of glucose residues in cellulose results in an insoluble crystalline structure. The nature of initial attack on such natural cellulose by cellulolytic enzymes is still obscure, but subsequent cleavage by β-1,4 glucanases and β-glucosidases follows a pattern analogous to the hydrolysis of starches and pectic substances.

Cellulolytic enzymes can be characterized as follows (King & Vessal, 1969).

(1) C_1 enzyme (unclassified) acting on highly oriented solid cellulose and exposing it to attack by other cellulases. This enzyme was first postulated as being formed by organisms which could grow with cellulose as sole carbon source in contrast to those which required a more assimilable form

such as carboxymethyl cellulose (Reese, Siu & Levinson, 1950). The C_1 component has been studied by a number of workers as a constituent of filtrates from *Trichoderma viride* and *T. koningii* and was isolated from *Penicillium furiculosum* (Selby, 1969). The nature of attack is still unknown.
(2) β-1,4 glucanases which will hydrolyse soluble cellulose derivatives such as carboxymethylcellulose and act synergistically with C_1 component on native cellulose.

These can be divided into exo-β-1,4 glucanases removing glucose units from the non-reducing end of the chains (sometimes placed in EC.3.2.1.21) and endo β-1,4 glucanases (EC.3.2.1.4) promoting hydrolysis of β-glycosidic links randomly within the chain.
(3) Cellobiases, acting on cellobiose and other β-dimers of glucose, are included within β-glucosidases (EC.3.2.1.21) although not all β-glucosidases act on cellobiose.

Most cellulolytic fungi, including *Penicillium* sp., *Aspergillus* sp., *Fusarium* sp., *Chrysosporium* sp. and members of the Basidiomycetes produce exo-β-1,4 glucanases, *Trichoderma viride* QM6a cultures are particularly active (Reese, 1969).

Endo-β-hydrolases of both β-1,4 and β-1,3 types have been reported in *Aspergillus niger* and *Trichoderma viride* by various workers and have been reviewed recently along with other β-glucanases (Barras, Moore & Stone, 1969).

INDUSTRIAL APPLICATIONS It appears that many potential uses of cellulases are not yet realized because preparations obtained so far, mostly from *Trichoderma viride* and *Aspergillus niger*, are not sufficiently active. Saccharification of cellulose by cellulases remains a laboratory process.

Nevertheless, cellulase preparations from European, American and Japanese manufacturers are available and Japan alone probably produces the better part of 100 tonnes annually. Their uses, including feed supplementation for cattle, digestive aids, extraction of green tea components, isolation of soya protein, modification of foods and production of single-cell preparations from vegetables are discussed by Tayama (1969). In some of these uses pectolytic and cellulolytic enzymes act together to degrade vegetable tissues. A recent United States patent cites the use of a mixture of cellulase, hemicellulase and pectinase for the conversion of food by-products to edible materials (Silberman, 1972).

The use of *Aspergillus niger* cellulase in a paper-making process was described in a patent (Bolaski & Gallatin, 1962), and the use of cellulases in preparation of coffee concentrates by Beckhorn *et al.* (1965).

Glucose oxidase

The term oxidase indicates an enzyme using oxygen as a hydrogen acceptor and forming usually water or hydrogen peroxide. Glucose oxidase (β-D-glucose: oxygen oxidoreductase, EC.1.1.3.4.) is a flavoprotein of this type and causes dehydrogenation of glucose at C_6 producing gluconic acid via the glucono-lactone and forming hydrogen peroxide with uptake of oxygen.

Unlike the enzymes hitherto discussed glucose oxidase is an intracellular enzyme and production entails the disruption of fungal mycelia.

Industrial extraction appears to involve grinding with an abrasive material and subsequent precipitation of enzyme from aqueous extract by addition of alcohol (Ward, 1967).

The enzyme was first isolated from *Aspergillus niger* which has been a major source but is produced by many other fungi, in particular the Penicillia. A process using *Penicillium vitale* grown with sugar-mineral medium for 7–9 days at 26–27°C was described as giving a very high enzyme yield (Pokrovskaya *et al.,* 1965). *P. amagasakaiense* has been claimed to give a product particularly free of mannose and galactose oxidases (Rosner & Foster, 1963). The name *notatin* has been used for glucose oxidase from *P. notatum.*

The pH optima for glucose oxidases are commonly between 5.5–5.8 and the enzyme shows a high specificity towards glucose as substrate. A purified *Aspergillus niger* enzyme oxidized mannose and galactose at rates which were 1% and 0.5% of the corresponding rate for glucose oxidation (Pazur & Kleppe, 1964).

While for practical uses the glucone-δ-lactone initially formed by the action of the enzyme might be assumed to form gluconic acid spontaneously, specific *lactonases* which catalyse this reaction are known (Spencer & Gorin, 1968).

APPLICATIONS Glucose oxidase has been used in a number of ways as an analytical reagent for glucose. It has, for example, along with peroxidase and gum guaiac, been compounded into test strips which change colour in the presence of glucose. This results from the action of peroxidase on the hydrogen peroxide produced causing oxidation of guaiac with consequent colour production (Rosner & Foster, 1963).

Several electrode systems incorporating glucose oxidase for measurement of blood glucose have been proposed. The nature of this application has recently been described (Guilbault, 1972).

In fermenter technology glucose oxidase systems can be used to measure oxygen transfer rates (Hsieh, Silver & Mateles, 1969).

In the food industry commercial applications have been found in desugaring egg products and in removing oxygen from mayonnaise, fruit juices and other products susceptible to deleterious oxidation (Reed, 1966). For uses such as removal of oxygen from the head-space of tinned food products, catalase is required also to destroy the hydrogen peroxide formed.

The proteases

The proteases or peptide hydrolases are enzymes hydrolysing the peptide links of proteins or peptides and like the amylolytic, pectinolytic and cellulolytic enzymes may be divided into endo-enzymes promoting attack within the chain of the macro-molecule and exo-enzymes hydrolysing off terminal amino-acids. However the position is more complex because of the heterogeneity of the amino-acid residues and the occurrence of specificity towards peptide links adjacent to particular amino-acids. This specificity is not absolute so that any strict systematic nomenclature is difficult to achieve.

The exo-enzyme types are α-amino-acyl-peptide hydrolases (E.C.3.4.1. enzymes) and peptidyl amino-acid hydrolases (E.C.3.4.2. enzymes) both splitting off terminal amino-acids, the former embracing aminopeptidases and the latter carboxypeptidases.

A separate class, dipeptide hydrolases, (E.C.3.4.3. enzymes) splits dipeptides and the endo-enzymes, peptidyl peptide hydrolases (E.C.3.4.4. enzymes) act on peptide links within the chain. Familiar animal proteases such as pepsin and trypsin are in this group, as in aspergillopeptidase A (E.C.3.4.4.17) which attacks bonds involving the carboxyl group of arginine or leucine.

The microbial proteases are also divided into three types according to their pH optima, which are pH 2–5 for the acid type, pH 7 for the neutral type and pH 9.5–10.5 for the alkaline.

The alkaline proteases have aroused much interest in recent years as detergent additives and are obtained mainly from strains of *Bacillus subtilis* or other bacteria. Commercial acid proteases on the other hand are mainly of fungal origin. The organisms which have been most studied and most used as sources are *Aspergillus saitoi, A. niger, A. oryzae, Trametes sanguineara* and *Mucor pusillus* (Keay *et al.*, 1972). Many other fungal acid proteases including those of *Rhizopus* spp. (Wang & Hesseltine, 1970) are known.

Despite the broad distinction between acid and alkaline protease sources made above, neutral and alkaline proteases also occur in the fungi and all three types are found in *Aspergillus oryzae*. The aspergillopeptidase A which has been crystallized from this organism seems to be the alkaline protease.

The neutral proteases which are metallo-enzymes, many containing zinc, are in general the least heat-stable of the three groups. The acid proteases from moulds show good heat stability at low pH values (Keay, 1971).

INDUSTRIAL APPLICATIONS The fungal proteases have found application mainly in the food industry and this is particularly true of the acid proteases which among other functions may substitute in the activities associated with pepsin, rennin and papain.

Papain has been used extensively for meat-tenderizing. Procedures have ranged from simply sprinkling the enzyme on meat to rehydration of the dried meat with enzyme solution and injection of proteases before slaughter. It appears that over 5% of United States beef is subject to tenderization procedures by packers. An account of current methods is given by Karmas (1970) who cites the use of *Aspergillus niger* and other microbial proteases.

Another use of proteases including those of fungal origin is in production of protein hydrolysates where attempts to control bitter flavours, probably due to peptide formation, have led to many studies on selection of suitable enzymes.

Fungal proteases are now being used increasingly in cheesemaking as a substitute for natural rennet. This enzymic material obtained from calf stomach is of limited supply and subject to fluctuating prices. The enzyme present, rennin, coagulates milk by hydrolysing off protective polypeptides from the casein micelle. The liberated casein forms a coagulum by interaction with calcium ions present and this is the primary event in the cheesemaking process.

The proteolytic action of the added enzyme may continue in the cheese and so influence taste by the formation of peptides. Many proteases will, in varying degrees, simulate rennin action but few are equally satisfactory for both clotting function and lack of subsequent production of undesirable flavours.

During the last few years, preparations from *Mucor miehei*, *Edothia parasitica* and *Mucor pusillus* have been successful enough to be marketed as rennets (Prins & Nielsen, 1970). Rennet activities of a number of other moulds have been discussed by Sardinas (1969).

Fungal proteases are used also in breadmaking and it has been stated that more than half the bread made in the United States is treated with enzymes derived from *Aspergillus oryzae*, the protease now being more important than the fungal amylase additives (Reed, 1966). The effect is on the visco-elastic properties of doughs, a reduction in viscosity and reduction in optimal mixing time being produced. This change in physical properties can be attributed to partial hydrolysis of the gluten-forming proteins.

Yet another use is in chill proofing bottled beer. This is subject to 'chill-haze', a precipitate formed by interaction of the protein and tannin present. Treatments with papain, bacterial protease and fungal protease have all been used. While papain appears to have been used most the acid proteases from fungi seem suitable for operation at the pH of beer which is about 4.5. A recent patent for a process for chillproofing malt beverages involves the use of insolubilized enzyme (Wildi & Boyce, 1972). It has been suggested by Beckhorn (1972) that, since some residual enzyme activity is required in the protease treatment for beer, ultimately a combination of soluble and insoluble enzyme may be used.

Minor applications of fungal enzymes

The applications of fungal enzymes already discussed are substantially those which are established. Other applications which are localized, restricted or only in tentative stages of development include addition of fungal lipases to cheeses, the use of naringinase to debitter citrus fruit juices, inclusion of microbial proteases in dentifrices and mouthwashes, and production of inosinic acid using 5′-phosphodiesterase. Japanese workers have been notably active in pioneering developments and a useful review is that of Sakaguchi *et al.* (1971).

Legal aspects

Since the bulk of fungal enzymes at present used are for food products, manufacturers must conform to food laws which differ from country to country. Care must be taken that the product is free of aflatoxins and pathogens. Fillers and additives must conform to normal standards, which in Britain, for example, will be as defined in the British Pharmacopeia.

In general since enzymes are normal constituents of uncooked foods their use as additives is widely accepted. *Aspergillus oryzae* proteases and carbohydrases, and *A. niger* carbohydrases, cellulases, glucose oxidase, pectic enzymes and lipases have all been permitted as additives in the

United States (Reed, 1966). Most enzyme additives will be inactivated before consumption and in any case would be expected to be digested with all the other dietary proteins. There appears to have been no evidence of harm from consumption of food enzymes in the diet.

It appears that concentrates of *Aspergillus oryzae* and *A. niger* enzymes will occur in a number of foods in quantities of up to 1000 p.p.m. (Beckhorn *et al.*, 1965).

Assay of industrial enzymes

The methods used for estimating potency of pure enzymes differ markedly from those used for crude ones. In the former case the International Union of Biochemistry (1965) have made recommendations that activity be measured in units, a unit being defined as the activity required to catalyse the transformation of one micromole of substrate in one minute, or, where more than one bond is attacked one micro-equivalent. Assays should be carried out at 30°C, the initial rate being measured at optimal pH and optimal substrate levels.

For crude enzymes where a mixture of enzyme of similar or synergistic function may exist as with the amylolytic or pectolytic types this approach is of little relevance. Here the major tendency is to adopt methods where measurement of enzyme potency can be carried out under conditions which are as close as possible to those under which the enzyme is to be used.

Over a number of years particular methods have gained general acceptance and a general appraisal of these and of the problem of assay has been made by Collier (1970).

10.4 Future Considerations

As pointed out previously, almost all of the fungal enzymes sold as industrial preparations are exo-enzymes, glucose oxidase being an exception.

It is becoming increasingly feasible to visualize large scale liberation of intracellular enzymes and it is interesting to speculate on whether a demand for them is likely to arise since a very small number of known enzymes are put to use.

All that can be said at present is that rising population will compel us in the search for food and avoidance of pollution to use the biosphere with greater precision and economy than in the past and this will involve a continued development of biotechnology at the molecular level.

New enzymic conversions will be in demand where specificity of chemical conversion cannot be readily achieved by non-enzymic methods as in the asparaginase treatment for leukemia.

Most of the intracellular enzymes however operate as entities within multi-enzyme systems and there still appears to be a large gap between the utilization of such systems outside the cell and our present capabilities.

10.5 References

ALBERSHEIM, P., NEWKOM, H., & DEUL, H. (1960). Über die Bildung von ungersättigten Abbauprodukten durch ein pectinabbaundes Enzym. *Helvetica Chimica Acta* 43, 1422–6.

AMOS, J.A. (1955). The use of enzymes in the baking industry. *Journal of the Science of Food and Agriculture* 6, 489–95.

ARIMA, K. (1964). Microbial enzyme production. In *Global Impacts of Applied Microbiology*, pp. 277–94. Ed. Starr, M.P. New York: Wiley.

ASCHENGREEN, N.H. (1969). Microbial enzymes for alcohol production. *Process Biochemistry* 4 (8), 23–5.

ASPINALL, G.O. (1970). *Polysaccharides.* Oxford: Pergamon Press.

BARKER, S.A. & EPTON, R. (1970). Water-insoluble enzymes. *Process Biochemistry* 5 (8), 14–19.

BARRAS, D.R., MOORE, A.E. & STONE, B.A. (1969). Enzyme-substrate relationships among β-glucan hydrolases. In *Cellulases and their Applications,* pp. 105–38. Ed. Gould, R.F. The American Chemical Society, Washington.

BECKHORN, E.J. (1967). Production of microbial enzymes. In *Microbial Technology*, pp. 366–80. Ed. Peppler, H. J. New York: Reinhold. Publishing Corporation.

BECKHORN, E.J. (1972). Speculations on the commercial future of immobilised enzymes. In *Enzyme Engineering*, pp. 355–9. Ed. Wingard, L.B. New York: Interscience Publishers.

BECKHORN, E.J., LABEE, M.D. & UNDERKOFLER, L.A. (1965). Production and use of microbial enzymes. *Journal of Agricultural and Food Chemistry* **13**, 30–4.

DE BECZE, G.I. (1970). Food enzymes. *Critical Reviews in Food Technology* **1** (4), 479–518.

BOLASKI, W. & GALLATIN, J.C. (1962). Enzymic conversion of cellulosic fibers. *United States Patent 3,041,246.*

CHIAN, E.S.K. & SELLDORFF, J.T. (1969). Ultrafiltration of biological materials. *Process Biochemistry* 4 (9), 47–51.

COLLIER, B. (1970). How active are commercial enzymes? *Process Biochemistry* 5 (8), 39–42.

CUATRECASAS, P. & AFINSEN, C.B. (1971). Affinity chromatography. *Annual Review of Biochemistry*, 259–78.

DAVIES, R. (1963). Microbial extracellular enzymes. In *Biochemistry of Industrial Micro-organisms*, pp. 68–150. Ed. Rainbow, C. & Rose, A.H. London and New York: Academic Press.

DENNIS, G.E. & QUITTENTON, R.C. (1962). Enzymes in brewing. *Canadian Patent 634,865.*

EDSTROM, R.D. & PHAFF, H.J. (1964). Purification and certain properties of pectin trans-eliminase from *Aspergillus fonsecaeus*. *Journal of Biological Chemistry* **239**, 2403–8.

ENKELUND, J. (1972). Externally added β-glucanase. *Process Biochemistry* 7 (8), 27–9.

FOGARTY, W.M. & WARD, O.P. (1972). Pectic substances and pectinolytic enzymes. *Process Biochemistry* 7 (8), 13–7.

GUILBAULT, G.G. (1972). Analytical uses of immobilized enzymes. In *Enzyme Engineering*, pp. 361–76. Ed. Wingard, L.B. New York: Interscience Publishers.

HARRISON, J.G. & ROWSELL, J.E. (1970). Glucoamylases in modern brewing. *Process Biochemistry* **5** (4), 37–9.

HSIEH, D.P.H., SILVER, R.S. & MATELES, R.I. (1969). Use of glucose oxidase system to measure oxygen transfer rates. *Biotechnology and Bioengineering* **11**, 1–17.

HURST, T.L. & TURNER, A.W. (1962). Method of refining amyloglucosidase. *United States Patent 3,047,471.*

ISHII, S. & YOKOTSUKA, T. (1971). Pectin trans-eliminase with fruit juice clarifying activity. *Journal of Agricultural and Food Chemistry* **19**, 958–61.

JOSLYN, M.A. (1962). The chemistry of protopectin. *Advances in Food Research* **11**, 2–107.

KARMAS, E. (1970). *Fresh meat processing.* Noyes Data Corporation, New Jersey.

KATCHALSKI, E., SILMAN, I. & GOLDMAN, R. (1971). Effect of micro-environment on the mode of action of immobilised enzymes. *Advances in Enzymology* **34**, 445–454.

KAY, G. (1968). Insolubilised enzymes. *Process Biochemistry* **3** (8), 36–9.

KEAY, L. (1971). Microbial proteases. *Process Biochemistry* **6** (8), 17–21.

KEAY, L., MOSELEY, M.H., ANDERSON, R.G., O'CONNOR, R.J. & WILDI, B.S. (1972). Production and isolation of microbial proteases. In *Enzyme Engineering,* pp. 63–92. Ed. Wingard, L.B. New York: Interscience Publishers.

KING, K.W. & VESSAL, M.I. (1969). Enzymes of the cellulase complex. In *Cellulases and their Applications,* pp. 7–25. Ed. Gould, R.F. American Chemical Society, Washington.

KOOI, E.R., HARJES, C.F. & GILKISON, J.S. (1962). Treatment and use of enzymes for the hydrolysis of starch. *United States Patent 3,042,584.*

LANGLOIS, D.P. (1959). Process for preparing starch syrups. *United States Patent 2,891,869.*

LILLY, M.D. & DUNNILL, P. (1972). Engineering aspects of enzyme reactors. In *Enzyme Engineering*, pp. 221–7. Ed. Wingard, L.B. New York: Interscience Publishers.

MCCREADY, R.M. & GEE, M. (1960). Determination of pectic substances by paper chromatography. *Journal of Agriculture and Food Chemistry* **8**, 510–3.

MILL, P.J. (1966). The pectic enzymes of *Aspergillus niger*. *Biochemical Journal* **99**, 557–65.

MILLER, B.S. & JOHNSON, J.A. (1954). Differential inactivation of enzymes. *United States Patent 2,683,682.*

MILLER, B.S., JOHNSON, J.A. & PALMER, D.L. (1953). A comparison of cereal, fungal and bacterial α-amylases as supplements for bread baking. *Food Technology* 7, 38–42.

NYIRI, L. (1968). Manufacture of pectinases, Part I. *Process Biochemistry* 3(8), 27–30.

NYIRI, L. (1969). Manufacture of pectinases, Part II. *Process Biochemistry* 4(8), 27–30.

PALMER, T.J. (1970). Glucose syrups in foods. *Process Biochemistry* 5(5), 23–4.

PAZUR, J.H. & KLEPPE, K. (1964). Oxidation of glucose and related compounds by glucose oxidase from *Aspergillus niger*. *Biochemistry* 3, 578–83.

POKROVSKAYA, N.V., OGANEZOVA, N.A., CHISTYAKOVA, E.A. & KISLYAKOVA, O.V. (1965). *Fermentinaya i Spirt. Prom.* 31, 22: *Chemical Abstracts* 56, 6482.

POMERANZ, Y. (1968). Relation between chemical composition and bread-making potentialities of wheat flour. *Advances in Food Research* 16, 335–455.

POMERANZ, Y., ROBINSON, R.J. & SHELLENBERGER, J.A. (1963). Evaluation of β-galactosidase (lactase) activity by paper chromatography. *Enzymologia* 25, 157–66.

PRINS, J. & NIELSEN, T.K. (1970). Microbial rennet. *Process Biochemistry* 5(5), 34–5.

REED, G. (1966). *Enzymes in food processing*. New York and London: Academic Press.

REESE, E.T. (1969). Estimation of exo-β-1,4,glucanase in crude cellulase solutions. In *Cellulases and Their Applications*, pp. 26–33. Ed. Gould, R.F. American Chemical Society, Washington.

REESE, E.T., SIU, R.G.H. & LEVINSON, H.S. (1950). The biological degradation of soluble cellulose derivatives. *Journal of Bacteriology* 59, 485–97.

ROBBINS, R.H. (1968). Clarification of fruit juices. *Process Biochemistry* 3(5), 38–40.

ROMBOUTS, F.M. & PILNIK, W. (1972). Research on pectin depolymerases in the sixties—a literature review. *Critical Reviews in Food Technology* 3(1), 1–26.

ROSNER, L. & FOSTER, R.O. (1963). Method for detecting the presence of glucose in cervical mucus. *United States Patent 3,116,223.*

SAKAGUCHI, K., VERNURA, T. & KINOSHITA, S. (1971). *Industrial aspects of fermentation*, Kodansha Ltd., Tokyo.

SARDINAS, J.L. (1969). New sources of rennet. *Process Biochemistry* 4(7), 13–21.

SELBY, K. (1969). The purification and properties of the C_1 component of the cellulase complex. In *Cellulases and Their Applications*, pp. 34–52. Ed. Gould, R.F. American Chemical Society, Washington.

SILBERMAN, H.C. (1972). Food by-products converted to edible material with an enzyme mixture of cellulase, hemicellulase and pectinase. *United States Patent 3,615,721.*

SPENCER, J.F.T. & GORIN, P.A.J. (1968). Microbiological transformations of sugars and related compounds. *Progress in Industrial Microbiology* 7, 178–220.

STENTEBJERG-OLESEN, B. (1971). Microbial enzymes in brewing. *Process Biochemistry* 6(4), 29–31.

THOMSON, A.R. (1971). Recent developments in separation methods. *Process Biochemistry* 6(9), 35–8.

TAYAMA, N. (1969). Applications of cellulases in Japan. In *Cellulases and Their Applications*, pp. 359–90. Ed. Gould, R.F. American Chemical Society, Washington.

UNDERKOFLER, L.A. (1969). Development of a commercial enzyme process: glucoamylase. In *Cellulases and Their Applications*, pp. 343–58. Ed. Gould, R.F. American Chemical Society, Washington.

WANG, M.L. & HESSELTINE, C.W. (1970). Multiple forms of *Rhizopus oligosporus* protease. *Archives of Biochemistry and Biophysics* 140, 459–63.

WARD, G.E. (1967). In *Microbial Technology*, pp. 208–23. Ed. Peppler, H.J. New York: Reinhold Publishing Corporation.

WIEG, A.J. (1969). Enzymes in washing powders. *Process Biochemistry* 4(2), 30–4.

WIEG, A.J., HOLLO, J. & VARGA, P. (1969). Brewing beer with enzymes. *Process Biochemistry* 4(5), 33–8.

WILDI, B.S. & BOYCE, D.C. (1972). Chillproofing of malt beverages. *United States Patent 3,597,219.*

CHAPTER 11

The Cultivation of *Agaricus bisporus* and other Edible Mushrooms

W. A. HAYES and N. G. NAIR

11.1 Introduction

Scepticism and prejudice have, for centuries, been associated with mushrooms, probably dating back to the time when man, on the basis of trial and error, began selecting food for his own use from the plants that grew in his surrounds. In the course of learning to distinguish those foods which sustained life and health from those which were poisonous, it can be assumed that many accidents occurred, some fatal. At a later stage of Man's history, the hieroglyphics of the Egyptians record legends showing belief that mushrooms were the plants of immortality, and the Pharaohs respecting the delicious flavour decreed they should never be touched by a commoner. Prolongation of life and aphrodisiac qualities, sometimes attributed to mushrooms, led Julius Caesar to issue an edict forbidding any of his troops other than the Captain of Cohorts to eat the plant. Mushrooms have even been associated with social distinction for many centuries; the epicures of Rome and the royalty of France and Britain permitted only the courts and palaces to serve them. In addition, civilizations from many parts of the world, Central America, Mexico, China, Siberia, Greece and Russia all practised mushroom rituals. Many believed that mushrooms contained properties which would confer the ability to find lost objects, to heal the sick, produce supernatural strength and aid the soul in reaching the realms of the Gods. These fascinations do not end with the past; a recently published book *The Sacred Mushroom and the Cross* by J. M. Allegro, an expert etymologist and scholar on Sumerian and Middle Eastern languages, argues that at the heart of the religion of fertility and nature worship of the Israelites and their inheritors, the Christians, was the 'cult of the sacred mushroom'.

It is therefore surprising that the artificial culture of mushrooms should ever be considered in an industrial context. Misapprehensions are however rapidly disappearing and in the last decade the art of mushroom cultivation has given way to a sophisticated technology, which like many other agri-industrial activities, has revolutionized outlooks and social attitudes. Where mushrooms were regarded as a luxury food, they are now being served regularly in the homes as well as in the restaurants. In most Western countries consumption rates per capita are increasing at the rate of ten per cent per annum. Consumption is in excess of local production in many countries such as West Germany and U.S.A., and with the general acceptance of the processed product in cans or in a dehydrated form, a world trade in mushrooms is rapidly expanding. This expansion in growing has happened not only in the traditional mushroom growing countries such as France, U.S.A. and Great Britain, but also in other countries new to growing notably Taiwan, Australia, New Zealand, Ireland and some Central European countries. Mushroom cultivation is also now being established in other regions of the world such as Korea, China, South America and certain Mediterranean countries. World production is estimated to be in excess of 300 000 metric tonnes per annum and an annual growth rate of at least ten per cent per annum is likely to continue world wide.

Most of the important edible species are members of the genus *Agaricus*, the type genus of the family Agaricaceae, class Basidiomycetes. Industrial exploitation is confined almost entirely to the species *Agaricus bisporus* (Lange) Sing. which according to Singer (1961) exists in the wild on soils rich in nitrogen, in greenhouses, near to manure heaps, on roadsides, in gardens and parks. It is known to exist in the wild in most European countries, Siberia and North Africa and its natural geographic area extends all over the Northern hemisphere outside the tropics and the arctic.

The basic life cycle of higher fungi is from spore to mycelium, to fruit body and following a meiotic division back to spore again. As the specific epithet *bisporus* indicates, the basidia bear only two spores, instead of the usual four of the Basidiomycetes, and as a result monokaryons are not formed (Fig. 11.1). In the Basidiomycetes the production of fruit-bodies is governed by genetic factors located in the nuclei, homothallic species possessing only one type of nucleus, while in heterothallic species the presence of two nuclear types with different incompatibility factors (A1 and A2) is essential for the production of fruit-bodies (Raper, 1966). The mushroom of cultivation, *Agaricus bisporus*, is heterothallic and from the cytological and genetic studies of Colson (1935), Evans (1959), Fritsche (1964), Raper & Raper (1972), and Elliot (1972) the behaviour of the nuclei throughout the life cycle can be postulated. In the vegetative tissue each cell is multinucleate and the nuclei are haploid. Two compatible nuclei are cut off in the terminal cells of the hymenium. These two nuclei fuse to form a short-lived diploid stage in the life cycle. Shortly after fusion, normal meiosis occurs resulting in four nuclei. Typically the meiotic nuclei migrate in pairs into each spore. If the spore contains factors A1 and A2, the mycelium produced from the germination of this spore is fertile; but if the spore contains two nuclei with only the A1 factor or the A2 factor, the mycelium formed by the germination of this spore is sterile.

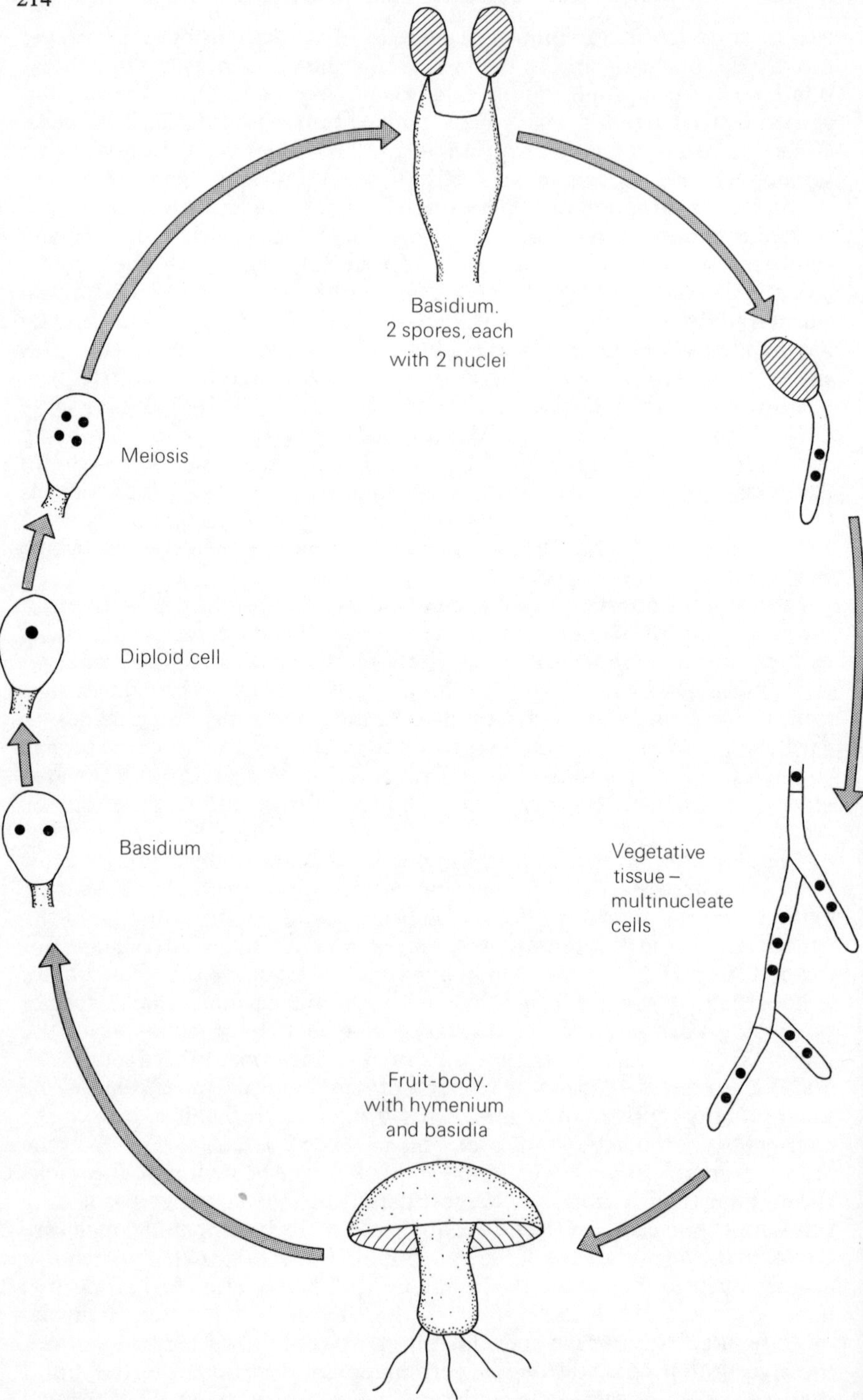

Fig. 11.1 Life cycle of *Agaricus bisporus*, the cultivated mushroom.

Nuclear control of the formation of the fruit-body, which is the crop for harvest and consumption, dictates the methods by which commercial strains (spawn) are obtained and propagated for cultivation. Spawn production is the first critical stage in the cultivation process, which provides the basis for the guarantee of edibility inherent in artificial culture, a feature which distinguishes the cultivated from those mushrooms gathered from the field. Single spore cultures are usually sterile and do not produce fruit-bodies unless a mating occurs through the fusion of two mycelia with compatible nuclei. Multiple spore cultures are usually fertile and are therefore used to establish mycelial cultures from which selections are made for propagation and establishment of strains used in commerce. Alternatively, mycelium isolated from the stipe, pilei or gills are used for selection and propagation purposes.

Pure cultures of *Agaricus bisporus* when propagated on to a suitable substrate remain in the vegetative stage for a minimum of 28 days before the onset of the reproductive or fruit-body stage which provides the crop for harvest. If the correct cultural conditions are maintained, fruit-bodies are produced in a series of *breaks* or *flushes* at approximately weekly intervals. In commerce the yield capacity of the culture diminishes with time and it is usual to terminate cropping after five or six weeks.

11.2 Early History of Cultivation

The first efforts of cultivating mushrooms are known to have taken place in France during the lifetime of Louis XIV of France (1683–1715). Underground caves in Paris provided a suitable environment and such caves are extensively used even in present times (Fig. 11.2). The natural self-heating properties of manure obtained from horse stables was found to be a good medium and was stacked into ridge beds in rows on the cave floors. These beds were inoculated with soil permeated with mycelium which was present in soils adjacent to old manure heaps or in situations where horses congregated. A popular source of inoculum was from the soil surrounding walk-tracks of horses used for working mills.

The art was soon to be exploited in England and Abercrombie in 1779 vividly described the uncertainties of culture in the 18th century. He reports that some of his contemporary growers 'make their beds under an airy covered shed, or barn or they erect a sort of awning of canvas; some also having a considerable range of glasshouses make them in these departments. I, however, have always found success in the open ground, and generally much better than when under any covering'. Despite Abercrombie's misgivings his contemporaries were originating the concept of protected mushroom culture in purpose built structures.

The importance of the mushroom as a delicacy to the 19th century aristocracy is revealed in the later writings of Callow (1831). He states 'the agreeable flavour of mushrooms makes them be in general request and to provide them throughout the year is an object of emulation amongst gardeners. Indeed such a supply is almost essentially necessary in every large establishment.' Callow further describes a house of a peculiar construction which was warmed by fire heat and 'well adapted to the growth of mushrooms throughout the year' and which was modified by himself to what is essentially a progenitor of what is now termed the shelf system of

Fig. 11.2 Print of mushroom growing and gathering for market in Paris caves during the early part of 19th century. (By Courtesy of R. Snetsinger, Pennsylvania State University.)

culture and is widely used even today. The shelves which carried the mushroom beds were arranged one above the other on brackets attached to the wall.

After the American Civil War, mushroom growing became established in the New World and it seems as if the techniques were introduced from Western Europe by English, French and Scandinavian gardeners employed by the wealthy citizens of New York and Philadelphia. Although many different kinds of buildings were used, greenhouses, usually thatched to provide insulation, were popular both in England and America in the early part of the 20th century. These early efforts of mushroom growing were unpredictable. Crop failures were common but quite acceptable since the return from an occasional successful crop fully compensated for such losses.

Many different locations and buildings are amenable to mushroom cultivation, *e.g.* disused railway tunnels, coke and limestone ovens, cellars, natural caves and man-made sheds *etc.* and as a result many diverse methods, systems and scales of growing operations have emerged.

The recent rapid change in the output and consumption of mushrooms has stemmed directly from improved and predictable methods of production. Even in the immediate post-war years, production methods were only a slight improvement on the hazardous techniques employed in the early part of the century, when crop failures were frequent. Over the last two decades the growth of the industry has been largely the result of intensification. While new knowledge has diversified techniques of growing to include simple low cost systems, which are being successfully applied in many countries new to mushroom cultivation, the large modern plants of today introduce a degree of precision into the production methods which is unique in horticulture and which are in many ways comparable to the production line methods employed in many manufacturing industries.

However, the base on which modern and predictable methods of culture are based, can be identified with improvements brought about during this century with many different facets of culture. These improvements are the result of the combined efforts of amateur enthusiasts, progressive growers and scientists.

11.3 Protected Cropping in Mushroom Growing Houses

Contrary to popular belief, darkness is not a prerequirement but mushrooms do require special conditions only provided by well ventilated caves or suitably constructed growing rooms. At about the end of the second decade of this century the mushroom growing community of Pennsylvania in U.S.A. adopted a standard house for mushroom growing. These buildings were constructed of wood or of hollow tile to provide insulation and conformed to a standard size of approximately 5.5 × 18.4 × 4.6 m high, sufficient to crop 464.5 m^2 of growing space in tiers of fixed shelves. These were called 'standard singles' and had a sloping roof and a natural system of ventilation. Somewhat later, growers began building two of these units under one roof, termed 'standard doubles'. The standard American mushroom growing house is now generally adopted in America and Canada and its design provided the basic principles for protected and continuous mushroom growing.

A typical small mushroom farm using the American style growing house consists of a series of singles or doubles and today is usually of cement block construction. The mushroom beds are about two metres wide with an aisle on each side and at the ends for picking, and watering. The beds are arranged in tiers of up to seven or eight and a wooden catwalk is provided for picking the upper beds. Heat is provided by circulated hot water and typically such houses are used for two crops a year and includes a mid-hot season break. The system of growing which has evolved with the American Standard is termed the 'shelf system' and is widely adopted throughout the world. Although successful it is generally regarded as being inflexible and because of the labour required to fill the fixed shelves it imposes a limit to the scale of the growing operation (Fig. 11.3).

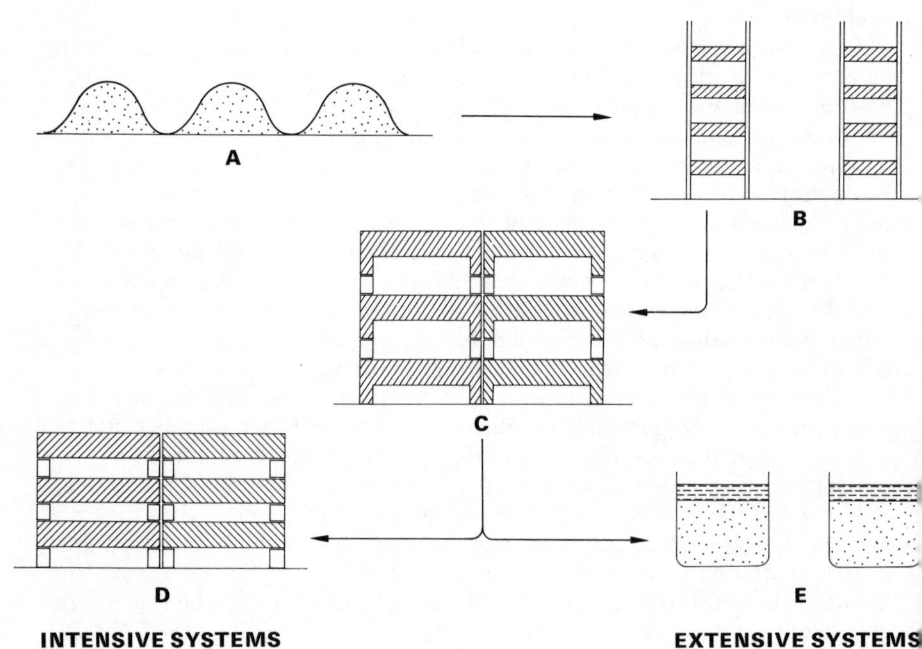

Fig. 11.3 The evolution of growing systems. A, Ridge beds in caves on the floors or on open ground. B, Shelf beds fixed to a purpose built mushroom growing house. Compost processing, spawn running, growing and harvesting *in situ*. C, Trays transportable from compost processing to spawn running and growing rooms. Sufficient space between trays for crop harvesting *in situ*. D, Tray transportable from compost processing to spawn running, growing rooms and to purpose built parlours for harvesting. E, Disposable containers. Compost processed in trays, filled into containers such as polythene bags and transported to purpose built growing rooms, old glasshouses, disused aircraft hangars or low cost houses constructed of polythene.

Shortly after the Second World War the tray system was developed. With the tray system, the bed is effectively split into movable portions contained in boxes which can be palletized and transported to the growing houses. This important development allows greater use of mechanization, a substantial increase in the potential size of a growing unit and a greater degree of capital investment. As a result, growing units cropping up to 20 000 m² of bed area are common and this system has directly contributed to the expansion of mushroom outputs in America and many European countries. Unlike the shelf method, containerized cropping in trays provides flexibility into the general system of growing and its application in commerce has tended to diversify general systems of growing. Today, a variety of houses of different construction and design exists to accommodate a variety of tray arrangements (Fig. 11.3).

A significant recent development is the use of plastic film materials both in the construction of growing houses and as containers for growing. Structures made from polythene film between an insulating layer of fibre glass or polystyrene are inexpensive and durable even in difficult climates. Furthermore, replacing the traditional wooden tray or shelf with cheap polythene containers which are disposed of at the end of crop minimizes the capital outlay required for mushroom growing.

11.4 Pure Culture Inoculum

None of the criteria normally recommended to distinguish poisonous from edible field-gathered mushrooms can be applied by an untrained person and the commercial exploitation of mushroom culture is clearly linked to the assurance provided by the methods developed for the culture of *Agaricus bisporus*. The production of a pure culture inoculum or spawn is a key operation in the success or otherwise of commercial culture. The original methods of obtaining spawn by digging up mycelium from meadows where wild agarics grew was unreliable and a spawn derived from pieces of colonized soil taken from established mushroom beds and known as 'flake spawn' was equally unreliable.

Improvements in spawn cultivation were however introduced by the 'brick spawn method' devised in England, and later adopted in America, in the early part of this century. The clay moulds, similar to those used for fire brick manufacture were used to shape finely ground and wetted manure before drying either artificially or more commonly by exposing them to the sun. This brick was inoculated with *Agaricus bisporus* and when colonized was smashed into many small pieces which were then used to sprinkle the beds.

At about the same time Ferguson (1902) described the process of spore germination in *Agaricus bisporus* and Duggar (1905) described how by removing a piece of tissue from a mushroom cap and culturing it on a suitable medium under sterile conditions, pure cultures of *A. bisporus* could be maintained to provide suitable inoculum for the bricks. However, despite the certainty of the inoculum these spawns were often contaminated by moulds, mites and insects and yields were consequently poor.

The use of milk bottles to produce a sterilized substrate for inoculation proved to be a significant development which eliminated the risk of contaminated spawn. The manure filled bottles were plugged with cotton wool and sterilized so as to kill harmful moulds, insects and bacteria. After the medium cooled, it was inoculated before incubation at about 21°C to encourage the growth of the inoculum.

The production of grain spawn according to the method of Sinden (1932) is now universally used as the means of vegetatively propagating mushroom mycelium. In this method grain to which water and chalk is added, is used as the base medium and pure cultures are maintained in the same way as that used for manure spawn. Rye is the most commonly used grain but wheat, millet and sorghum are equally satisfactory. The production of grain spawn is now a highly mechanized and specialized process and the application of modern aseptic techniques have eliminated sources of contamination which were common when manure was used as the base medium. Grain spawn can be safely distributed throughout the world from any one spawn manufacturer. It is claimed that the reservoir of nutrient available in the grain accelerates the growth of mycelium in the beds and its granular nature allows it to be mixed throughout the compost instead of being planted on the surface. This practice known as through spawning has become standard procedure in most modern production units.

11.5 Substrates and Pasteurization

The nutrients for mushroom growth are provided in composts which traditionally are prepared from horse manure and wheat straw. Spawn will develop mycelium in unfermented horse manure but satisfactory development is prevented by the self-generation of heat and competition from other micro-organisms. It is therefore necessary to compost the manure in order to produce a medium which will remain stable and in which the quantity of readily available nutrients for competing organisms is considerably reduced. Preparing such composts is probably the most difficult procedure in mushroom growing and consistently satisfactory results are obtained only after much experience.

Preparation of composts and microbial activity

Composting techniques currently employed stem directly from the trial and error approach adopted by growers and experimenters in the early decades of this century. But, before even the basic principles of the composting process were in any way appreciated, the availability of horse manure to the developing American industry became critical and the replacement of horses by cars, trucks and tractors during the 1930's led to a search to secure a replacement mix. A successful mix which is frequently used in America even today, consists of corn-cobs, legume hay, gypsum, ammonium nitrate, muriate of potash and dried brewers grains (Sinden, 1946). Other synthetic formulations devised later have been used successfully in other countries especially a mix described by Edwards (1949) consisting of wheat straw and dried blood as major ingredients.

However, in the post Second World War years increased numbers of pleasure horses and the expanding numbers of race tracks have increased the availability of horse manure. While synthetic mixtures are used to supplement local or seasonal shortages, most farms rely on at least part horse manure as an ingredient, although other animal manures, *e.g.* chicken manure, pig manure, and sheep manure are sometimes used as ingredients of straw based composts.

A variety of techniques and formulations exist for composting, all of which incorporate the base principles outlined by Lambert (1941). He identified the major physical and environmental variables which contributed to successful composting. In stacks of composting mixtures he found four main areas in which different conditions existed, which arose largely from the effect of aeration or a lack of it on microbial activity. At the base of the stack an acid mixture resulted, in which the carbon dioxide concentration was at its highest and the temperature at its lowest (37–55°C). In the hottest areas (65–80°C) decomposition was rapid and produced a neutral or alkaline compost. The outer layer, exposed to the air, had the highest oxygen content, but was generally cool (45°C), with very little decomposition taking place. Lambert found that compost produced at temperatures between 50–60°C was the most suitable for mushroom growing providing conditions favoured aerobic fermentation. The anaerobic core and the areas of extreme heat produced poor quality compost, but if subsequently subjected to aerobic fermentation at 50–60°C the quality improved. Thus Lambert concluded that a procedure known in the industry as peak-heating during which the entire compost is kept at 50–60°C under aerobic conditions, must be an integral part of the composting process.

The stacks investigated by Lambert were 3.7 m or more wide and about 1.2 m high. In order to avoid having an anaerobic core the width of the stacks can be reduced and some growers use a system of ventilation at the base.

Techniques were perfected by Sinden & Hauser (1950*a*, 1954) in a procedure known as the short method of composting by which the zoning described by Lambert was almost eliminated, by using stacks of smaller width (about 1.8 m). These workers laid great stress on the importance of the size, shape and compactness of the stacks having found that stacks could be made larger with loose long material than with dense substances, and that the stacks should be smaller in summer than in winter. In the short method, Lambert's concept of peak-heating is also applied and an economy of materials is claimed. The principles outlined in the method are now widely adopted.

In present day composting, although techniques vary in detail, they conform to a general schedule involving two stages. Stage I is normally done out of doors where the straw and manure are arranged in stacks and water as well as nitrogenous supplements are added to initiate the fermentation. Stacks are regularly turned to maintain aerobic conditions and to ensure thorough mixing of the ingredients. During this stage high temperatures are obtained in the stack, the result of intense microbial activity, which in turn selects for specific groups of thermophilic micro-organisms (Fig. 11.4).

Stage II which represents the peak-heat stage, is done in specially constructed buildings (called heat rooms) if trays are used for growing, or in the growing rooms where mushrooms are grown in shelves fixed to the building. Aeration and temperature control are important features of the Stage II process.

Although Sinden & Hauser (1954) recommended for their system of short composting a nitrogen content of the initial mixture of about 1.5% on a dry matter basis and in Edwards' (1949) formula, based on dried blood and wheat straw the level of nitrogen recommended was 2%, little attention was given to the detail of ingredient formulation. Pizer (1937) and Pizer & Thomson (1938) attached importance to the dispersion of the various constituents of compost and found that when the colloids were dispersed, growth of mushroom mycelium was poor. The addition of calcium sulphate to flocculate the colloids encouraged a strong growth of

Fig. 11.4 Composting for modern mushroom production on covered yards. (By courtesy of Burcros Limited, Camberley, Surrey).

mycelium. As a result of adding calcium sulphate (gypsum) to compost a more granular structure is obtained and its ability to retain water is increased.

Evidence obtained from a number of workers, Waksman & Nissen (1932), Burrows (1951), Gerrits, Bels-Koning & Muller (1967), and Hayes (1969*a*), shows that energy giving (containing carbon) materials are primary nutrients in composts for mushroom production. Nitrogen, although important, is secondary, a basic feature which contrasts with the soil/green plant-relationship in which nitrogen is the major nutrient required for growth.

During both Stages I and II of composting many changes in the composition and numbers of the natural micro-flora have been shown to occur (Hayes, 1969*a*). Mesophiles flourish at ordinary temperatures, but soon give rise to a predominantly thermophilic flora which is specific to mushroom composts and other heat-generating fermentations. Thermophilic bacteria, which utilize simple forms of energy, like sugars and amino acids, outnumber the Actinomycetes, a group which can utilize more complex forms of energy, like cellulose and hemicellulose. These Actinomycetes, commonly referred to as *fire fang*, flourish in the later stages of composting, especially during Stage II. Fungi, which are also active as decomposers of complex sources of energy, remain at a constant level throughout Stage I, but their numbers increase slightly in Stage II.

These changes are orderly and consistent, but variations from the general pattern occur according to (a) the nutrients available for microbial exploitation during composting, and (b) temperature conditions prevailing in the composting mass.

Concurrently with the build up of vast microbial populations some of the chemical components are transformed into microbial tissues which together with what remains of the mixture, forms the subsequent substrate for mushroom growth. Much nutriment is lost, however, in gaseous by-products—carbon dioxide and ammonia—and much energy is released as heat. This results in an overall loss of dry matter during composting which is accounted for, firstly by the breakdown of simple and most readily available forms of energy *e.g.* sugars and secondly, by the removal of lignins, cellulose and hemicellulose fractions of straw (Gerrits *et al.*, 1967; Hayes & Randle, 1968).

This understanding of the components which function in composting led to experiments designed to shift the pattern of the microbiological sequence in the composting fermentation in favour of the organisms which are unable to break down cellulose and hemicellulose (Hayes, 1969*a*). Increasing the levels of soluble carbohydrates by sucrose supplementation favoured high populations of bacteria at the expense of Actinomycetes, with corresponding increases in the levels of lignin, cellulose and hemicellulose fractions, a conservation of nutrients which was associated with increased mushroom yields. A net conservation of nitrogen also resulted, less nitrogen being released as ammonia, a feature which was also exploited in the time taken to complete Stage II.

Composting mixtures for mushroom production was seen by Hayes & Randle (1969) as a wasteful but necessary energy consuming process and while the concept of conserving the nutrients of value to the mushroom was emphasized by their work, its achievement in practice required more precise definition of the chemical composition of mixture ingredients and the balancing of the primary nutrient, carbon with that of nitrogen which previously was the only component which was adjusted by supplementation. The schedules for composting outlined by Hayes & Randle (1969; 1970) exploit the use of cheap and plentiful by-products of the sugar industry, *e.g.* molasses to minimize losses of cellulose and hemicellulose, nutrients which are important for mushroom growth. Alternatives include sugar beet pulp, leafy hay, apple and grape pumice or other energy rich materials which are easily broken down by the composting microflora such as fats and vegetable oil (see Schisler, 1967).

Ready-mixed compost

Preparing a satisfactory compost requires much experience and skill; in particular, the many chemical variations inherent in the composition of mixtures based on animal manures have to be taken into account. Every mixture is peculiar to itself and for each the routine of supplementation should ideally be modified to allow the course of the fermentation to conform to a predictable pattern, so that factory production line management can be applied.

A relatively recent trend which has had a profound effect on farms of small size in the U.S.A., U.K. and Holland is ready-mixed or custom compost prepared by specialist composters in bulk quantities before delivery to the farm, with the grower taking over responsibility for the final preparations in Stage II. This trend is likely to continue and Hayes (1969*b*) suggested that a logical development of this specialist activity would be to supply such growing units with a substrate which requires no further treatment before spawning or even to supply compost already permeated with mushroom mycelium. This has yet to be applied to commercial cultivation although this concept is now widely adopted in the marketing of small containers for amateur enthusiasts.

The casing layer—a substrate

The familiar mushroom sporophores are formed only after the compost is covered with a layer of casing material which can be top or sub-soil, loam, sand, gravel, ash, or peat made alkaline with calcium carbonate. Loam soil is widely used for this purpose in America but in the U.K. and other European countries its use has been superseded by peat/chalk mixtures following the investigations of Edwards (1949) and Edwards & Slegg (1953).

In contrast to the compost layer, the casing layer is a dilute medium for growth, but its function as part of the total substrate has only recently been considered. The investigations of Eger (1961) showed that micro-organisms which naturally colonize the casing layer are implicated in this change from vegetative to reproductive growth. The activities of *Pseudomonas putida* isolated from casing soils (Hayes, Randle & Last 1969; Arrold, 1972) was linked by Hayes (1972, 1973) to its ability to solubilize iron, an essential nutrient for the growth and development of primordia but concluded that iron availability in a casing soil could be influenced by many factors other than those related to microbial activity such as the formation of acids.

Also on the basis of their experiments, Hayes *et al.* (1969) postulated

VEGETATIVE PHASE

A

B

REPRODUCTIVE PHASE

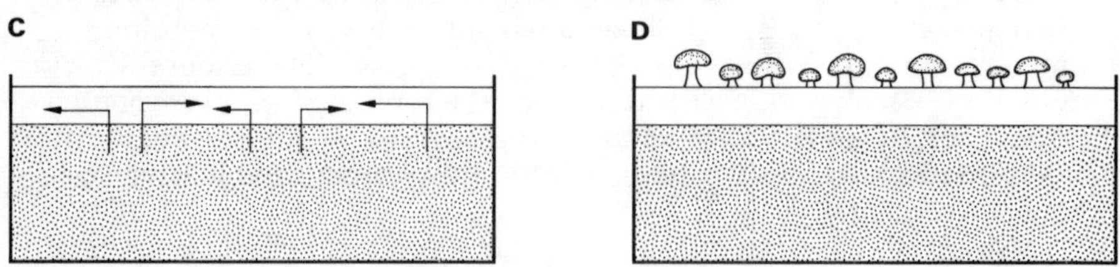

Fig. 11.5 Suggested relationship between compost, the casing layer and the formation of fruit-bodies. A, Mycelium growing on substrate. B, Volatile by-products such as carbon dioxide, ethanol, acetaldehyde, acetone, ethyl acetate. C. Volatiles which accumulate in soil layer select for required bacteria. D, Fruit-bodies form as a result of microbial action in the soil layer.

that the occurrence and activity of *Pseudomonas putida* in casing soil is the result of the environment created by the growing mycelium in the compost (Fig. 11.5). Thus the two substrates required for the commercial culture of mushroom sporophores is intimately linked to the activity of other micro-organisms which generally operate by affecting the availability of nutrients. The procedures involved in preparing a compost, in casing and the management of the culture to initiate sporophore formation are fundamentally concerned in establishing those micro-organisms which contribute to the growth process of the mushroom and are beneficial, at the expense of those which are natural competitors or are pathogenic.

Control by pasteurization

A form of selective control over the range of micro-organisms which contribute to the composting process was introduced by Lambert (1941) and by Sinden & Hauser (1950*a*) in describing their method of composting which included a pasteurizing phase in which a moderate level of heat is applied to the mass to eliminate nematodes, insect pests and other pathogens. At the time of inoculation with spawn, therefore, the compost is nutritionally and biologically selective to favour the growth of the inoculated *Agaricus bisporus* culture.

Heat treatment by steam or dry heat is also essential for the eradication of pests from soil used for casing mushroom beds and temperatures of 76°C for 30 minutes are generally regarded as minimal for this purpose. Peat/chalk mixtures are relatively pathogen free and pasteurization is not frequently practised.

Heat treatment by steam although widely adopted is difficult to control and is not selective in the organisms it destroys. In addition it is costly and causes damage to materials and structures. The same principles of pasteurization were applied by Hayes & Randle (1970) in using a gas, methyl bromide, instead of steam which is used in a composting procedure known as G.C.R.I. Formula 3. The improved yields obtained by these workers by fumigation is attributed to the destruction of Actinomycetes and fungi which utilize valuable cellulosic foods (Hayes, 1969).

In describing the unique selective properties of methyl bromide Hayes (1970) and Smith & Hayes (1972) showed that pasteurization of peat/chalk casing soils by this method did not cause the reduction in yield which normally results from steam treatment. This was considered to be a reflection on the low toxicity of methyl bromide to the bacteria previously mentioned as being important in stimulating the formation of mushrooms.

A technique for treating loam soil at lower temperatures employing the principles of aerated steam has recently been introduced in the U.S.A. but because of the expense involved in constructing both trays and a soil pasteurization house there has not been a widespread adoption of this method.

It is now widely accepted that for the production of disease-free crops it is necessary at the end of cropping to disinfect the growing structures and trays (or shelves) with both substrates, compost and casing layer remaining *in situ*. This is known as terminal disinfection or cook-out and has traditionally been done by steam to raise the levels sufficient to kill all pests and pathogens which accumulate during cropping (see later). Especially important in this connection are the insect pests and viruses which are transmitted by the tissues of the mushroom. Contamination is maximal at the end of cropping and because of this, together with the difficulties of heat penetration and distribution throughout the structure, it is necessary to maintain a compost bed temperature of at least 76°C for 12 hours.

Methyl bromide is also toxic to many of the pests and pathogens which build up during normal cropping (Hayes, 1970) if applied at a dosage rate of 4 lb/1000 ft to achieve a minimum CTP of 600 oz/h/1000 ft^3. Its use is now well established in terminal disinfection procedures despite contrary findings by van Dieleman (1971) who found poor toxicities to methyl bromide at lower rates of application.

The impact of pasteurization and its application to mushroom growing has been profound and the provision of an alternative to steam in the three major pasteurizations of the process offers an alternative method of cultivation which is applicable to low cost systems of growing (see Hayes, 1970).

11.6 Mechanization and Environmental Control

Mechanical aids for use in mushroom growing have resulted from the seemingly innate abilities of growers to improvise and it is only in recent years that machinery designed specifically for the mushroom industry have been marketed. Up to about two decades ago the laborious task of turning compost stacks was done by hand, but mechanical turning equipment designed to mix and aerate compost are now standard items.

The shelf or the fixed bed system of growing which predominates in the U.S.A. and Holland imposes major constraints to sophisticated mechanization but the tray system provides a means by which units of compost can be readily transported by fork-lifts over considerable distances. Large growing units are thus more flexible (Fig. 11.6). The introduction of the tray system, provided a stimulus for the phasing of culture procedures into, peak-heating, spawning and growing, and this could in turn provide the means by which spawning and casing could be mechanized. On new modern plants, tray handling lines consisting of devices for destacking, tipping, spawn metering and mixing, compressing of the compost, casing and automatic destacking are standard. These arrangements can be viewed as a preliminary to automated mushroom production.

Recently, a number of farms in England have extended the principle of line operation to harvesting the crop. This involves transporting the crop from the growing area to a conveyor line with pickers on either side. Although it is interesting to speculate that this approach may be a forerunner of a mobile system of growing, perhaps on a continuous line, the so called picking line readily lends itself to a fixed device for mechanical harvesting—a logical next step on these highly mechanized farms. A prototype harvester developed by Persson (1972) which uses oscillating knives to cut mushrooms immediately above the casing layer, demonstrates its feasibility.

However, a major limitation to the application of mechanical aids to harvesting stems from the variables of the growth process. Control over growth in order to minimize the wastage which is inevitable when the stage of crop maturity is spread over several days, is currently a major

Fig. 11.6 Automated tray handling line installed at test and demonstration facility, Pennsylvania State University.

objective of research. Reflecting this need, there has been a general trend in mushroom engineering towards the development of mechanical equipment to provide more precise environmental control, so as to create the best possible conditions for mushroom growth.

Nineteenth century growers realized the need for some fresh air for good crops, but the real importance of climate was not fully realized until Lambert (1933) demonstrated the harmful effects of carbon dioxide tensions on the development of sporophores. Later Middlebrook & Storey (1950) compared the seasonal cropping patterns in three different types of mushroom house. They concluded that the ratio of air space to bed area was a very important factor in determining yield and if this ratio was below 0.12 m^3 per m^2 of bed area, yield was reduced.

Tschierpe (1959) confirmed the earlier findings on carbon dioxide concentration and showed that a concentration of 0.5% carbon dioxide in the air overlapping the casing inhibited fruit formation and later Tschierpe & Sinden (1964) showed that concentrations of carbon dioxide over the range 0.03–0.1% were optimal. Similar optima were recorded by Long & Jacobs (1968). These researches also emphasize that the exact critical level varies with strain and that above optimal concentrations cause abnormally long stalks, but once sporophores are formed high concentrations can be tolerated. An endogenous supply of carbon dioxide is required for satisfactory growth of mycelium through the casing and during the early stage of sporophore development.

The effect of temperature has been extensively studied by Flegg & Gandy (1962) and Flegg (1968, 1970). For maximum yield and earliest cropping the air temperature during the period of fruiting after casing should be about 15–18°C while during cropping it should be about 15–16°C. Higher or lower temperature than the optimum during the two weeks after casing delays the onset of fruiting, the extent of the delay being variable and ranging from about one to seven days for a difference in air temperature of about 5.5°C (Flegg, 1968).

The relative humidity of the air immediately above the casing is also known to be an important environmental variable. Flegg & Gandy (1962) showed that at high relative humidities (80–95%) the onset of fruiting tended to be earlier and the weight of mushrooms greater than at low humidities (40–50%). Evidence produced by Storey (1965) suggests that at very high humidities, when evaporation from the casing layer is much reduced and water may condense from the air on to the casing layer cropping is delayed and yields reduced.

With this knowledge on the relationship between environmental factors and the growth process, simple systems of ventilation have given way to the concept of air conditioning which allows a greater degree of control of temperature and humidity throughout the year. Electronic monitoring and control devices are now being introduced to automatically provide the optimum temperature, humidity and of fresh and recirculated air required in the various phases of mushroom production. While the economics of these sophistications have not yet been established, the trend in these intensive units contrasts sharply with some of the newer concepts of extensive culture where costs are minimized at the expense of precision.

11.7 Pests, Pathogens and Competitors

The mushroom, like any other cultivated crop, is subject to attack by pests and pathogens. Before the days of protected cropping, pests were known to be troublesome and Loudon (1850) suggests that 'A toad kept in a mushroom house will eat the vermin, snails and slugs mentioned, and also worms, and ants and other insects; but to most people the idea would be disgusting of a toad crawling over anything intended for the table'. This interesting observation of a 19th century gardener is an example of the control of pests by biological means, an approach that is only now, over a century later, being considered as potentially valuable in modern systems of culture (Hussey, 1969; Nair & Fahy, 1972).

With intensification and the perfection of techniques of culture, much attention has been given to the general biology and epidemiology of a range of pests (insects and nematodes) and pathogens (fungi, bacteria and viruses) which reduce yield. Although the introduction of pasteurization procedures in substrate preparation and in disinfection radically checks the establishment of disease, its occurrence is inevitable in a continuous system of cropping.

In addition, other non-pathogenic micro-organisms can establish at the expense of the mushroom and cause reduction in yields. These competitors are not pathogenic and are usually the result of imperfections in composting or some other cultural practice. However, the unique nature of mushroom cultivation based on closely manipulated environmental conditions provides an opportunity for controlling most of the diseases and excluding the competitive organisms with greater ease than in the case of other plant crops.

Pests associated with mushroom crop

PHORIDS Species of the phorid *Megaselia* are one of the most important insect pests of the cultivated mushroom. The larvae tunnel into the sporophore tissues. Light-trap studies have shown that *M. nigra* flies from early June to December (Hussey *et al.*, 1969), and eggs are laid only where daylight falls on mushroom beds. In modern mushroom growing units light intensity is low enough to inhibit oviposition. The biology of phorids has been studied by Hussey (1959).

SCIARIDS *Lycoriella solani* Winnertz, *L. auripila* Winnertz and *Bradysia brunnipes* Meigen threaten a crop after a large population has developed (Hussey *et al.,* 1969). There is evidence that they may be attracted to composts containing nitrogenous supplements (Hussey, 1972). Sciarids show less distinct seasonal activity than phorids and they seem to require a certain type of microflora in the casing for efficient breeding (Hussey, 1972). Sciarids transmit mites and spores of *Verticillium* spp. According to Hussey *et al.* (1969) 'the association between Lycoriids and the virus disease "die-back" seems too frequent to be entirely accidental'.

CECIDS Two genera *Heteropeza* and *Mycophila* have been shown to be pests of the cultivated mushroom. Early infestations of *Heteropeza* have been shown to reduce the yield of mushrooms by about 14% (Hussey, 1972). *Heteropeza* is able to enter into a resting stage and thus survive unfavourable conditions. Early attacks of *Mycophila* are also quite damaging to the crop.

MITES Tyroglyphid mites are saprophagus but they could establish high populations on sporophores. They occur in straw, chicken manure, cotton seed meal *etc.* and are introduced into the mushroom crop through substandard composts.
The tarsonemid mite *Tarsonemus myceliophagus* Hussey feeds on mushroom mycelium. It may also destroy the mycelial strands at the base of sporophores. The mites can survive after-crop sterilization and thus may transmit mushroom viruses (Hussey *et al.*, 1969).
Pygmephorus spp. (the red pepper mites) are also known to infect mushroom beds (Hussey & Gurney, 1967). They are apparently associated with *Cheatomium* sp. Whereas *P. sellnicki* completes its life cycle on fungi such as *Trichoderma, Monilia* or *Humicola, P. mesembrinae* breeds only on *Trichoderma* (Hussey *et al.,* 1969).

NEMATODES Three types are known to be associated with mushroom growing—saprophagus, predacious and pathogenic nematodes. Of the pathogenic nematodes, *Ditylenchus myceliophagus* Goodey and *Aphelenchoides composticola* Franklin occur most commonly and severe infestations by these can result in significant loss of mushrooms. Their modes of feeding differ but result in damage to mycelium leading to secondary infection of the cell contents by bacteria. They are capable of a rapid rate of multiplication, in the order of 100 000-fold in 4 weeks are common

(Hussey *et al.*, 1969). Their damaging effects are dependent both on the level of population and the stage of crop development at infestation. Contamination of the mushroom crop is mainly through casing soil or is the result of inefficient pasteurization.

Pathogens associated with mushroom crop

FUNGAL PATHOGENS *Verticillium malthousei* Ware causes severe losses of mushroom crops. It appears to be an endemic disease (Sinden, 1971). In most countries where mushrooms are grown the loss from this disease is greater than from any other disease except virus (Sinden, 1972) and probably bacterial blotch in Australia (Nair, 1969). Infection of sporophores results in deformed pilei and the formation of brown spots and spheres called dry bubbles. A greyish white mouldy growth is seen on the pileus. Inoculation experiments showed that severity of the disease was affected more by inoculum concentration than time of infection (Gandy, 1972). Fekete (1967) and Cross & Jacobs (1969) demonstrated human transmission of *V. malthousie*; however, Gandy (1972) pointed out that although pickers could spread the infection, immediate removal of the picked mushrooms prevented any significant transmission of the disease in this manner. Splash dispersal of the conidia of *V. malthousie* (Cross & Jacobs, 1969) has now been shown to be only over short distances (Gandy, 1972) but is of sufficient significance to spread the disease. Conidia of *Verticillium* species are readily spread by phorid flies (Hussey, 1972) and this is another source of transmission of the disease. The major source of contamination, however, appears to be debris and dust of floors of mushroom growing units. No resting spore is known for *V. malthousei* although mycelium from dried bubbles can survive for long periods.

Another species, *Verticillium psalliotae* Treschow is also known to infect the mushroom sporophore. Treschow (1941) pointed out that *V. psalloitae* did not form as conspicuous an aerial mycelium over the infected pileus as *V. malthousei*. Dayal & Barron (1970) and Barron & Fletcher (1970) have demonstrated parasitism of *Rhopalomyces elegans* Corda by *V. psalliotae in vitro*. Spores of *R. elegans* could thus be sources of infection of a mushroom crop as well as agents of transmission of the disease.

Mycogne perniciosa Magnus causes a disease commonly known as Wet Bubble. Infection of the sporophores results in malformed stipe and pileus; in severe cases the sporophore ceases to develop resulting in a shapeless mass of hyphae. The surface is covered with mycelium producing two kinds of spores—hyaline conidia on verticillate conidiophores and two-celled, brown, chlamydospores. The diseased sporophores finally show the typical symptom of dripping brown liquid with a foul odour. The chlamydosphores are known to survive in soil for years (Smith, 1924; Lambert, 1930; Chaze & Sarazin, 1936) and thus form an important source of infection. Barron & Fletcher (1972) showed that conidiophores, vesicles and conidia of *Rhopalomyces elegans* were parasitized by *M. perniciosa*. *R. elegans* is a coprophilous fungus (Ellis, 1963) and if it is widely distributed its spores may not only form a source of the disease but also transmit *M. perniciosa*.

Other fungal pathogens of relatively less importance associated with the mushroom crop are *Hypomyces auranteus* (Pers. ex Fr.) Tul., *H. rosellus* (Alb. and Schw.) Tul., *Dactylium dendroides* (Bull.) Fr., *Trichoderma koningi* Oud. and *T. viride* Pers.

BACTERIAL PATHOGENS *Pseudomonas tolaasii* Paine causes brown blotch or bacterial blotch of the cultivated mushroom. The disease produces brown, slightly sunken spots and blotches on the maturing pileus. The browning appears to involve only the superficial mushroom tissues. In severe instances of the disease brown lesions develop on the stipe. *P. tolaasii* can also cause considerable damage to mushrooms in storage and in transit. There have been several instances where mushrooms have been cropped unblemished, but in storage at low temperature (1.6°C) have turned brown with blotched pilei. The bacterium is present in soil and, in several cases, in water used for mushroom growing. It has not been isolated from peat used for casing the mushroom beds, possibly due to low pH of the peat. The disease can become quite severe under conditions of high humidity and temperature.

Lelliott *et al.* (1966) in a study of the fluorescent groups of the genus *Pseudomonas* suggested that *P. tolaasii* could be regarded as a normal constituent of the microflora of mushroom beds which, under certain conditions, produce a metabolite toxic to mushrooms. Nair & Fahy (1973) have provided evidence for the production of toxin by *P. tolaasii* and demonstrated the ability of this toxin to cause browning of sporophores actively growing under conditions similar to those in commercial mushroom growing units. The observations of Nair & Fahy (1973) that the toxin as well as cells of *P. tolaasii* inhibit mushroom mycelial growth is relevant to commercial growing as *P. tolaasii* may inhibit the growth of mycelium at the spawn growing stage of mushroom cultivation.

From what is known about the etiology of the brown blotch disease we could probably conclude that

$$\text{Disease severity} = \underset{(P.\ tolaasii)}{\text{inoculum potential}} + \underset{(P.\ fluorescens)}{\text{adaptation to pathogenesis}}$$

because *P. tolaasii* may be closely related to *P. fluorescens* complex of the mushroom bed. *P. tolaasii* and *P. fluorescens* closely resemble each other in their physiological characteristics except that the former is pathogenic to the mushroom. *P. fluorescens* can become pathogenically adapted to the mushroom and according to Stolp (1961) pathogenic mutants can be selected directly from a statistically avirulent population. The disease severity would therefore depend on the inoculum potential, that is the total number of *P. tolaasii* plus the number of *P. fluorescens* which have become pathogenically adapted to the mushroom.

Pseudomonas sp. causes a disease known as mummy. Tucker & Routein (1942) have described this disease in the United States. Diseased mushrooms have thin stipes and the pilei fail to reach normal size. The sporophores are often tilted and open prematurely. Secondary infection by the bacteria sets in once the sporophores become necrotic. Due to the infectious nature of the disease it was thought for a long time that this disease was caused by a virus. Schisler *et al.* (1968) isolated a species of *Pseudomonas* from mushrooms afflicted by mummy disease. One of the characteristics of the disease is that it appears at one point on the mushroom bed and then spreads fast at the rate of a foot a day. The mycelium of the bed is also infected and Schisler *et al.* (1968) reported intracellular occurrence of the bacterium. The movement of the bacterium along the hyphae aided by protoplasmic streaming probably results in the rapid spread of the disease. There is no information on the transmission of the disease and improved methods of isolation of the pathogenic organism will help us in understanding the etiology of the mummy disease.

VIRUSES Since the first description of an infectious disease of the cultivated mushroom by Sinden & Hauser (1950*b*) in Pennsylvania, U.S.A., mushroom disorders of a similar nature have been observed in countries where mushrooms are extensively grown. Various names like La France (Sinden & Hauser, 1950*b*) Brown disease and Watery Stipe (Gandy, 1960), X-disease (Kneebone, Lockard & Hager, 1962) and Die-back disease (Gandy & Hollings, 1962) were given to these disorders due to the variety of symptoms observed and the difficulty in determining their causes. It is difficult to diagnose the disease on the basis of sporophore symptoms such as drum stick-like mushrooms and premature opening of the veils because similar symptoms can be caused by certain environmental and cultural factors. Virus infections may even be symptomless (Nair, 1972). Reduction in yield of mushrooms is perhaps the most reliable symptom. One other symptom that is commonly associated with an infected crop is the slow and depressed growth of mycelium isolated from infected mushrooms. Gandy (1960) showed that mycelial isolates of diseased mushrooms grew more slowly forming buff coloured instead of white colonies. However, there have been reports of diseased isolates showing intermediate growth rate, *i.e.* growth rate falling between slow and normal types (Hager, 1969; Hollings & Stone 1969; Nair, 1972); the rate of growth being inversely related to the number of virus particles present (Last, Hollings & Stone, 1967). Virus-infected cultures could be cured by heat treatment at 32–33°C for 1–3 weeks (Gandy, 1960; Hollings, 1962; Hollings & Stone, 1969; Rasmussen, Mitchel & Slack, 1972; Nair, 1973).

Six viruses have been isolated from diseased mushrooms; five of these are polyhedral virus particles 19, 25, 29, 35 and 50 nm diameter and one is a bacilliform particle 19 × 50 nm (Hollings & Stone, 1969, 1971; Hollings, 1972; Nair, 1972). They can occur singly or more commonly in combinations. The intracellular occurrence of virus particles has been reported by Dieleman-van Zaayen & Igesz (1969) and Nair (1972). It has not been possible to associate any particular type of virus with any specific symptom.

The observations of virus transmission through mushroom spores by Schisler *et al.* (1967) and through mushroom spawn by Nair (1972) were vital factors in our understanding of the etiology of mushroom virus disease. Phorid larvae and tarsonemid mites are known to transmit viruses (Hussey, 1972). Certain species of field mushroom (*e.g. Laccaria laccata*) have been suspected as natural reservoirs of mushroom viruses.

Fungal competitors associated with the mushroom crop

Several fungi have been isolated from mushroom compost at different stages in its preparation and use in mushroom cultivation. Many of these fungal competitors or weed-moulds interfere with the growth of mushrooms. Some of these fungi, by rapid utilization of the substrate, have an ecological advantage over the mushroom and establish themselves in the compost without inhibiting the mycelial growth and formation of mushroom sporophores. Others may inhibit the growth and development of mushrooms through microbial excretions or biologically generated toxins. To date, the molecules involved in this inhibitory process in mushroom compost have not been characterized.

Three soil-borne fungi *Diehliomyces microsporus* (Diehl and Lamb.) Gil., *Chrysosporium luteum* (Cost.) Carm. and *Geotrichum* sp. (= *Sporendonema purpurescens* (Ben.) Mason and Hughes) are known to inhabit mushroom compost. *D. microsporus* is known to cause the truffle disease and *C. luteum* the mat disease. Since the adoption of high standards of hygiene such as preventing soil contamination *etc.* during Stage I or outdoor composting, contamination of the compost by these fungi has been limited. Over-composted substrates as a result of a relatively long Stage I are colonized by fungi like *Papulospora byssina* Hotson and *Scopulariopsis fimicola* (Cost. and Matr.) Vuill. often referred to as the brown plaster mould and white plaster mould respectively. With the introduction of short composting methods for Stage I, these fungi have largely been controlled. All these fungi appear on mushroom beds during cropping as extensive patches of colonies, and inhibit the growth of mushroom mycelium.

Certain conditions during Stage II enable the mushroom compost to be colonized by fungal competitors. Insufficient amounts of available carbohydrates in Stage II lead to incomplete conversion of ammonia and amines and their accumulation in the compost. This condition provides an opportunity for fungi like *Coprinus fimetarius* (L.) Fr., *Oedocephalum* sp. and *Thielavia thermophile* Fer. and Sinden to colonize the compost. If aeration of the compost is insufficient and the level of carbon dioxide is high during Stage II *Chaetomium olivaceum* Cooke and Ellis, a widely distributed coprophilous fungus is found to grow in it. This fungus inhibits the growth of mushroom mycelium.

Strategy in disease control

Three avenues of control are open to the mushroom grower.

1. Chemicals toxic to the pathogen, pest or competitive organism may be added during the process. Use of fungicides in controlling mushroom diseases poses special problems since in several instances both the host and the pathogen are fungi. Likewise, pesticide usage brings with it several problems connected with distribution of these substances and development of resistance towards them.
2. The physical factors of the micro- and macro-environment of a mushroom crop such as temperature, humidity and air movement can be manipulated, and is referred to by Sinden (1971) as ecological control. The term ecological control used here also takes into account modifications in farm design that lead to the prevention of diseases.
3. Several biological methods of control have been proposed and include the use of micro-organisms antagonistic to a pathogen (Nair & Fahy,

1972), the use of aerated steam (Baker, 1957) and methyl bromide (Hayes, 1970) to pasteurize the casing soil in order to avoid a 'biological vacuum', or by establishing a population of natural enemies or predators of insect pests as shown by Hussey (1969, 1972). An example of an integrated scheme for mushroom disease control is illustrated in Fig. 11.7.

CHEMICAL CONTROL Incorporation of pesticides such as diazinon, malathion and thionazin in the compost or the casing is the most common method of controlling phorids, sciarids and cecids (Hussey & Wyatt, 1959; Hussey *et al.*, 1969; Wyatt, 1970). Adult flies may be killed by using aerosols of diazinon, malathion and pyrethrin. Efficient after-crop sterilization and Stage II should eliminate some of the insect pests such as cecids and mites. Methyl bromide fumigation appears to be a better proposition than steam for after-crop sterilization. Nematode infestation of the mushroom beds has been controlled by nematicide treatment (Hesling & Kempton, 1969). Cayrol & Ritter (1972) have reported an effective control of *Ditylenchus myceliophagus* with thionazin.

Fungal pathogens associated with the mushroom crop have been controlled largely by the use of fungicides. Of the dithiocarbamates used zineb, has been most commonly sprayed on mushroom beds; mancozeb and cufraneb have also been used more recently (Feket & Kuhn, 1965, 1966; Newman & Savidge, 1969). A relatively new fungicide, benomyl, has been shown to control *Verticillium malthousei* (Gandy, 1971; Wuest, 1971). The likelihood of *V. malthousei* developing resistance towards benomyl should not be overlooked (Gandy, 1972). Mutagenic effects of benomyl on fungi have been recorded (Hastie, 1970). Benomyl has a disadvantage in that it can cause severe allergy reactions in some people. This chemical has also been found effective in controlling certain fungal competitors of the mushroom bed such as *Trichoderma* spp. (Peake, 1972).

Bactericides such as chlorinated water (Lambert, 1938), combined halogen compounds such as chlorine and bromine (Nair & Milham, 1970) and hexachlorophene (Stoller, 1968) have been used to control the bacterial pathogens of the cultivated mushroom.

No chemical methods have yet been found for controlling the virus disease complex of the cultivated mushroom; however, fumigation of the mushroom compost *in situ* after the completion of cropping with methyl bromide is a good procedure in the prevention of virus disease.

ECOLOGICAL CONTROL This method of control may be achieved either by manipulation of the physical factors of the crop environment or by modification in farm design. For instance, *Verticillium malthousei* may be controlled by reducing the relative humidity to 80–85% and the temperature to 14°C (Sinden, 1971); however, the use of this method may result in the lowering of quality and quantity of mushrooms.

Similarly, bacterial blotch can be controlled by environmental manipulation (Sinden, 1971). The optimum conditions necessary for infection are brought about by slight changes in temperature at the surface of the mushroom (Fig. 11.7). A rise in temperature of not more than 0.5°C upsets the balance between the surface of the mushroom and the ambient air resulting in sufficient condensation of moisture on the surface of the mushroom for infection. Fluctuation in temperature and relative humidity

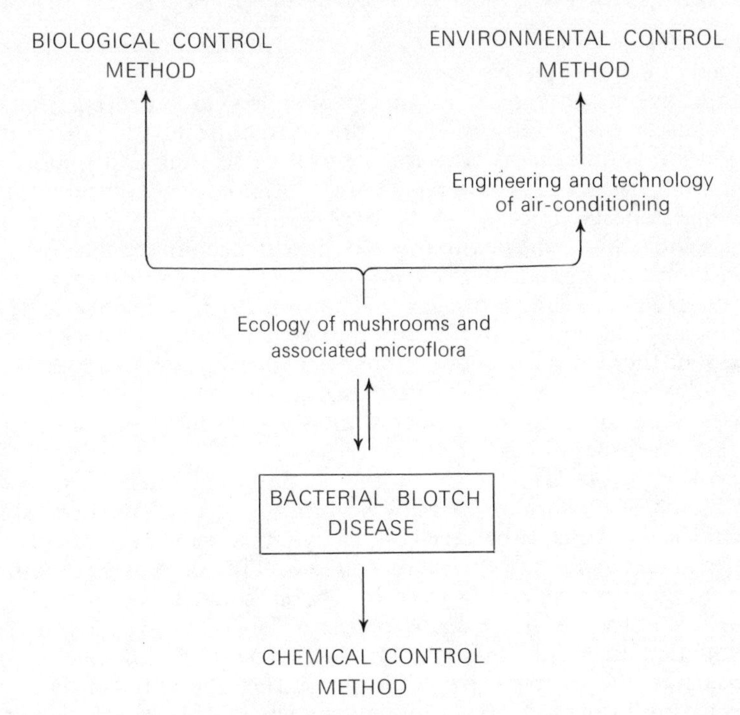

Fig. 11.7 Methods of control for bacterial blotch disease of cultivated mushroom *Agaricus bisporus*.

within a mushroom growing unit must, therefore, be kept to a minimum.

Farm design is a major factor in virus disease control. Ganney (1972) has described farm layouts aimed at reducing chances of infection at all stages of cropping and maintaining a high standard of hygiene. Control of mushroom viruses also includes filtration of the air introduced into the spawn-running rooms, the spawning and casing areas and the growing rooms.

BIOLOGICAL CONTROL The application of biological control to mushroom cultivation is comparatively recent. Efficient distribution of chemicals in compost and casing has always been the most difficult aspect of chemical methods of controlling mushroom diseases. Resistance shown by the pests and pathogens to chemicals poses another serious problem in this field and as Hussey (1972) remarks, 'it is tempting to think of biological control by living organisms which are capable of moving away from the limited introduction sites achieved by the spawning-machines'.

Normally the mushroom crop is grown in hygienic environment without any pathogens or pests. When a pathogen or pest is introduced accidentally into this environment, its numbers increase at a maximum rate; however, the environment can also favour the development of an antagonist or an enemy of the pathogen or pest. If the population of the antagonist or enemy can be maintained at a high level, the increase in numbers of the pathogen or pest can be greatly inhibited. This is one of the principles on which the programmes of biological control of certain mushroom insect pests and bacterial blotch disease are based.

In mushroom cultivation this concept of biological control was first used against insect pests. Hussey (1969) drew attention to the possibility of using entomophilic nematodes for controlling sciarids and phorids and recent work has revealed several examples of biological control of mushroom insect pests (Hussey, 1969, 1972; Hudson, 1972). More work is necessary to exploit this possibility and make it a commercial proposition. For instance, the nematode *Bradynema* sp. and *Tetradonema* sp. attacking phorids (Hussey, 1965) and sciarids (Hudson, 1972) respectively, have to be artificially cultured in sufficient numbers to achieve biological control. At present there is not a sound system for mass-producing these nematodes. Similarly, techniques have to be developed for preparing mass-cultures of the predator red gamasid mites to prevent pepper mites from infesting mushroom beds.

Nair & Fahy (1972) have shown the possibility of controlling the bacterial blotch of mushroom by biological means. They have been able to prevent the incidence of bacterial blotch by introducing peat inoculants of bacteria antagonistic to *Pseudomonas tolaasii*. The three bacteria with the ability to suppress blotching were *P. cepacia* isolated from soil, *P. fluorescens*, and a strain of *Enterobacter aerogenes* isolated from peat. There was no reduction in yield of mushrooms as a result of introducing these bacteria into the casing layer. This shows that the antagonists had no apparent inhibitory effects on organisms such as *P. putida* known to stimulate the formation of sporophores (Hayes, Randle & Last, 1969; Hayes, 1972; Hume & Hayes, 1972). Full commercial application of this method depends on commercial production of peat cultures of bacterial antagonists.

The technique of mushroom cultivation lends itself to the application of such chemical, ecological and biological methods for disease control. It is difficult to single out any one of these three methods as more applicable to mushroom growing because a disease is hardly ever controlled by any single method. A combination of either the chemical and biological methods or the chemical and environmental manipulation (ecological method) offer the best means of controlling the diseases of the cultivated mushroom. The integrated pest control scheme in the field of entomology is a typical example of this mode of approach.

11.8 Cultivation of Other Edible Fungi

A number of fungi within a wide range of taxonomic groups are edible, but most are Basidiomycetes, the notable exceptions being the edible species in the genus *Tuber* (truffles) and *Morchella* (morels) which are Ascomycetes. Despite the popularity of *Agaricus bisporus* cultivation, techniques for the mass cultivation of other edible mushrooms have not as yet been fruitful. However, a ready market is found for field gathered exotic species, especially in many of the European countries such as France and Italy, a demand which dates back to the nineteenth century. According to Last (1970) the records of the Halles market in Paris show that a punnett (maniveau) of cultivated mushrooms (champignon ordinaire) cost 35 cents in April 1828, a pound (livre) of truffles 14 francs and a small basket (petit panier) of morels 2 francs. However, some species are cultivated especially *Volvariella volvacea* (Bull ex Fr.) Sing. family: Amanitaceae, *Lentinus edodes* (Berk) Sing., in the Orient; *Pleurotus ostreatus* (Jacqu. ex Fr.) Kunsmer (family Tricholomataceae) and *Tuber melanosporum* Vitt. are commonly cultivated in Europe.

The padi-straw mushroom—Volvariella volvacea

A full account of the systematics of this genus is given by Singer (1961). Although it has been cultivated by the Chinese for many centuries, large scale commercial cultivation is only now being considered possible.

The traditional method of growing the padi-straw mushroom involves the use of ridge beds under partial shade or in the open field. The ridge bed is made on a soil base 70–80 cm wide. Although the most common substrate is rice straw other materials such as the residues from tapioca manufacturing and the straw of *Bombax melaboricum* are known to be suitable. The methods of preparing the substrate are simple; rice straw is soaked in water by complete immersion in tanks, before being folded and arranged into a stack on a soil base. The stack is set up in layers, each layer being spawned with pure culture *Volvariella volvacea* grown on sterilized straw at points about 15–20 cm apart. The procedure is repeated with each layer gradually decreasing in width until the fifth or the last layer is reached at about 80 cm. After three days, temperature in the beds reaches about 40–45°C and the beds are usually watered after five days. Fruiting takes place about twelve days after preparing and spawning the bed without the application of a casing layer. Employing this conventional method yields are not predictable and 6–7 kg of mushrooms per 100 kg is considered to be satisfactory.

Recently, however, Ho (1972) described a method of growing *Volvariella volvacea* in plastic houses where environmental factors such as temperature, humidity and ventilation are controlled. Peak-heated or pasteurized compost was used and the mushroom beds were cased with a 2 cm thick layer of clay loam. This new method of indoor cultivation of *V. volvacea* resulted in twice as much yield as obtainable by the outdoor method of cultivation and clearly shows the potential of this tropical species.

Shiitake—Lentinus edodes

Shiitake is cultivated in Eastern Asia and Japan and because of its natural lignicolous habit is grown on wood. Cultivation of Shiitake starts with the preparation of the inoculum which is a pure culture of *Lentinus edodes* on pieces of oak wood. Logs cut from hardwood trees such as oak (*Quercus* spp.), Chestnut (*Castanea crenata*), certain species *Carpinus, Alnus* and *Acer*. Holes are drilled into the log substrate and the solid pieces of inoculum placed inside.

After inoculation the bed logs are placed on a site called the laying yard (Fig. 11.8). The site should not be excessively moist as faulty moisture conditions may lead to loss of yield. The logs are placed in an oblique upright manner with single logs laid crosswise to facilitate aeration. The optimum temperature for the growth of mycelium is 24–28°C. After 7–8 months, when the mycelium has grown through the wood, the logs are transferred to another site called the raising yard. This site is shaded

Fig. 11.8 The cultivation of the Shiitake mushroom. a (*left*). Spawned logs in laying yard. b (*top right*). Logs in oblique upright positions showing heavy mushroom production. c, (*bottom right*). Mushroom formation.

and the moisture level is high. Formation of sporophores takes place during early spring or late autumn at a temperature between 12 and 20°C and the logs are kept moist by frequent watering. The minimum cropping time is about three years and the maximum is six years.

The oyster mushroom—Pleurotus ostreatus

Traditionally this fungus is cultivated on beech trunks and logs of deciduous trees in a manner similar to that for Shiitake and recently many growing operations new to the cultivation of this fungus have been established in Europe. Toth (1970) described a method of cultivating *Pleurotus ostreatus* on crushed corn cobs, but under sterile conditions. Another Bulgarian researcher Gyurkó (unpublished results) modified the method under non-sterile conditions again using corn cobs as substrate.

Solid spawn is used as in the cultivation of other mushrooms but Kostadinov, Turev & Rantcheva (1972) describe a method of producing *Pleurotus ostreatus* mycelium in submerged culture which may be used as inoculation material (spawn) for the production of sporophores on a substrate of crushed corn cobs.

Unlike *Agaricus bisporus* light is obligatory for the formation of *Pleurotus* sporophores (Gyurkó, 1972). In complete darkness the stipe remains thin and the cap rudimentary, while the formation of primordia is hindered by light over 40 lux. But below 40 lux primordia form normally. Light of short wavelength is said to be most effective.

Truffles—Tuber melanosporum

Unlike the mushrooms, the fruiting structures of the Ascomycete truffles are found underground. In the direct method of truffle cultivation cultures are introduced into the earth while in the indirect method the fungus is allowed to colonize the prepared habitat by natural means. Truffles grow in association with roots of trees, such as *Quercus, Carpinus, Fagus, Olea, Pinus, Cedras,* but for commercial growing, *Quercus* (Oak) is most commonly cultivated. Plantations are established in alkaline and well aerated soil and frequently artificial and natural irrigation systems are employed. Although truffle production may be continued in the same grove for several years, a relatively long interval of time, from six to ten years is required and as far as is known, yield of fruit-bodies varies according to season, but even so Gray (1970) estimated that production of truffles in the 1968–69 season was in excess of 2 million pounds.

The successful industrialization of growing other edible mushrooms depends on many factors, not the least of which is the nature of the growth habit of the fungus being cultivated. Of these currently being cultivated, it would seem that *Pleurotus ostreatus* is the only species which displays a simplicity of growth habit which suggests that culture methods can be evolved to allow commercial methods of production to advance on lines similar to those adopted for *Agaricus bisporus*. Knowledge of the basic physiology of growth, however, would aid this development. Indeed it may be possible to exploit other known edible mushrooms known to form mycorrhizal associations with roots, such as *Boletus edulis* Bull. ex Fr., and *Tricholoma matsutake* (S. Ito and Imai) Sing. The highly prized morel mushrooms, *Morchella hortensis* Bond., *M. esculenta* Fr. and *M. costata*

Vent. seemingly can be grown on open beds made from residues of apples and paper waste (Singer, 1961) although it has been suggested that these are mycorrhizal.

11.9 Future Prospects

Mushrooms are generally valued for adding flavour and zest to other foods and because of this they are usually projected as a commodity with little or even no food value. In a comprehensive review by Worgan (1968) on the nutritional value of higher fungi, it can be seen that although much of the evidence which relates to the chemical composition of the cultivated species is contradictory, the mushroom ranks favourably with other low calorie vegetables in the same group as peas, green beans, cauliflower, tomatoes, asparagus and lettuce. Relative to the mentioned vegetables, mushrooms are a good source of the following nutrients: protein, fat, phosphorous, iron, thiamin, riboflavin and niacin (Morgareidge, 1958). Further research is needed to quantify accurately the nutritional value and this is becoming increasingly urgent in view of the general increase in mushroom consumption. In many countries the *per capita* consumption of mushrooms is increasing steadily; where mushrooms are considered as part of the national diet, *e.g.* Canada and some European countries, consumption exceeds 1.5 kg per person per annum. In the U.K. it is estimated that the figure is about 1 kg and in the U.S.A. about 0.5 kg per person per annum.

Several factors appear to contribute to the increasing *per capita* consumption of mushrooms, many of which are interrelated. Increases in the general standard of living influences the consumption of beef steaks, away-from-home eating and the increasing use of convenience foods. The general availability of mushrooms irrespective of season, improved methods of marketing and processing, industry promotion and price stability have also aided increased consumption.

Food value also assumes importance in regard to the increasing interest shown by industrial concerns and public agencies in the safe recycling of wastes which pollute the environment. The possibility of converting industrial and agricultural wastes, urban refuse, wood waste *etc.* into a product which is directly edible by humans is attractive, especially since many wastes can readily be manipulated to provide a complete substrate.

However, irrespective of public demand and food values, the future prospects for continued expansion and mass production largely concerns the economics of production methods. For the present at least there appears to be little indication that methods of growing are evolving to one or even a few ideal systems. The tendency in recent years has been to intensify production through mechanization and sophisticated controls and the high capital costs involved must be offset by consistent high yields to remain profitable. The impact of some new low cost systems of culture, which require, by comparison, small amounts of capital outlay and can be operated as family units orientated to local quality markets, have yet to be realized by the industry generally. It is from the modern intensive units that the concept of mass production originates. For their growth and continued expansion, production methods are required which can accommodate not only the sophisticated engineering which is now being introduced, but also the fundamental complexities of the growth processes of *Agaricus bisporus* and other edible fungi, which at best are only partially understood.

The aim of techniques employed in the commercial culture of *Agaricus bisporus* is to produce a pure culture of the organism on a large scale, which is similar in concept to many other production processes involving micro-organisms, fermentation techniques and even routine laboratory culture techniques. The procedures of spawning (inoculation), control of temperature in establishing the different phases of growth (incubation) and terminal disinfection (sterilization of cultures and vessels before disposal and for re-use) can be readily identified with normal laboratory and fermentation practice. However, in the laboratory culture of micro-organisms, the substrate is first sterilized by using pressurized steam in autoclaves, but in mushroom growing the substrate is fermented to achieve a chemical and biological balance which favours the mushroom at the expense of other organisms. This fermentation is specific to mushroom cultivation and involves the sequential establishment of different groups of micro-organisms, the biomass of which partially contributes to the nutritional status of the substrate. The nature of the substrate and its preparation, more than all other aspects of growing, dictates the method by which mushrooms are grown.

In addition, the activities of other micro-organisms in the casing layer are known to be involved in the transition from vegetative to reproductive growth. These organisms must be preserved while other organisms which are pathogenic must be excluded by physical and/or chemical means. Furthermore, biological techniques of disease control require the establishment of specialized groups of micro-organisms in the immediate environment of mushroom growth.

The associations formed between the mushroom and other micro-organisms in commercial production have been shown to be significant in attempts to devise culture methods which are more comparable to the pure culture techniques of the fermentation industries. A procedure devised by Till (1962) in which the ingredients of chopped straw, peat, soya-bean meal, lucerne meal and calcium carbonate are autoclaved and the culture maintained sterile to the stage of applying the casing layer, was later modified by Hunké & von Sengbusch (1968) to include a fermentation (composting) to facilitate applicability. In the modified procedure the sterile phase ends after the sterilization of the Till substrate which according to these authors inhibits and prevents 'the subsequent development of rival organisms'. This procedure however has not been applied commercially.

More recently Smith & Hayes (1972) in describing a simple laboratory system for experimental work using a liquid substrate (hydroponics) found it necessary to use inert carrier materials to provide physical support for full mushroom development which imposed a major limitation on the yield potential. However, better yields were obtained by replacing the carrier materials with a partially inert sphagnum peat and supplementing with solids. Despite encouraging results from these methods, the high capital expenditure involved in the installation of pressurized equipment is prohibitive for commercial cultivation purposes.

The above methods, together with laboratory methods of cultivation in Petri-dishes devised by Hume & Hayes (1972), have provided evidence of the minimal nutritional requirements which relate to all stages in the growth cycle of *Agaricus bisporus* (Hayes, 1972). The requirements for mycelium growth and the formation of fruit-bodies were found to be different. For the production of mycelium, a defined medium composed of simple sugars and salts supplemented with biotin and thiamin were adequate. For the formation of primordia a liquid medium composed of a mixture of chemical elements requires solidification with agar or must be carried in a vermiculite pumice mixture. In addition to those nutrients required for mycelium production, the medium must be supplemented with a source of acetate. This requirement for acetate is absolute for the formation of primordia and consequently fruit-bodies. It is provided in trace amounts of ethylacetate or sodium acetate at 0.1 g dm^{-3} of medium. Other more complex chemicals which yield acetate can be used in place of the simple acetates, *e.g.* fatty acids. A number of vegetable oils rich in fatty acids have been found by Schisler (1967) to stimulate yield when applied to spawned compost at casing time.

The mushroom is also especially demanding in its requirements for the major nutrients, carbon and nitrogen, in order to complete its growth cycle and this underlines the need for a precisely controlled method of substrate preparation. This fact was substantiated by Smith & Hayes (1972) in laboratory culture. Ratios of carbon: nitrogen outside the optimum of 17:1 caused substantial reductions in yield. For the production of primordia in Petri-plates a mixed source of carbohydrates which must include dextrin is essential and there is a specific requirement for a hydrolysed protein such as casein hydrolysate.

In the preparation of composts for commercial growing it is necessary to ensure that all the food requirements are contained in the ingredients of the compost, and are made available to the mushroom by a thorough and carefully controlled composting procedure. This is a demanding and difficult exercise when variable and bulky materials such as animal manures are used. Since the whole economic structure of growing largely stems from the nature of the substrate a greater degree of precision is required in techniques in order to achieve the predictability and control over mushroom output which is standard for other industrial fermentation processes.

In France, Laborde & Delmas (1969) have worked extensively on a method of composting which accelerates the fermentation and involves careful preparation of the ingredients and includes a mechanical breakdown of the vegetable matter. From the time of obtaining the mixture, composting is a one stage process done in trays and in a structure akin to a traditional peak-heat room. This same concept is extended a stage further in what is known as controlled environment composting developed by Randle & Hayes (1972) in which as much control as possible is introduced to limit both chemical and environmental variables. This concept is in many ways similar to that of accelerated composting but the means by which substrates are prepared differs. Composting is done in a cylindrical drum capable of being rotated and aerated simultaneously, to provide the ideal conditions of aeration, temperature control and mixing instead of a conventional heat room. Composting in drums offers the opportunity to explore new methods and systems of growing, with enough flexibility for composting, spawning and the spawn running stage to be done in one stage in a closed system, free of contamination from pests and pathogens. This can be viewed as the first stage of a continuous production line.

The discovery of the involvement of micro-organisms, notably bacteria of the genus *Pseudomonas*, in the casing layer with the development of mushroom fruit-bodies, presents yet another dimension to the complex of factors known to be important in controlling growth. To what extent these bacteria and the environmental factors, carbon dioxide concentration and temperature are related is not yet known but could provide the basis for an objective approach to controlling the growth and maturity of the mushroom fruit-body. Even with sophisticated environment controls, the stage of maturity is spread over several days requiring each mushroom to be selected for harvesting by hand. This requires much skill and is costly.

The application of mechanical aids to harvesting demands a uniform crop. To achieve this further improvements to the culture system are required. Lately, attention has been given to control of the number of primordia which form on the surface of mushroom beds and in experiments using agar media in Petri-plates designed to simulate the chemical environment of the casing soil Hayes (1972) demonstrated that by manipulating chemical factors which inhibit growth (*e.g.* EDTA) and factors which are required for the formation of primordia (*e.g.* Fe^{2+}) control over the numbers of primordia which form was possible. Since the active *Pseudomonads* which are located in the casing layer are known to be involved in the solubilization and availability of iron, control of the numbers of primordia under commercial conditions can best be achieved by manipulating the occurrence and activity of the active *Pseudomonads*. This, in addition to close control over the physical environment, offers the possibility of achieving the control that is required for mechanical harvesting which would not only accelerate the complete industrialization of the large modern tray plants with centralized harvesting facilities, but would also further the concept of a continuous system of culture.

A report published in 1973 by the National Economic Development Organization on U.K. farming and the Common Market indicates that labour is a major single item of cost in the production of mushrooms in the U.K. which accounts for 40% of the total costs, half being for labour for harvesting and packing. Composts represents the next major cost item which amounts to 20–25% of the total production costs. This suggests that there exists much scope for the continued evolution of culture systems to include improvements in composting and harvesting which may not only contribute to standardization and control but also to the economics of culture.

A further development currently under investigation which could also drastically alter the economics of growing stems from the work of Gerrits *et al.* (1967) who demonstrated that the mushroom in its growth only partially utilizes the substrate. This offers the possibility of replenishing the nutrients which are depleted during cropping and recycling the base compost. Unpublished work by Dr W.S. Murphy of the Campbell Soup Company in the U.S.A. demonstrates its feasibility. In many countries, however, growing cereals, composting the straw and growing the mushroom as a cash crop, to return the compost to the soil as a fertilizer is probably as economical and advantageous as recycling the compost for further use.

In its short history—mushroom science—a composite science of many different microbiologically orientated disciplines and specialities has contributed a great deal to the world-wide growth of the industry. Much of the research effort in the past has been devoted to pathology. Lately, more attention has been given to the fundamental aspects of nutrition and substrates, but the benefits of this research has yet to be fully realized. The overriding constraint to the ultimate pairing of biology and engineering and therefore complete industrialization, hinges on the fact that the mushroom is a living and variable micro-organism and knowledge of the fundamental processes which govern its growth is scant. Furthermore, a range of other micro-organisms equally as variable are essential components of its environment in commerce. Advances in the technology of mushroom cultivation would benefit from a more balanced approach by researchers to include genetics, breeding and selection of strains, and physiology and biochemistry aimed at obtaining a better understanding of the biological processes governing growth. Unlike many other foods of fungal origin, of which some are being canvassed as protein substitutes in the diet, mushrooms have been accepted by man for centuries past. This is perhaps a safeguard to the future and the continued development of the industry. It may also provide the necessary stimulus to mushroom science to perfect and further the industrialization of mushroom growing.

11.10 References

ABERCROMBIE, J. (1817). *Abercrombie's practical gardener, or improved system of modern horticulture*, 2nd ed. revised by J. MEAN. London: Cadell and Davies.

ALLEGRO, J.M. (1970). *The sacred mushroom and the cross.* London: Hodder and Stoughton.

ARROLD, N.P. (1972). Confirmation of the ability of *Pseudomonas putida* to cause fruiting of the cultivated mushroom. *Bulletin of the Mushroom Growers Association* **269**, 200.

BAKER, K.F. (1957). The U.C. system for producing healthy container grown plants. *California Agriculture Experimental Station Manual No. 23.*

BARRON, G.L. & FLETCHER, J.T. (1970). *Verticillium albo-atrum* V. *dahliae* as mycoparasites. *Canadian Journal of Botany* **48**, 1137–9.

BARRON, G.L. & FLETCHER, J.R. (1972). *Rhopalomyces elegans* Corda, a host of *Mycogne perniciosa* Magn. *Mushroom Science* **8**, 383–6.

BURROWS, S. (1951). The chemistry of mushroom composts. *Journal of the Science of Food and Agriculture* **3**, 395–410.

CALLOW, E. (1831). Observations on the methods now in use for the artificial growth of mushrooms, with a full explanation of an improved mode of culture, by which a most abundant supply may be procured and continued throughout every month in the year, with a degree of certainty that has in no instance failed. London: Fellows.

CAYROL, J.C. & RITTER, M. (1972). Nematodes et nematicides en myciculture. *Mushroom Science* **8**, 867–79

CHAZE, J. & SARAZIN, A. (1936). Nouvelles données biologiques et expérimentales sur la Mole, maladie du champignon de couche. *Annales des Science Naturelles (Botanique)* **18**, 1–86.

COLSON, B. (1935). The cytology of the mushroom *Psalliota campestris* Quel. *Annals of Botany (London)* **49**, 1–18.

CROSS, M.J. & JACOBS, L. (1969). Some observations on the biology of the spores of *Verticillium malthousei. Mushroom Science* **7**, 239–44.

DAYAL, R. & BARRON, G.L. (1970). *Verticillium psalliotae* as a parasite of *Rhopalomyces. Mycologia* **62**, 826–30.

DIELMAN-VAN ZAAYEN, A. (1971). Methyl bromide fumigation versus other ways to prevent the spread of mushroom virus disease. *Journal of Agricultural Science Netherlands* **19**, 154–67.

DIELMAN-VAN ZAAYEN, A. & IZESZ, O. (1969). Intra-cellular appearance of mushroom virus. *Virology* **39**, 147–52.

DUGGAR, B.M. (1905). The principles of mushroom growing and mushroom spawn making. *Bulletin of the United States Bureau.of Plant Industries*, No. 55.

EDWARDS, R.L. (1949). M.R.A. report on synthetic composts. *Bulletin of the Mushroom Growers Association* **15**, 84–8.

EDWARDS, R.L., & SLEGG, P.B. (1953). Cropping experiments on casing soil. *Annual Report Mushroom Research Station*, 1952, 20–9.

EGER, G. (1961). Untersuchungen uber die Funktion der Deckschicht bei der Fruchtkoerperbildung des Kulturchampignons, *Psalliota bispora* Lge. *Archiv für Mikrobiologie* **39**, 313–34.

ELLIOT, T.J. (1972). Sex and the single spore. *Mushroom Science* **8**, 11–8.

ELLIS, J.J. (1963). A study of *Rhopalomyces elegans* in pure culture. *Mycologia* **55**, 115–35.

EVANS, H.J. (1959). Nuclear behaviour in the cultivated mushroom. *Chromosoma* **10**, 115–35.

FEKETE, K. (1967). Uber morphologie, biologie und bekampfung von *Verticillium malthousei*, einem parasiten des kulturchampignons. *Phytopathologische Zeitschrift* **59**, 1–32.

FEKETE, K. & KUHN, J. (1965). Bekampfung von *Verticillium* und *Mycogne*. (Vourlaufige Mitteilung). *Mushroom Science* **6**, 495–506.

FEKETE, K. & KUHN, J. (1966). Vertomyc, ein neues fungizid zur bekampfung von *Verticillium* und *Mycogne*. *Champignon* **6**, 16–22.

FERGUSON, M. (1902). A preliminary study of the germination of the spores of *Agaricus campestris* and other basidiomycetous fungi. *Bulletin of United States Bureau of Plant Industries* No. 16.

FLEGG, P.B. (1968). Response of the cultivated mushroom to temperature at various stages of crop growth. *Journal of Horticultural Science* **43**, 441–52.

FLEGG, P.B. (1970). Response of the cultivated mushroom to temperature during the two-week period after casing. *Journal of Horticultural Science* **45**, 187–96.

FLEGG, P.B. & GANDY, D.G. (1962). Controlled environment cabinets for experiments with the cultivated mushroom. *Journal of Horticultural Science* **37**, 124–33.

FRITSCHE, G. (1964). Versuche zur frage der Merkmalsubertragung beim Kulturchampignon. *Der Zuchter* **34**, 76–93.

GANDY, D.G. (1960). 'Watery stipe' of cultivated mushrooms. *Nature, London* **185**, 482–3.

GANDY, D.G. (1971). Experiments on the use of benomyl ('Benlate') against *Verticillium. Bulletin of the Mushroom Growers Association* **257**, 184–7.

GANDY, D.G. (1972). Observations on the development of *Verticillium malthousei* in mushroom crops and the role of cultural practices in its control. *Mushroom Science* **8**, 171–81.

GANDY, D.G. & HOLLINGS, M. (1962). Dieback of mushrooms; a disease associated with a virus. *Annual Report of the Glasshouse Crops Research Institute*, 103–8.

GANNEY, G.W. (1972). Critical observations of mushroom virus problems in the United Kingdom. *Mushroom Science* **8**, 739–53.

GERRITS, J.P.G., BELS-KONING, H.C. & MULLER, F.C. (1967). Changes in compost constituents during composting pasteurisation and cropping. *Mushroom Science* **6**, 225–43.

GRAY, W.D. (1970). *The use of fungi as food and in food processing*. London: Butterworths.

GYURKÓ, P. (1972). Die rolle der belichtung bei dem anbau des austermseitlings (*Pleurotus ostreatus*). *Mushroom Science* **8**, 461–9.

HAGER, R.A. (1969). An investigation of X-disease. *Mushroom Science* **7**, 205–11.

HASTIE, A.C. (1970). Benlate induced instability of *Aspergillus diploids. Nature, London* **226**, 771.

HAYES, W.A. (1969*a*). Microbial changes occurring when differently treated mixtures of wheat straw and horse manure are composted. *Mushroom Science* **7**, 173–186.

HAYES, W.A. (1969*b*). New techniques with mushroom compost. *Span*, **12**, 162–6.

HAYES, W.A. (1970). Fumigation—its application to commercial mushroom growing. *Bulletin of the Mushroom Growers Association* **257**, 213.

HAYES, W.A. (1972). Nutritional factors in relation to mushroom production. *Mushroom Science* **8**, 663–74.

HAYES, W.A. (1973). The emergence of mushroom science and technology—its impact on present and future methods of production. *AMI Mushroom News 2*, 11–26.

HAYES, W.A. & RANDLE, P.E. (1968). The use of water-soluble carbohydrates and methyl bromide in the preparation of mushroom composts. *Bulletin of the Mushroom Growers Association* **218**, 81–102.

HAYES, W.A. & RANDLE, P.E. (1969). Use of molasses as an ingredient of wheat straw mixtures used for the preparation of mushroom composts. *Report of the Glasshouse Crops Research Institute*, 142–7.

HAYES, W.A. & RANDLE, P.E. (1970). An alternative method of preparing composts using methyl bromide as a pasteurising agent. *Report of the Glasshouse Crops Research Institute,* 166–9.

HAYES, W.A., RANDLE, P.E. & LAST, F.T. (1969). The nature of the microbial stimulus affecting sporophore formation in *Agaricus bisporus* (Lange) Sing. *Annals of Applied Biology* **64**, 177–87.

HESLING, J.J. & KEMPTON, R.J. (1969). The control of nematodes in mushroom compost. *Proceedings of the 5th British Insecticide and Fungicide Conference* **1**, 185–8.

HO, MING-SHU, (1972). Straw mushroom cultivation in plastic bags. *Mushroom Science*, **8**, 257–63.

HOLLINGS, M. (1962). Viruses associated with a die-back disease of cultivated mushroom. *Nature, London* **196**, 962–5.

HOLLINGS, M. (1972). Recent research on mushroom viruses. *Mushroom Science* **8**, 733–8.

HOLLINGS, M. & STONE, O.M. (1969). Viruses in fungi. *Science Progress* Oxf. **57**, 371–91.

HOLLINGS, M. & STONE, O.M. (1971). Viruses that infect fungi. *Annual Review of Phytopathology* **9**, 93–118.

HUDSON, E.K. (1972). Nematodes as biological control agents. *Mushroom Science* **8**, 527–32.

HUME, D.P. & HAYES, W.A. (1972). The production of fruit-body primordia in *Agaricus bisporus* (Lange) Sing. on agar media. *Mushroom Science* **8**, 527–32.

HUNKÉ, W. & VON SENGBUSH, R. (1968). Champignonanbau auf nicht kompostiertem nährsubstrat. *Mushroom Science* **7**, 405–9.

HUSSEY, N.W. (1959). Biology of mushroom phorids. *Mushroom Science* **4**, 160–70.

HUSSEY, N.W. (1965). Observations on the association between *Bradynema* sp. (Nematoda: Allantonematideae) and the mushroom infesting fly *Megaselia halterata* (Diptera: Phoridae). *Proceedings of the XII International Congress of Entomology* (1964) 752 pp.

HUSSEY, N.W. (1969). Biological control of mushroom pests—Facts and fantasy. *Bulletin of the Mushroom Growers Association* **238**, 448–65.

HUSSEY, N.W. (1972). Pests in perspective. *Mushroom Science* **8**, 183–92.

HUSSEY, N.W. & GURNEY, B. (1967). *Pygmephorus* spp. (Acarina: Pyemotidae) associated with cultivated mushrooms. *Acarologia* **9**, 353–8.

HUSSEY, N.W. & WYATT, J.J. (1959). Cecid control by incorporation of insecticides in compost. *Mushroom Science* **4**, 280–86.

HUSSEY, N.W., READ, W.H. & HESLING, J.J. (1969). *The pests of protected cultivation. The biology and control of glasshouse and mushroom pests.* London: Edward Arnold.

KNEEBONE, L.R., LOCKARD, J.D. & HAGAR, R.A. (1962). Infectivity studies with X-disease. *Mushroom Science* **5**, 461–7.

KOSTADINOV, I., TURÉV, A. & RANTCHEVA, T. (1972). Some aspects of the production of *Pleurotus ostriatus. Mushroom Science* **8**, 253–6.

LABORDE, J. & DELMAS, J. (1969). La préparation express des substrates. *Bulletin Federation National Syndicates Agricole Cultivateurs de Champignons* **184**, 2093–109.

LAMBERT, E.B. (1930). Studies on the relation of temperature to the growth, parasitism, thermal death point and control of *Mycogne perniciosa. Phytopathology* **20**, 75–83.

LAMBERT, E.B. (1933). Effect of excess carbon dioxide on growing mushrooms. *Journal of Agricultural Research* **47**, 599–608.

LAMBERT, E.B. (1938). Principles and problems of mushroom culture. *Botanical Reviews* **4**, 397–426.

LAMBERT, E.B. (1941). Studies on the preparation of mushroom compost. *Journal of Agricultural Research* **62**, 415–422.

LAST, F.T. (1970). Mushroom cultivation. Mystique or method? The changing scene. *Bulletin of the Mushroom Growers Association* **246**, 259–73; **247**, 312–22; **248**, 370–9.

LAST, F.T., HOLLINGS, M. & STONE, O.M. (1967). Some effects of cultural treatments on virus diseases of cultivated mushroom *Agaricus bisporus. Annals of Applied Biology* **59**, 451–62.

LELLIOTT, R.A., BILLING, E. & HAYWARD, A.C. (1966). A determinative scheme for the fluorescent plant pathogenic *Pseudomonads. Journal of Applied Bacteriology* **29**, 470–89.

LONG, P.E. & JACOBS, L. (1968). Some observations on carbon dioxide and sporophore initiation in the cultivated mushroom.

Mushroom Science 7, 727–31.

LOUDON, J.C. (1850). *An encyclopaedia of gardening; comprising the theory and practice of horticulture, floriculture, aboriculture and landscape gardening*. 2nd Ed. Longman, Brown, Green and Longmans, London.

MIDDLEBROOK, S. & STOREY, D.I.F. (1950). Observation on cropping in three types of mushroom house. *Mushroom Science* **1**, 26–34.

MORGAREIDGE, K. (1958). The nutritional value of mushrooms. *American Mushroom Institute Report*, Kennett Square, Pennsylvania.

NAIR, N.G. (1969). Two diseases of cultivated mushrooms. *Agricultural Gazette, New South Wales* **80**, 638–9.

NAIR, N.G. (1972). Observations on virus disease of the cultivated mushroom *Agaricus bisporus*, in Australia. *Mushroom Science* **8**, 155–70.

NAIR, N.G. (1973). Heat therapy of virus-infected cultures of the cultivated mushroom *Agaricus bisporus*. *Australian Journal of Agricultural Research* **24**.

NAIR, N.G. & FAHY, P.C. (1972). Bacteria antagonistic to *Pseudomonas tolaasii* and their control of the brown blotch of the cultivated mushroom *Agaricus bisporus*. *Journal of Applied Bacteriology* **35**, 439–442.

NAIR, N.G. & FAHY, P.C. (1973). Toxin production by *Pseudomonas tolaasii* Paine. *Australian Journal of Biological Sciences* **26**, 509–12.

NAIR, N.G. & MILHAM, P.J. (1970). Brominated chemicals in mushroom cultivation. *Agricultural Gazette, New South Wales* **81**, 419.

NEWMAN, R.H. & SAVIDGE, M. (1969). Mancozeb dust—a breakthrough in mushroom disease control. *Bulletin of the Mushroom Growers Association* **232**, 161–162.

PEAKE, R. (1972). Benlate—The fungicide for mushrooms. *Bulletin of the Mushroom Growers Association* **267**, 127–9.

PERSSON, S.P.E. (1972). Mechanical harvesting of mushrooms and its implications. *Mushroom Science* **8**, 115–23.

PIZER, N.A. (1937). Investigations into the environment and nutrition of cultivated mushroom *Psalliota campestris*. I. Some properties of composts in relation to the growth of the mycelium. *Journal of Agricultural Science* **27**, 349–75.

PIZER, N.H. & THOMPSON, A.J. (1938). Investigations into the environment and the nutrition of the cultivated mushroom *Psalliota campestris*. II. The effect of calcium and phosphate on growth and productivity. *Journal of Agricultural Science* **28**, 604–17.

RANDLE, P.E. & HAYES, W.A. (1972). Progress in experimentation on the efficiency of composting and compost. *Mushroom Science* **8**, 789–95.

RAPER, J.R. (1966). *Genetics and sexuality in higher fungi*. New York: Ronald.

RAPER, J.R. & RAPER, C.P. (1972). Life cycle and prospects for interstrain breeding in *Agaricus bisporus*. *Mushroom Science* **8**, 1–9.

RASMUSSEN, C.R., MITCHELL, R.E. & SLACK, C.I. (1972). 'Heat treatment' of cultures from apparently healthy and virus-infected mushrooms and subsequent effects on cropping yields. *Mushroom Science* **8**, 239–51.

SCHISLER, L.C. (1967). Stimulation of yield in the cultivated mushrooms by vegetable oils. *Applied Microbiology* **15**, 844–50.

SCHISLER, L.C., SINDEN, J.W. & SIGEL, E.M. (1967). Etiology, symptomatology and epidemiology of a virus disease of cultivated mushrooms. *Phytopathology* **57**, 519–26.

SCHISLER, L.C., SINDEN, J.W. & SIGEL, E.M. (1968). Etiology of mummy disease of cultivated mushrooms. *Phytopathology* **58**, 944–8.

SINGER, R. (1961). *Mushrooms and truffles*. London: Leonard Hill.

SINDEN, J.W. (1932). Mushroom spawn and method of making same. *U.S. Patent, 1,869,517*.

SINDEN, J.W. (1946). Synthetic compost for mushroom growing. *Bulletin of the Pennsylvania Agricultural Experimental Station* **482**.

SINDEN, J.W. (1971). Ecological control of pathogens and weed-molds in mushroom culture. *Annual Review of Phytopathology* **9**, 411–32.

SINDEN, J.W. (1972). Disease problems in technologically advanced mushroom nurseries. *Mushroom Science* **8**, 124–30.

SINDEN, J.W. & HAUSER, E. (1950). The short method of composting. *Mushroom Science* **1**, 52–9.

SINDEN, J.W. & HAUSER, E. (1950). Report on two new mushroom diseases. *Mushroom Science* **1**, 96–100.

SINDEN, J.W. & HAUSER, E. (1954). The nature of the composting process and its relation to short composting. *Mushroom Science* **2**, 123–31.

SMITH, F.E.V. (1924). Three diseases of cultivated mushroom. *Transactions of the British Mycological Society* **10**, 81–97.

SMITH, J.F. & HAYES, W.A. (1972). Use of autoclaved substrates in nutritional investigations on the cultivated mushrooms. *Mushroom Science* 8, 355–61.

STOLLER, B.B. (1968). The use of hexachlorophene to prevent bacterial spotting of mushrooms. *Bulletin of the Mushroom Growers Association* 224, 398–401.

STOLP, H. (1961). Neue Erkenntnisse über phytopathogene Bakterien und die von ihnen verursachten Krankheiten. I. Verwandtschaftsbeziehungen zwischen phytopathogenen Pseudomonas—Arten und saprophytischen Fluoreszenten auf der Grundlage von Phagenreaktonen. *Phytopathologische Zeitschrift* 42, 197–262.

STOREY, I.F. (1965) The cropping problem in 1964. *Mushroom Growers Bull.* 184, 154–160.

TILL, O. (1962). Champignonkultur auf Steriliziertem nährsubstrat und die wiederverwendung von abgetragenem kompost. *Mushroom Science* 5, 127–33.

TOTH, E. (1970). Sterile method for production of *Pleusotus ostreatus*. *Gradinarstwo* 6-*Sophie*, 42–4.

TSCHIERPE, H.J. (1959). Die Bedeutung des Kohlendioxyds fur den Kulturchampignon. *Gartenbau Wissenschaft* 24, 18–75.

TSCHIERPE, H.J. & SINDEN, J.W. (1964). Weitere Untersuchungen Kohlendioxyd fur die Fruktifikgation des Kulturchampignons *Agaricus campestris* var. *bisporus* (L) Lge. *Archiv für Mikrobiologie* 49, 405–25.

TRESCHOW, C. (1941). The Verticillium diseases of cultivated mushroom. *Dansk Botanical Archiv* 11, 1–31.

TUCKER, C.M. & ROUTIEN, J.B. (1924). The mummy disease of the cultivated mushroom. *Research Bulletin of the Missouri Agricultural Experimental Station,* 358.

WAKSMAN, S.A. & NISSEN, W. (1932). On the nutrition of the cultivated mushroom and the chemical changes brought about by this organism in the manure compost. *American Journal of Botany* 19, 514–37.

WORGAN, J.T. (1968). Cultivation of higher fungi. *Progress in Industrial Microbiology* 8, 73–139.

WUEST, P.J. (1971). A new mushroom fungicide is discovered. *Mushroom News* 19, 1–4.

WYATT, I.J. (1970). The control of paedogenic cecid larvae in mushroom beds by the use of thionazin. *Annals of Applied Biology* 66, 497–504.

CHAPTER 12

Submerged Culture Production of Mycelial Biomass

G. L. SOLOMONS

12.1 Introduction

The World's fermentation industry has been producing mycelial biomass for many years, but only as a method of obtaining certain desired secondary metabolites. The use of *Penicillium chrysogenum* to produce penicillin, *Rhizopus nigricans* to convert progesterone to 11-hydroxy-progesterone and *Aspergillus niger* to produce citric acid are three major examples of this 'secondary' biomass production, but in no case is the organism itself required. It is, in fact, a considerable nuisance, consuming valuable carbohydrates and other nutrients to grow, and poses some difficult disposal problems when its useful work is done. Therefore, it is commercial sense to *restrict* the amount of organism produced to the absolute *minimum* consistent with the production of the maximum amount of secondary metabolite. Nevertheless, much valuable information has been obtained with regard to the microbiology, the biochemistry and the biochemical (or microbiological) engineering of these existing fermentation processes, which can be applied to the problems of producing mycelial biomass as an end product.

In considering submerged culture production of fungi (considering only the Eumycetes, but excluding the yeasts), two distinct areas of interest are apparent. The first of these is the production of mushrooms for flavour and structure to replace mushrooms grown in the conventional manner. The second is the production of mycelium largely for its protein content, as another contender in the SCP (Single Cell Protein) field; the title of SCP being a strict misnomer in this context, since fungi usually grow in multi-cellular form.

12.2 The Production of Mycelium for use as Flavouring Agents

The wide acceptance of the fungus *Agaricus bisporus* has led to large scale commercial cultivation using a type of surface-culture based on compost (Smith, 1969). This is a lengthy and complex process, and for basidiocarp formation requires a process termed 'casing', which involves the use of a soil layer. It has been shown that the soil contains bacteria which produce compounds that trigger basidiocarp formation (Park & Agnihotri, 1969; see also Chapter 11).

Because there is a large commercial market for mushrooms, work was undertaken to see if they could be grown in submerged culture. Humfeld (1948) carried out the first commercially orientated researches and showed that spherical pellets of mycelium could be obtained with a remarkably high yield factor of 0.6 (g of dry cell/g of carbohydrate consumed). Since then, many other workers have examined the problem using a wide range of fungi. When compared to many other species *Agaricus bisporus* scores low marks for flavour. The problems encountered fall into three categories: growth, growth rate and yield; growth form, *i.e.* pellets or filamentous mycelium (see Fig. 12.1); and flavour. Since (so far) basidiocarp

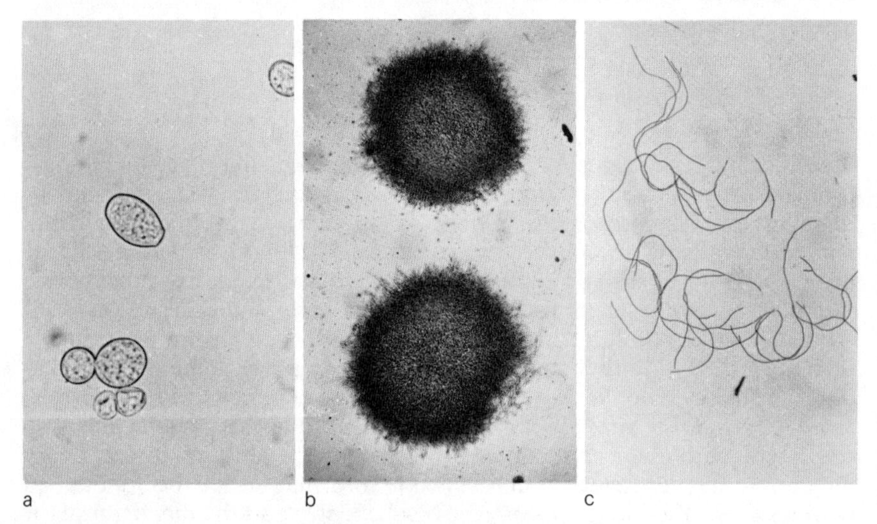

Fig. 12.1 Forms of fungal growths (a) single cell growth; (b) pelleted growth, and (c) filamentous growth.

formation in submerged culture does not occur, mushrooms cultivated in this manner, must be used for selected purposes. The main use envisaged was for food manufacturing, when its form as a powder would not necessarily be detrimental. It could be used as a flavour enhancer or in the form of comminuted mushroom soup.

Growth, growth rate and yield

In a review of mushrooms viewed as a potential source for protein production, Litchfield (1968) gives a comprehensive list of the growth as g dm^{-3} dry weight obtained by investigation of 13 different fungi grown in many different culture media, both simple and complex. Most of the data obtained shows that growth of only a few (2–8) g dm^{-3} were obtained.

Exceptionally, 20–30 g dm^{-3} were reported for two fungi *Agaricus blazei* and *Morchella hydrida* when grown on corn-steep liquor and beet molasses media respectively. Humfeld (1952) claimed a growth of 30.7 g dm^{-3} using *Agaricus campestris* (white variety) grown on glucose/salts medium—a remarkable value when only 50 g dm^{-3} of carbohydrate were supplied as carbon source. Heinemann (1963) only claimed around 7–8 g dm^{-3} for species of *Agaricus, Coprinus, Helvella* and *Morchella*. Growing species of *Morchella*, Szuecs (1956) obtained around 18 g dm^{-3} but, although not reported, the ash content could be expected to be unusually high, since 2.4 g dm^{-3} of calcium carbonate were used in the medium and this was 'digested'.

Growth rate values in any of the work discussed so far are not given in terms of μh^{-1} or $\mu_{max} h^{-1}$. Cirullo, Hardwick & Seeley (1960) give values of dry weight *v* time, which allow calculation of a doubling time of 2.6 h and $\mu_{max} = 0.26$ h^{-1}. It is therefore not possible to give any true idea of how fast most of these organisms will grow, but, as in the question of growth, times of growth to reach a given dry weight vary considerably.

As with amount of growth, Humfeld (1952, 1954) with *Agaricus campestris* has reported the shortest incubation times for mycelium production, 24 h to reach 30.7 g dm^{-3} and 25.5 g dm^{-3} in 33 h, also with the same organism. Heinemann (1963) quotes about 96 h to achieve 7–8 g dm^{-3} dry weight; whilst Szuecs (1956) states that in 72 h *Morchella esculenta* reached 18.35 g dm^{-3} dry weight.

In the literature, the yield of cells is not usually expressed in standard terminology. If we accept yield factor Y = g cell dry weight/g carbohydrate consumed, some unbelievably high values are given. Thus Litchfield (1968) quotes Y values of up to 0.861 and a number in the range 0.60–0.575. This is difficult to accept at its face value, since Y_{max} of around 0.5 for carbohydrates is a reasonable value, if the carbon content of cells are in the range of 45–50%, as is the case with most micro-organisms examined. However, in complex medium there can be many other sources of carbon besides the carbohydrate content (*e.g.* amino-acids, lipids *etc.*) and yield factors expressed only on the carbohydrate content instead of assimilable carbon content can be misleading. Values given by Litchfield, Overbeck & Davidson (1963) for three species of *Morchella* grown on glucose, maltose and lactose are 0.43–0.49, a range which is in keeping with other micro-organisms grown on the same substrates. In addition, there is some evidence to suggest that mushrooms can have exceptionally high ash contents, Robinson (1952) quotes ash values of from 4.67 to 18.60%. Under these circumstances, of course, yield factors can achieve higher values than would usually be expected, since in most micro-organisms ash is of the order of 5–8%. At 18% ash, a yield factor of 0.60 would be equivalent at 8% ash to 0.54.

Growth form

It has been generally accepted that the desirable form of fungal growth for submerged mushroom mycelium production is as spherical pellets. As will be discussed in some detail in a later section, the growth form of fungi is of considerable importance to the overall physico-chemical environment within the fermenter. One undeniable advantage of pellet growth is the ease of harvest, a simple sieve is all that is required. Moreover, the complex structure of pellets, which usually involve an autolysed centre and growth only at the periphery of the pellet, might also be important for taste and flavour. In the literature concerning mushroom production there appears to be no knowledge of the work by Camici, Sermonti & Chain (1952), who showed that pellet or mycelial type of culture in shake flasks could be controlled by inoculum level, which in turn depended upon the type of culture medium used. Simple medium requires smaller inoculum levels to avoid pellet formation than complex medium using, for example, corn-steep liquor.

The fermenters used for mushroom culture have employed slow speed agitation, a feature which in itself tends to favour pellet formation (Szuecs, 1956, 1958). Agitation besides simply suspending particles also transfers oxygen from the supplied air stream into the culture medium. Litchfield, Overbeck & Davidson (1963) reported that for a *Morchella* sp. an oxygen transfer rate of 4.8 mMole O_2/l/h gave the highest yield of pellet forms and over a rate of 9.0 mMole O_2/l/h filamentous growth occurred with a reduction in yield. This is somewhat surprising since the pellets are usually anaerobic in the centre and oxygen limitation occurs when pellet size of *Penicillium* sp. is only 0.2 mm in diameter (Phillips, 1966). On the other hand, once filamentous mycelium is formed, the oxygen transfer rate falls rapidly and this may account for the lower yield found. In order to ensure pellet formation Szuecs (1956) added insoluble compounds such as calcium carbonate or calcium sulphate to his medium. He termed these 'particulate support material', which no doubt acted in a similar manner to 'seed' crystals in a crystallation procedure, in allowing a nucleate point of growth.

Flavour

Since mushrooms are used as a garnish, the flavour is of utmost importance. The development and assessment of flavour is a somewhat subjective analysis and it is, therefore, not surprising to find rather conflicting evidence presented in the literature.

It seems generally accepted, however, that flavour is not growth associated, but appears in its most intense form, after the carbohydrate has been consumed. Humfeld (1952) provides an example by showing that although carbohydrate had been used up at 24 h, the product was not harvested until 93 h. In this example the dry weight had fallen from 30.7 g dm^{-3} at 24 h to 22.8 g dm^{-3} by the time of harvest. This certainly suggests an autolytic process contributing to flavour, a case well established with meat from both animals and poultry, which are virtually tasteless when completely fresh and only become acceptable when they are 'hung' for a short period of time.

Cultural conditions also appear to affect flavour. Heinemann (1963) claiming that the incorporation of skim milk in the range 1–10% substantially improved the flavour of a number of species. Humfeld & Sugihara (1952) investigated the fungus *Agaricus campestris* and found that the medium requirements of nitrogen, phosphorus, potassium and sulphur for the organism to develop maximum flavour were 1.5–2.0, 6.0–8.0, 1.0–3.0, and 4.0 times respectively, the concentrations required for maximum growth.

In a more recent review, Litchfield (1967) discusses some of the chemical compounds possibly responsible for the flavour of morel mushroom mycelium.

12.3 The Production of Protein by Mycelial Biomass

Before considering specific problems associated with the use of mycelial biomass for protein production, we must consider general problems of SCP production, which the fungi have in common with yeasts, bacteria and algae.

The first essential is to be clear as to the purpose of producing protein, that is, to what use will your product(s) be put. Essentially, the protein may be used as food for direct human consumption (Spicers, 1973) or as animal feedstuff (Forss *et al.*, 1972; Imrie, 1973). The choice of which market to aim for is critical to subsequent operations. Whilst off-specification material intended for foodstuffs can perhaps be disposed of in feedstuffs, all the initial process costings will have been geared to the more expensive product and its disposal via the less lucrative route is only used to save complete loss. Some of the implications of choosing to market a foodstuff commodity include: use of higher grade substrate and associated chemicals; a higher standard of microbiological and chemical control; more severe nutritional and particularly toxicological testing; more sophisticated and therefore expensive food formulation; and more expensive marketing operations.

Having decided which market one should aim for, some general requirements of any organism chosen are: good nutritional quality of protein (*e.g.* high NPU/PER values, Duthie, 1973); non-toxic to humans, animals, fish and plants; high protein content; satisfactory growth rate; high yield factor; ability to grow in a reasonably simple culture media and almost certainly the ability to use ammonia or ammonium as nitrogen source; ability to grow in submerged culture to a cell density of 10–40 g dm^{-3} dry weight; should not present additional problems of recovery other than size (*e.g.* should not have a slimy capsule, *etc.*); and in the case of microfungi in particular, should be able to grow in a suitable growth form, *i.e.* filamentous and not in pellet or single cell form.

When we consider the use of fungal mycelium as a potential source of biomass, we find that in spite of the now huge amount of literature on the single cell protein subject, little work has been done with fungi. Perhaps the reasons for this are the commonly held views (not of course supported by many mycologists) that fungi: are often 'poisonous' (look at toadstools!), only grow slowly, are difficult to grow, have a low protein content, and what protein they contain is of poor quality (see for example Litchfield, 1968). The case for the use of fungi really consists of a detailed rebuttal of these points and providing actual experimental evidence against them.

Why fungi?

Although fungi, both micro and macro, are used on a world wide basis for or in foods, *e.g.* mushrooms, mould inoculated cheeses (Stilton, Roquefort, Gorgonzola), mould inoculated staples such as Meso and Tempeh, great play is made of the toxicity of many fungi. Whilst it is true that a considerable number of toxic compounds are produced by fungi, a large number are produced by bacteria and plants. Gray (1970), in a masterly review of the fungi, makes this point forcibly.

If the choice of the potential manufacturer of SCP is for foodstuffs, many arguments for the use of fungi, as opposed to other micro-organisms, can be made; for example, they are used in existing foodstuffs and their naturally occurring texture (Fig. 12.1) makes them particularly suitable for human foodstuffs (Fig. 12.2). It cannot be stressed enough that people eat food not nutrition, so that acceptable new foods have to meet this primary requirement. It is of interest that of the many fungi screened in our laboratory, only one, an *Alternaria* sp., produced acute toxicity in rats and this organism is a well-known toxin producer.

Growth rate and yield factor

When single cell organisms multiply they do so in a manner which can be expressed by the equation:

$$\mu = \frac{1}{x}\frac{dx}{dt} \qquad \text{equation } (1)$$

where x is weight of cells
t is time
μ is specific growth rate, h^{-1}.
This on integration leads to:

$$\mu = \frac{\log_e 2}{td} = \frac{0.693}{td} \qquad \text{equation } (2)$$

where μ is specific growth rate, h^{-1}
td is doubling time (h).

When growth is not limited by an external nutrient (*e.g.* at the beginning of a batch culture) then $\mu = \mu^{max}$, that is the maximum specific growth rate in that particular medium, under the particular environmental conditions of temperature, pH *etc.* Since micro-fungi do not multiply by simple fission, it was held by a number of mycologists that fungi could therefore only grow at a linear rate and not at an exponential rate. This assumption has been shown to be wrong since fungi may grow at each end of the filament of mycelium and can branch (Trinci, 1969). It can be shown that fungi will in fact grow exponentially until they form such a structure that there is not the space available for this type of growth and nutrient diffusion becomes limiting, when they form 'pellet' types and these then grow according to a cube law. Nevertheless, we have been able to measure true exponential growth up to 10–15 g dm^{-3} dry weight. We have also found that shake flask cultures can be used for accurate growth rate experiments, provided a small inoculum is used and measurements are made over a limited range of dry weights up to 150–200 mg/100 cm^3. It is of interest to note that the experiments can be carried out using optical density measurements as well as dry weights, in a fashion analogous to bacterial growth experiments. Similar experimental techniques have been reported by Trinci (1972).

Fig. 12.2 One type of food dish prepared from mycelium protein.

If one turns for information to the literature on fungal growth rates, one is immediately struck by the paucity of data. Many papers refer to x g dm^{-3} obtained after y hours of batch growth, but detailed information is very difficult to obtain. Amongst those fungi which have been examined for maximum growth rates are *Aspergillus niger* 0.20 h^{-1} (Carter & Bull, 1969; Fencl & Novák, 1969) and *Graphium* sp. 0.187 h^{-1} (Volesky & Zajic, 1971). Trinci (1969) gave the following values at 25°C: *Aspergillus nidulans*, 0.148 h^{-1}; *Penicillium chrysogenum*, 0.123 h^{-1}; and *Mucor hiemalis*, 0.099 h^{-1}. The effect of temperature was shown by values for *Aspergillus nidulans*: 25°C, 0.148 h^{-1}; 30°C, 0.215 h^{-1}; 37°C, 0.360 h^{-1}. Trinci (1972) also reported the fastest fungal growth rate that we are aware of, a *Geotrichum candidum* with a doubling time of 1.10 h, which is equivalent to a growth rate of 0.61 h^{-1} at 30°C.

In examining a large number of micro-fungi, the fastest growth rate we have found is that of *Neurospora sitophila* which has a μ_{max} of 0.40 h^{-1}, which represent a cell doubling time of 1.73 h. This is a fast rate and is in fact faster than a number of bacteria. The organism we have examined in greatest detail is a strain of *Fusarium graminearium* Schwabe (Solomons & Scammell, 1970) which has a μ_{max} at 30°C of 0.28 h^{-1}, a doubling time of 2 h 25 min. Of course the maximum growth rate is not by itself of great relevance in SCP production by continuous culture, but assumes great importance if batch culture is to be employed. The relative merits of these two methods of cultivation will be examined in detail in a later section, but it is probably true to state that provided an organism will grow and utilize 98% plus of its limiting substrate (usually carbon source for SCP) at a dilution rate D of $0.1-0.2$ h^{-1}, its maximum rate is not important.

From a production standpoint, the yield factor (Y) is of great importance. Clearly it is necessary to use as little of the limiting substrate (carbon source for SCP) as possible since this is almost always the most expensive component supplied. Even this statement has to be interpreted with caution since a high value for Y can be misleading, although such a caution usually applies to bacteria rather than fungi. Nevertheless, it has been shown by Herbert (1959) that some organisms do produce considerable quantities of polysaccharide, which while appearing in dry weight determinations of cell production, greatly lower the protein content of the product even if the protein content of the cells themselves remains constant. Therefore, we must take this type of growth into consideration and require a high value of Y, but at a constant (or nearly so) value of N% of cell dry weight.

Based on glucose (or other monosaccharides) a yield factor of 0.48–0.50 is considered achievable, this of course requires that carbon source is the only limiting nutrient. Failure to supply sufficient oxygen for aerobic glycolysis usually leads to a reduction in yield as carbon is wasted via the anaerobic CO_2 producing pathways.

Another factor causing lowering of yield is the necessity to ensure that as well as oxygen supply, the pH of the medium needs controlling. Too many authors do not provide data of the final culture pH when quoting optimum incubation times. Shuklar & Datta (1965) growing a species of *Rhizopus* stated that eight days was required for maximum growth of their organism, achieving 36 g dm^{-3} in shake flask at 30°C. They provided 100 g dm^{-3} of sucrose and this would (assuming total sugar utilization) give a $Y=0.36$. The main point however is that most *Rhizopus* spp. are very fast growing organisms and eight days is an excessive incubation time. The Takata's medium they employed possibly had insufficient buffering capacity to cope with 100 g dm^{-3} of sucrose and probably the pH of their culture fell so low that the growth rate declined. We have grown a *Rhizopus* in single cell form (see later) to 40 g dm^{-3} dry weight in 18 h, but of course pH control and aeration were adequate.

Other examples of yield factors which appear to be extraordinarily high are given by Falanghe, Smith & Rackis (1964). Growing *Boletus indecisus* on soya bean whey, they achieved a yield factor of 0.74 on the carbohydrate. Again, whey contains other assimilable carbon sources and the elemental analysis of this organism is not given, so caution must be exercised in interpreting this data.

Some interesting results that clearly show this need for caution were provided by Reusser, Spencer & Sallans (1958). This experienced group of workers carefully provide data on terminal pH's and compare simple and complex culture medium. Growing *Tricholoma nudum* in stirred fermenters on glucose salts/medium they obtained the results shown in Table 12.1.

Table 12.1 Effect of nitrogen source in a glucose/salts medium on yield of mycelium (Data from Reusser, Spencer & Sallans (1958))

Nitrogen source	Terminal pH	Carbohydrate utilization %	Yield factor
Ammonium nitrate	3.4	100	0.376
Potassium nitrate	7.5	100	0.376
Ammonium sulphate	2.0	14.9	0.532
Ammonium chloride	1.8	12.7	0.443
Ammonium tartrate	2.2	83	0.334
Urea	3.0	100	0.418

When grown on a complex medium the results shown on Table 12.2 were obtained. This table shows the effects of increased substrate concentration on dry weight and yield factor. At low levels of sugar, yield factors are high 0.54–0.66, whereas at high substrate levels, the yield factor fell to 0.296. This fall is almost certainly due to the oxygen limitation, since the fermenter does not appear to be able to support more than around 35 g dm^{-3} dry weight.

In comparing the results in Tables 1 and 2 it can be seen that the yield factor for a glucose/salts medium with good substrate utilization was 0.38–0.42. From complex medium the values were up to 0.66, showing the effects of assimilable non-carbohydrate sources.

Our own work has shown that many microfungi, when grown under conditions that provide carbohydrate limitation of batch growth, have a yield factor of 0.48–0.52. It is in fact a difficult determination to carry out

Table 12.2 Effect of substrate concentrate in complex medium on yield factor. (Data from Reusser, Spencer & Sallans (1958))

Medium	Total sugars g dm^{-3}	Sugars used %	Mycelium g dm^{-3}	Yield factor
Molasses/tartrate	24.8	94.3	12.7	0.539
	42.0	95.7	21.6	0.538
	59.6	96.1	35.2	0.614
	81.2	96.2	38.8	0.497
	101.0	95.8	38.5	0.398
	120.0	91.9	32.6	0.296
Sulphite liquor/ Sulphate	21.3	78.9	11.1	0.662
	23.2	77.5	11.6	0.640

accurately, since the usual method employed with micro-fungi of filtration, washing and drying leads to small errors not easily avoided.

Problems in growing micro-fungi in submerged culture

PROVISION OF INOCULUM The work-up of most bacterial and yeasts cultures to form an inoculum for fermenter studies is usually very easy. Agar cultures usually readily and uniformly suspend in Ringer solution and simple optical density measurements allow reproducible cell levels to be used. By using young agar cultures the lag phase can often be reduced to 1–3 h. A number of micro-fungi will readily allow uniform spore suspensions to be produced *e.g. Aspergillus niger, Penicillium notatum etc.* and whilst it is easy to provide dense suspension on a small scale, two problems result. Firstly, being in spore form a long lag phase results, usually 12–18 h, and secondly the preparation of large numbers of spores is not readily carried out with agar cultures. To provide large numbers of spores, two methods are often suitable (see also Chapter 9). In the first instance, spores are produced by inoculating the fungus on to pearl barley (30 g in a 500 cm^3 Erlenmeyer flask) which has been autoclaved at 15 p.s.i.g. for 2 h (Whiffen & Savage, 1946). Spores from an agar slope are suspended in a solution of 3.0% w/v glycerol, 0.1% L-asparagine and sterilized 15 p.s.i.g. for 15 min and 10 cm^3 of this spore suspension used to inoculate a flask of the pearl barley. The contents of the flask should be mixed by gently tapping the bottom of the flask against the palm of one's hand. Incubation at a suitable temperature, usually 25–30°C for most fungi, will often result in a very dense spore production which can then be suspended in a very dilute sterile surface active agent. Suitable solutions can be prepared from Tween 80 or Lubrol W at concentrations of 0.05%–0.10% w/v.

Spores can also be produced in submerged culture. We have found that the method described by Capellini & Peterson (1965) works well with a number of fungi. The method consists essentially of a culture medium with low nutritive value and uses carboxy-methyl cellulose which presumably acts as a means of providing, in shake flask culture, conditions of low oxygen tension.

Because spores as an inoculum entails considerable delay in the fermentation, due to the long lag phase for spore germination, use is often made of grown mycelium as inoculum. This can be produced from surface mycelium cultures by homogenizing the 'felt' in a sterile blender fitted with high speed cutting blades. An alternative method is to use germinated spores as inoculum. This is often a good method, but for maximum convenience in use, the period of germination is usually very important; with too short a germination period the lag phase is still long and with too long a germination period the inoculum has grown too much, which leads to a poor inoculum because of physiological changes in the organism. If inoculum flasks have too high a cell density of mycelium, usually above 1–2 g dm^{-3} dry weight, oxygen becomes limiting due to viscosity changes in the culture (see later). To prevent formation of long filamentous strands of mycelium which lead to this condition, some cultures can tolerate the presence of a few small glass beads in the flask. By their rolling action these glass beads can often produce a culture which is more single cell in appearance and can form an excellent inoculum, as it provides high cell densities and a large number of growing points.

TYPES OF MYCELIAL GROWTH In submerged culture, fungi can grow in three forms, although there is often not a sharp differentiation between them; they are single cell, pelletal and filamentous (see Fig. 12.1). Single cell forms resemble yeast cultures in that the cells are usually spherical or ellipsoidal measuring about 5–12 μm. These forms are not spores but true single cells; they are often produced as a result of high shear conditions in a fermenter. Single cell cultures can also be produced in shake flasks by the introduction of glass beads into the flasks.

The formation of pellets is a balance between inoculum type and levels, culture medium and culture conditions (Camici *et al.*, 1952; Galbraith & Smith, 1969). For a number of secondary metabolite fermentations the formation of pellet type cultures is desirable *e.g.* citric acid (Steel, Martin & Lentz, 1954). Essentially they produce a culture which is easily stirred and the fact that the centre of the pellet autolyses and finally forms a hollow shell does not necessarily affect the formation of some metabolites. It has been shown by Phillips (1966) that *Penicillium chrysogenum* pellets above 0.2 mm in diameter are anaerobic at the centre due to diffusion resistance to oxygen. Pirt (1966) also considered that oxygen is the rate limiting nutrient in pelletal growth.

Filamentous growth is characterized by strands of interlocking mycelium which is usually branching and the culture appears to be 'thick' with an appearance which can be most aptly described as resembling porridge. Even without forming true pellets as such, most filamentous cultures form what we have termed 'agglomerates'. These structures are not permanent like pellets, but can be seen to form, disperse and reform depending on the local shear rate in the fermenter. They can usually be dispersed by simple dilution. Even with true filamentous growth, the actual length of the hyphae can vary considerably due to changes of culture medium and conditions such as pH. With a strain of *Penicillium chrysogenum*, Pirt & Callow (1959) showed that comparatively small changes in pH considerably altered the length of the hyphae.

EFFECT OF GROWTH FORM ON THE RHEOLOGICAL PROPERTIES OF THE CULTURE AND ITS CONSEQUENCES It is difficult to exaggerate the importance of growth form of fungal cultures. From changes in morphology a whole range of biochemical changes result, some no doubt due to inherent changes within the cell which merely reflect the changes of form (Marks, Keller & Guarino, 1971). In addition, a considerable number of changes are brought about due simply to the alteration of the physical properties of the culture. In particular, the mass transfer rates are influenced by the apparent viscosity of the culture and this in turn can 'trigger off' biochemical pathways which reflect, for example, the degree of anaerobiosis of the culture; also the growth rate of the organism is affected. Single cell cultures, because of the small size of the individual cells, result in a low viscosity broth which does not depart too significantly from a Newtonian fluid. This property allows maximum mass transfer rates, especially oxygen, with the result that cell growth rates are at maximum and a fermenter can support a fairly high cell density and still meet the peak oxygen demand. We have already quoted the example of a *Rhizopus* sp. attaining a cell dry weight of 40 g dm^{-3} in 18 h when growing as single cells. The whole fermentation proceeds very efficiently from a thermodynamic point of view with fast growth rate, excellent yield factors and high protein content of cells. However, due to the growth form, cell recovery requires the use of solid bowl centrifuges as used for yeast production. The product so obtained is difficult or impossible to incorporate into foodstuffs at high concentrations and there is therefore little or no advantage to be gained in using that type of culture compared to growing true yeasts. Pellet growth, with sizes varying from 0.1 mm to 1 cm remain thin Newtonian fluids with what is the approximate equivalent of 'solid' bodies suspended in the fluid. The oxygen transfer to the fluid is high, but diffusion rates into the pellets are minimal. Because, as Phillips (1966) has shown, pellets are anaerobic in the centre; growth only proceeds at the periphery of the pellet and is therefore cubic in nature. Also due to the anaerobic condition within the pellet, various biochemical pathways are induced and most products of these types of metabolism lead to poor yield factors and protein content. Recovery of this type of growth is very easy, a simple nylon mesh being all that is required. Due to poor growth rate and yield, as well as the poor physical properties of the final product, this type of growth is not usually economic for protein production.

Filamentous growth leads to the formation of culture broths which are mostly pseudoplastic in nature; that is, the viscosity of the fluid is dependent upon the shear rate. The effects of this change of viscosity were first investigated in detail by Deindoerfer & Gaden (1955). Basically, what happens is that as cell density increases or hyphal length increases, the apparent viscosity rises, the turbulence of the system falls and the heat and mass transfers fall. The apparent viscosity values reached by some culture broths can be very high indeed, we have found values of up to 2000 cP, that is two thousand times the viscosity of the original culture medium. Shear due to the impellor $\dot{\gamma} = \text{r.p.s.} \times 11.5$ (Metzner *et al.*, 1961) and shear due to the induced flow of fluid elements in the fermenter act to lower the apparent viscosity as shown in Fig. 12.3. The effects of different apparent viscosities on oxygen transfer rates were described by Solomons & Weston (1961). This work showed, that with small fermenters at least, very high viscosity broths could not be adequately aerated. A further consequence of the high viscosity characteristics of mould culture broths is that the apparent critical oxygen concentration (*i.e.* that concentration of oxygen above which respiration is independent of oxygen concentration) rises (Steel & Maxon, 1966). This effect is brought about by the cell 'agglomerates' requiring a high level of external oxygen to provide the driving force for diffusion into the cells in the centre.

Protein content and protein quality

In conventional foods, protein content can be determined with a reasonable degree of accuracy by the determination of total nitrogen by the Kjeldahl method and multiplying the value by 6.25.

With SCP this method is totally inadequate, since the proportion of non-protein nitrogen (NPN) of the total nitrogen (TN) is much higher than in conventional foods. The main reason for this large NPN is accounted for by the much higher levels of RNA in SCP. This of course is a consequence of high protein content and fast growth rate. Levels of RNA

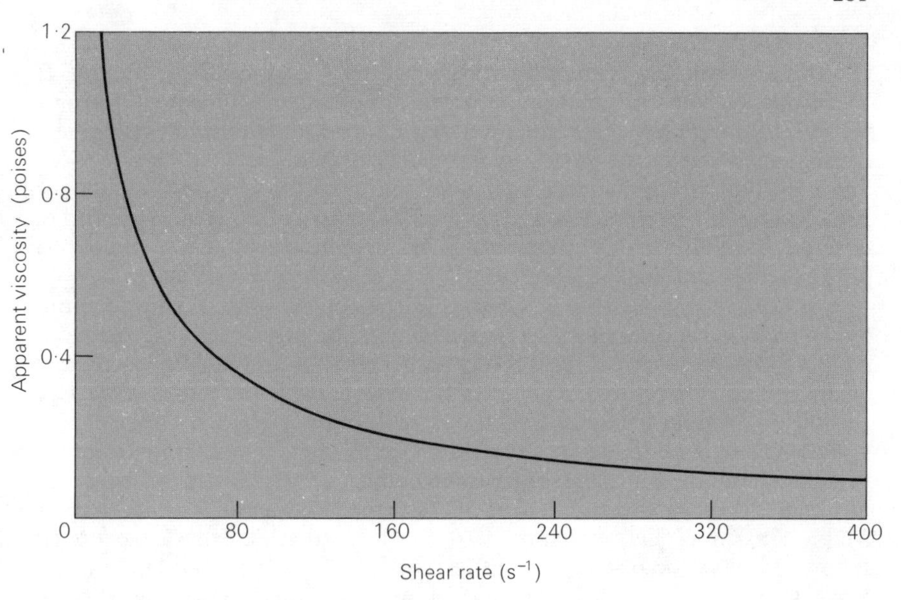

Fig. 12.3 Effect of shear rate on apparent viscosity of a filamentous mycelium culture broth.

vary chiefly due to differences in growth rate, but in micro-fungi the highest values we have found are around 15% of the cell dry weight. In bacteria the figure can be nearly 25% (Herbert, 1959). In addition to the levels of RNA, micro-fungi have a considerable amount of nitrogen occurring in the *n*-acetyl glucosamine of chitin, which constitutes some 10% of fungal cell walls. This latter value varies considerably, but in a number of micro-fungi we have found the NPN constitutes from 20–50% of the TN (Anderson *et al.*, 1973). This means great care must be exercised in judging protein content of micro-fungi by TN analysis. Indeed, we have of necessity developed a total amino acid nitrogen analytical method to establish a more reliable method of assessment. This method is similar to that developed by Gehrke & Wall (1971). According to our methods, fungi can vary from around 3%–7.5% α-amino nitrogen, corresponding to 19–47% of protein.

An accepted method of measuring protein quality is to use the rat assay, the NPU (Net Protein Utilization) of Miller & Payne (1961). In this method a diet is made up to contain 10% protein, that is 1.6% N. We feel that for SCP products this 1.6% should be based on α-amino nitrogen rather than TN. The basis of this opinion is that chitin-N is non-digestible and RNA-N does not appear to take part in the normal N metabolism of rats. If only TN assays are used, we feel that the true protein value of SCPs will be considerably underestimated.

Using these techniques we have found a wide spread of NPU (based on α-AN) values ranging from about 30 to nearly 90. This latter value reflects the exceptionally high sulphur amino acids values that some micro-fungi possess which put them into a class of proteins normally considered to belong to animal protein products, *e.g.* milk protein, meat *etc.*

RNA levels

On the basis of experiments carried out by Edozien, Udo, Young & Scrimshaw (1970) on human volunteers, the Protein Advisory Group of the WHO/FAO have recommended that humans should limit their intake of single cell nucleic acids to 2 g/day (PAG, 1972).

Most SCP products currently available contain around 50–60% of protein and 10–15% of RNA. If we take a figure of 35 g of protein/day derived from SCP, this means that the percentage of RNA in such a product must not exceed 2.9–3.3%.

It therefore follows that RNA must either be prevented from forming or removed subsequent to the fermentation. The first possibility seems to have only limited scope since RNA is the protein manufacturing machinery of the cell and therefore we could only achieve low RNA values if we can accept low protein levels and/or low growth rates.

The removal of RNA by treatment after the fermentation offers a feasible method, the real requirement being that protein losses must be minimal. Two methods have been described in detail, one is that of Maul, Sinskey & Tannenbaum (1970), who used a heat shock treatment followed by two incubation stages, so that the RNAases could act on the RNA and the products, nucleotides and nucleosides could then diffuse from the cell. The method would seem to depend on the heat shock to selectively inactivate the cell proteases, which are more heat labile than the RNA'ses. The other was recently described by Akin & Kwei (1972) and consists of treating cells at pH 9.0–10.5 with NH_4OH and heating for 10 min at 120°C. They claim RNA hydrolysis, without damage to proteins. Other methods appear to have been developed (Decerle, Franckowiak & Gatellier, 1969), but no details have been given.

The existing problem is that as well as removing RNA, cell pools and other small molecular weight compounds must also be removed and therefore there is substantial loss in cell weight, up to 30% being lost. If this material could be recovered economically it could be used in animal feeds. Another possibility is recycling the material back into the fermenter.

Batch and continuous culture

All of the existing fermentation processes involving micro-fungi use batch culture methods. This is because the fermentations are designed to produce secondary metabolites and high yields of these cannot be presently achieved in extended continuous culture. The reason for this is basically that these fermentations employ highly mutated and genetically unstable strains which revert during continuous cultivation. Biomass production on the other hand is only interested in growing cells and therefore continuous culture is not only feasible, but from an economic standpoint essential. If we take it that a large industrial fermenter requires 36 h to complete a cycle of sterilizing, inoculating, growing, emptying and washing, in the same period the chemostat will have produced, at a dilution rate of 0.1 h^{-1}, 3.6 times as much product, other considerations being equal.

In addition, continuous culture offers the possibility of controlling, at least to some extent, the composition of the cell by the control of growth rate. With batch culture no such control is feasible. In particular with continuous culture the concentration of RNA can be reduced by as much as 50% by alteration of growth rate (Herbert, 1959).

Growing a strain of *Penicillium chrysogenum*, Pirt & Callow (1959) showed the feasibility of using continuous culture for micro-fungi and we have certainly confirmed their findings. Provided that all the normal precautions are taken, we have been able to operate chemostats of up to 1300 dm^3 capacity for periods of up to 3000 h on stream. There appears to be no reason why a single fermentation should not last for, say, 8000 h, equivalent to a year's running when the plant would probably have to be shut down for normal annual overhaul.

Continuous culture also makes possible the continuous recovery of cells. This is important because it is most desirable to keep recovery times down to a very short period, since this prevents the degradation of proteins by the cell's own autolytic processes. In conclusion, we may say that virtually all of the factors both scientific and economic now point to continuous culture as the only feasible production route for single cell protein.

The author wishes to thank the American Society for Microbiology for permission to use the data presented in Tables 12.1 and 12.2.

12.4 References

AKIN, C. & KWEI, C.C. (1972). Improvements of nutrient properties of single-cell protein material used as food products. *German Patent 2,158,261.*

ANDERSON, C., LONGTON, J., MADDIX, C., SCAMMELL, G.W. & SOLOMONS, G.L. (1973). The growth of micro-fungi on carbohydrates. Paper presented at Second International Symposium on SCP, Boston, Massachusetts, May 29–31st.

CAMICI, L., SERMONTI, G. & CHAIN, E.B. (1952). Observations on *Penicillium chrysogenum* in submerged culture. *Bulletin of the World Health Organisation* **6**, 265–76.

CAPELLINI, R.A. & PETERSON, J.L. (1965). Macroconidium formation in submerged culture by a non-sporulating strain of *Gibberella zeae Mycologia* **57**, 962–6.

CARTER, B.L.A. & BULL, A.T. (1969). Studies of fungal growth and intermediary carbon metabolism under steady and non-steady state conditions. *Biotechnology & Bioengineering* **11**, 785–804.

CIRILLO, V.P., HARDWICK, W.A. & SEELEY, R.D. (1960). Fermentation process for producing edible mushroom mycelium. *U.S. Patent 2,928,210.*

DECERLE, C., FRANCKOWIAK, S. & GATELLIER, C. (1969). How I.F.P. makes food yeast. *Hydrocarbon Processing* **48**, (3) 109–12.

DEINDOERFER, F.H. & GADEN, JR., E.L. (1955). Effects of liquid physical properties on oxygen transfer in penicillin fermenters. *Applied Microbiology* **3**, 253–7.

DUTHIE, I.F. (1973). Animal feeding trials with a microfungal protein. Paper presented at Second International Symposium on SCP. Boston, Massachusetts, May 29–31st.

EDOZEIN, J.C., UDO, U.U., YOUNG, V.R. & SCRIMSHAW, N.S. (1970). Effects of high levels of yeast feeding on uric acid metabolism of young men. *Nature, London* **228**, 180.

FALANGHE, H., SMITH, A.K. & RACKIS, J.J. (1964). Production of fungal mycelial protein in submerged culture of soyabean whey. *Applied Microbiology* **12**, 330–4.

FENCL, Z. & NOVÁK, M. (1969). Prediction of the course of continuous fermentation on the basis of analysis of the batch process. *Folia Microbiologica, Praha* **14**, 314–21.

FORSS, K.G., GADD, O.G., LUNDELL, R.O., WILLIAMSON, H.W., & KAUNIAINEN, D. (1972). *German Patent 2055306.*

GALBRAITH, J.C. & SMITH, J.E. (1969). Filamentous growth of *Aspergillus niger* in submerged shake culture. *Transactions of the British Mycological Society* **52**, 237–46.

GEHRKE, C.W. & WALL, L.L. (1971). Automated trinitrobenzene sulfonic acid method for protein analysis in forages and grain. *Journal of the Association of Official Analytical Chemists* **54**, 187–91.

GRAY, W.D. (1970). The use of fungi as food and in food processing. *Critical Review of Food Technology* **1**, 225–329.

HEINEMANN, B. (1963). Process and composition for growing mushroom mycelium in submerged fermentation. *U.S. Patent 3,086,320.*

HERBERT, D. (1959). Some principles of continuous culture. In *Recent Progress in*

Microbiology, pp. 381–96. Ed. Tunevall, G. Stockholm.

HUMFELD, H. (1948). The production of mushroom mycelium (*Agaricus campestris*) in submerged culture. *Science* **107**, 373.

HUMFELD, H. (1952). Production of mushroom mycelium. *U.S. Patent 2,618,900.*

HUMFELD, H. (1954). Production of mushroom mycelium by submerged culture in a liquid medium. *U.S. Patent 2,693,665.*

HUMFELD, H. & SUGIHARA, T.F. (1952). The nutrient requirements of *Agaricus campestris* grown in submerged culture. *Mycologia* **44**, 605–20.

IMRIE, F.K.E. (1973). The production of fungal protein from carob in Cyprus. *Journal of Science of Food & Agriculture.* **24**, 639.

LITCHFIELD, J.H. (1967) Morel mushroom mycelium as a food-flavouring material. *Biotechnology & Bioengineering* **9**, 289–304.

LITCHFIELD, J.H. (1968). In *Single-Cell Protein*, pp. 309–29. Ed. Mateles, R.I. & Tannenbaum, S.R., M.I.T. Press. London and Cambridge.

LITCHFIELD, J.H., OVERBECK, R.C. & DAVIDSON, R.S. (1963). Factors affecting the growth of morel mushroom mycelium in submerged culture. *Journal of Agricultural & Food Chemistry* **11**, 158–62.

MARKS, D.B., KELLER, B.J. & GUARINO, A.J. (1971). Growth of unicellular forms of the fungus *Cordyceps militaris* and analysis of the chemical composition of their walls. *Journal of General Microbiology* **69**, 253–9.

MAUL, S.B., SINSKEY, A.J. & TANNENBAUM, S.R. (1970). New process for reducing the nucleic acid content of yeast. *Nature, London* **228**, 181.

METZNER, A.B., FEEHS, R.H., RAMOS, H.L., OTTO, R.E. & TUTHILL, J.D. (1961). Agitation of viscous Newtonian and non-Newtonian fluids. *American Institution of Chemical Engineers' Journal* 7, 3–9.

MILLER, D.S. & PAYNE, P.R. (1961). Problems in the prediction of protein values of diets. The influence of protein concentration. *British Journal of Nutrition* **15**, 11–19.

P.A.G. Bulletin, 13, (FAO/WHO). Vol. 2. No. 1, p. 2 (1972).

PARK, J.Y. & AGNIHOTRI, V.P. (1969). Sporophore production of *Agaricus bisporus* in aseptic environments. *Antonie van Leewenhoek* **35**, 523–8.

PHILLIPS, D.H. (1966). Oxygen transfer into mycelial pellets. *Biotechnology & Bioengineering* **8**, 456–60.

PIRT, S.J. (1966). A theory of the mode of growth of fungi in the form of pellets in submerged culture. *Proceedings of the Royal Society Series B* **166**, 369–73.

PIRT, S.J. & CALLOW, D.S. (1959). Continuous flow culture of the filamentous mould *Pencillium chrysogenum* and the control of its morphology. *Nature, London* **184**, 307–310.

REUSSER, F., SPENCER, J.F.T. & SALLANS, H.R. (1958). *Tricholoma nudum* as a source of microbiological protein. *Applied Microbiology* **6**, 5–8.

ROBINSON, R.F. (1952). Food production by fungi. *The Scientific Monthly* **75**, 149–54.

SHUKLA, J.P. & DUTTA, S.M. (1965). Production of fungal protein from *Rhizopus* sp. *Indian Journal of Experimental Biology* **3**, 242–4.

SMITH, J. (1969). Commercial mushroom production. *Process Biochemistry* May 43–6, 52.

SOLOMONS, G.L. & SCAMMELL, G.W. (1970). Production of edible protein containing substances. *British Patent 1346061.*

SOLOMONS, G.L. & WESTON, G.O. (1961). The prediction of oxygen transfer rates in the presence of mould mycelium. *Journal of Biochemical & Microbiological Technology & Engineering* **3**, 1–6.

SPICER, A. (1973). Proteins from carbohydrate. *Chemistry in Britain*, **9**, No. 3, 100–3.

STEEL, R., MARTIN, S.M. & LENTZ, C.P. (1954). A standard inoculum for citric acid production in submerged culture. *Canadian Journal of Microbiology* **1**, 150–7.

STEEL, R. & MAXON, W.D. (1966). Dissolved oxygen measurements in pilot and production-scale novobiocin fermentations. *Biotechnology & Bioengineering* **8**, 97–108.

SZUECS, J. (1956). Mushroom culture. *U.S. Patent 2,761,246.*

SZUECS, J. (1958). Method of growing mushroom mycelium and the resulting products. *U.S. Patent 2,850,841.*

TRINCI, A.P.J. (1969). A kinetic study of the growth of *Aspergillus nidulans* and other fungi. *Journal of General Microbiology* **57**, 11–24.

TRINCI, A.P.J. (1972). Culture turbidity as a measure of mould growth. *Transactions of the British Mycological Society* **58**, 467–73.

VOLESKY, B. & ZAJIC, J.E. (1971). Batch production of protein from ethane and ethane–methane mixtures. *Applied Microbiology* **21**, 614–22.

WHIFFEN, A.J. & SAVAGE, G.M. (1947). The relation of natural variation in *Pencillium notatum* to the yield of pencillin in surface culture. *Journal of Bacteriology* **53**, 231–240.

CHAPTER 13

Oriental Food Fermentations

B. J. B. WOOD and YONG FOOK MIN

13.1 Introduction

Because of the great popularity which Chinese restaurants have enjoyed in recent years, most people in the West are, by now, fairly familiar with soya sauce. What is less generally realized is that this is only one of a wide range of fermented food products to be found in Oriental cuisine. To the Occidental, three features of these products will be particularly impressive, *viz.* their range, their antiquity and their use of mould fermentations. As to diversity, they range from condiments, through a kind of cheese, red rice, snacks and even a fermented beverage, the so-called rice wine, to tempeh, a main meal component. From Chinese records, it is clear that such arts as the preparation of soya sauce were already of considerable antiquity a thousand years B.C. To the Westerner, the development of mouldiness is normally associated with deterioration of foodstuffs, and in only a few products, such as English Blue Stilton and comparable foreign cheeses, is the development of a mould an essential part of the process. In the Orient, on the other hand, moulds serve the function filled by malt in the West, and are also highly prized for their ability to increase digestibility and palatability of the rather bland and indigestible vegetables which form an important source of protein in the frequently vegetarian or near-vegetarian diets. These products have commercial potential in the West both because of the greater current awareness of, and interest in, foreign foods, and also because scarcity and high prices of traditional animal proteins are forcing consideration of alternative dietary protein sources.

13.2 Koji

The Koji is the central feature of most fungal food preparations. The term tends to be applied to both the inoculum and the main fermentation. For clarity, we shall call the inoculum a seed-koji and reserve the unqualified term Koji for the main mash.

For the majority of fermentations, the Koji involves strains of *Aspergillus*, notably *A. oryzae* and *A. soyae*. Typically, the stages in the production of a Koji will be as follows. The desired strain of mould is maintained as a culture on rice agar slopes. Under these conditions, the mould will sporulate freely. When a fresh seed-koji is required, the spores are washed off a slope with sterile water, and the spore suspension is used to inoculate steamed rice. When satisfactory mould growth has taken place, between 1–10% of seed-koji is added to the main substrate. For laboratory-scale investigations, we have found it perfectly satisfactory to use the spore suspension directly to inoculate the appropriate substrate.

Classically, the Koji is incubated in baskets made of woven bamboo, thus providing very good aeration. We have found that expanded aluminium makes a very convenient, readily available, easily fabricated, cleaned and sterilized replacement for rattan. The trays or baskets are filled to a depth of two or three inches and incubated in rooms maintained at 25–35°C. Under these conditions, mould growth is rapid, and considerable heat is produced. Temperatures in excess of 40°C are normally considered to be harmful, and during the incubation it is usually necessary to cool the Koji from time to time by mixing it thoroughly.

During this period, the mould is growing vegetatively, utilizing low molecular weight materials present in the substrate, and releasing a range of extracellular enzymes appropriate to the conditions. Thus, *Aspergillus oryzae* growing on soya beans will liberate proteinases and amylases; we have additionally found invertase and lipase (Yong, 1971) and also cellulase (Goel, S.K. and Wood, B.J.B., unpublished results).

The length of time for which the Koji is allowed to develop is very important. Clearly, too short an incubation time will lead to insufficient mould growth and inadequate production of extracellular enzymes. If the Koji is cultured for too long a time, the mould will start to sporulate, and this is normally undesirable since it causes mouldy off-flavours in the finished product. In soya-koji there is also a considerable release of ammonia (presumably from deamination of amino acids) if incubation is continued for too long a time. This latter phenomenon is of considerable interest to the microbial biochemist since it may well indicate that the mould is using this as a rapid means to obtain carbon compounds needed for anaplerotic pathways of metabolism associated with sporulation. However, from a practical point of view it is very undesirable since it adversely affects the flavour of the finished product. In practice, about three days is normally the optimum period for development of the Koji.

The subsequent treatment of the mass of mould-covered material will depend upon the particular product being manufactured, but in general, the aim is to inhibit further growth or development of the mould, while permitting the extra-cellular enzymatic breakdown of the proteins, polysaccharides, *etc.*, present in the substrate. This severely limits the options open to the manufacturer, and the provision of anaerobic conditions is the favoured method, since this completely inhibits mould growth, while allowing the continuation of enzyme activity, and the development of a secondary microbial flora to continue. It might be expected that incubating protein-rich materials such as partially-degraded soya beans under anaerobic conditions would lead to the rapid establishment of a flora rich in food-poisoning organisms such as *Botulinus*. However, considerable subtlety has gone into the provision of conditions such that desirable organisms are encouraged, while undesirable ones are completely suppressed. When we consider that the role of microbes in food poisoning has been known for less than a century, yet these foodstuffs have been utilized for more than three millennia, it is impossible not to be impressed with the skill of the people who discovered and perfected these arts. Similarly, the first reaction of any mycologist familiar with the current work on aflatoxin and similar mycotoxins, when the oriental uses of moulds in food is described to him, is to express profound concern at the dangers inherent in such processes. Yet all work on these methods, and on the moulds employed, fails to reveal the presence of any mycotoxins. This is all the more remarkable when it is considered that the bulk of the soya sauce produced outside Japan, was until recently made by the individual householder or by small cottage industries. How non-toxic moulds were selected out of the untold numbers of spores present in the air of humid sub-tropical and tropical regions is a complete mystery. In practice, however, infections of soya-koji by unwanted moulds seem to be limited to an occasional outbreak of *Rhizopus nigricans* if the atmosphere of the koji-chamber becomes super-saturated with moisture. Indeed it has been suggested (Smith, 1949) that the general occurrence of the soya sauce mould throughout the small Japanese city of Noda (where soya sauce has been made by the Kikkoman Shoyu Company since 1764, and where 23 million gallons were produced in 1948) suppresses other microbial life, and maintains a high health standard for the city.

13.3 Soya Sauce

Among Oriental foodstuffs made from the soya bean, soya sauce or Shoyu is the one best known in the Occident. In terms of tonnes of soya beans consumed, it ranks second to Tofu (q.v.) in Japan (Shibasaki & Hesseltine, 1962).

It is not our intention to present a detailed review of the extensive literature pertaining to soya sauce; the reader who desires this is referred to our recent survey (Yong & Wood, 1973) which incorporates a very selective but extensive bibliography. It is, however, necessary to pay tribute to the outstanding contributions of Dr. C.W. Hesseltine and his co-workers. He has certainly done more than any other person to inform the Western scientist about this and other fermentation arts practised in the East, and to demonstrate their suitability for production and use in the Occident (see, for example, Hesseltine, 1965, 1966; Hesseltine & Wang, 1967, 1968; Wang & Hesseltine, 1970).

While much fascinating research remains to be performed in unravelling the details of the microbiology and biochemistry of the Shoyu fermentation, we can now be reasonably certain of the sequence of events in a process which was already well known as an art in the time of Confucius, since in the book of Chau Lai (one of the Thirteen Classics of Confucius) it is recorded that the King's cook used twenty jars of soya sauce for the ceremonial rites of the Chau Dynasty (Groff, 1919). Despite this great antiquity, even the sequence of micro-organisms involved was, until recently, a matter for considerable controversy (Yong & Wood, 1973).

In practice, two distinct stages—Koji and Moromi—are recognized. First a Koji is produced, and secondly in the Moromi stage the Koji is mixed with brine, and a mixed lactic acid and yeast fermentation takes place. The filtrate from this second stage is the soya sauce of commerce.

Koji

The starting material is a mixture of soya beans and wheat. The beans are first soaked in cold water for about 12 h. It is necessary to use running water, or to drain the beans and add fresh water at frequent intervals, since otherwise bacterial fermentations will develop, and give rise to undesirable flavours in the final product. The beans are then boiled or autoclaved until soft. This heat treatment softens the beans, destroys proteinase inhibitors which are normally present in the bean, and kills bacteria growing on the surface of the bean.

In modern Japanese practice, the use of whole beans seems to have been largely displaced by defatted soyabean meal (Shibasaki & Hesseltine, 1962). The process of preparation by soaking and cooking seems to be much the same as with whole beans, although a much shorter soaking time is required; indeed, it is probably sufficient simply to moisten the meal prior to steaming or autoclaving. Whether beans or meal are used, the cooked material must be cooled as rapidly as possible and, to minimize bacterial growth, blended at once with the wheat and seed-koji.

Meanwhile, the wheat is prepared by roasting the grains, then crushing them to give a mixture of powder and larger pieces. Alternatively, wheat bran and flour may be used; in this case, they are steamed, rather than being roasted. The treatment accorded to the wheat will have a considerable influence on the properties of the soya sauce, roasted wheat giving a darker product of more robust flavour, due to higher concentrations of phenolic compounds such as vanillin, vanillic acid, ferulic acid and 4-ethylguaiacol.

These two raw materials contribute rather different things to the Koji. Soyabeans typically contain about 38% protein, 15–20% fat, and 15–19% sucrose (Yokotsuka, 1960). These analyses are based on a water content of 10–13% in the beans as received. Kawamura (1967) has reported that carbohydrate represents about 34% of the dry matter of the bean. The soft wheat normally employed contains much less protein (around 8–9%) but those present are rich in glutamic acid. The starches present in soft wheat represent essential precursors for fermentations occurring in the Moromi stage.

The wheat and beans or bean-meal are mixed together to give a dryish, open-textured material. In Japanese practice, roughly equal weights of beans and wheat are used, but the Chinese apparently prefer to use soyabeans alone or in excess over the wheat. These differences give rise to finished products which have rather different flavour, colour, aroma, and nitrogen and sugar contents (Smith, 1949; Yokotsuka, 1964).

A three- to five-day old seed-koji is mixed in with the beans and wheat. In Japan, the seed-koji is produced by growing *Aspergillus soyae* or *A. oryzae* on steamed, polished rice, while in China, a mixture of wheat bran and soyabean flour is the preferred substrate. Usually 1–2% w/w of the seed koji is added to the main mass of beans and wheat. In our own studies, we were anxious to avoid complicating the picture with respect to changes in enzyme levels during development of the Koji, and we therefore employed a suspension of washed fungal spores as inoculum. We found this to be a very easy and satisfactory method of dispersing the mould evenly through the Koji. The inoculated Koji is then incubated for approximately 72 h at 30°C, with occasional mixing.

The changes occurring during development of the Koji have recently been intensively studied (Table 13.1) (Yong, 1971). In an attempt to mimic

Table 13.1 Changes in composition of koji during the soya sauce fermentation (Yong, 1971)

Time (H)	pH	Temp.	Moisture % w/w	Reducing sugars as glucose g % dry wt	Total soluble nitrogen g % dry wt	Amino nitrogen g % dry wt	Ammonia nitrogen g % dry wt
0	6.55	28.5	47.0	0.3	0.57	0	0.02
18	6.49	28.5	49.0	2.2	0.63	0.02	0.04
22	6.44	30.5	48.0	2.8	0.62	0.09	0.03
29.5	6.48	41.0	49.0	3.7	1.04	0.34	0.07
42.0	6.74	—	47.0	2.5	1.07	0.49	0.11
47.0	6.86	39.0	42.0	2.4	1.25	0.29	0.20
52.5	6.90	36.0	39.0	1.7	1.30	0.26	0.30
66.0	7.08	32.0	36.0	2.1	1.44	0.39	0.34
73.0	7.34	31.5	34.0	2.5	1.59	0.32	0.40
90.5	7.48	—	35.0	2.1	1.59	0.33	0.39
96.5	7.50	30.0	34.0	1.7	1.58	0.33	0.39

industrial conditions as closely as possible, while retaining strict control over the microbial flora, we incubated the Koji in shallow trays made of expanded aluminium. Aeration was with a continual stream of moistened, filtered air delivered to the underside of the trays (Fig. 13.1). Detailed reports of the results will be published elsewhere; suffice it to say here that in trials with three different strains of *Aspergillus oryzae*, invertase was always the first carbohydrate-hydrolysing enzyme to be detected, followed by α-amylase later in the fermentation. β-Amylase was not detected in the Koji, although the steady increase in the level of reducing sugars towards the end of the fermentation strongly suggests that this type of enzyme activity is present. Experiments with liquid cultures of soya sauce strains of *A. oryzae* (Goel, S.K. and Wood, B.J.B., unpublished results) indicate that β-amylase levels may rival those of α-amylase under these conditions; our studies have not yet progressed far enough to permit us to offer any explanation of these differing results.

Total soluble nitrogen increases fairly steadily between 20–70 h after incubation, but amino-nitrogen first increases rapidly, then declines somewhat and fluctuates around a somewhat lower level. Free ammonia increases slowly during the first part of the fermentation, but starts to increase very rapidly after about 40–50 h, coinciding with the onset of sporulation.

After about 72 h, the Koji is given a final mix, then blended with about twice its own volume of brine containing 20% salt.

Fig. 13.1 Flow sheet for laboratory production of soya sauce (Yong, 1971).

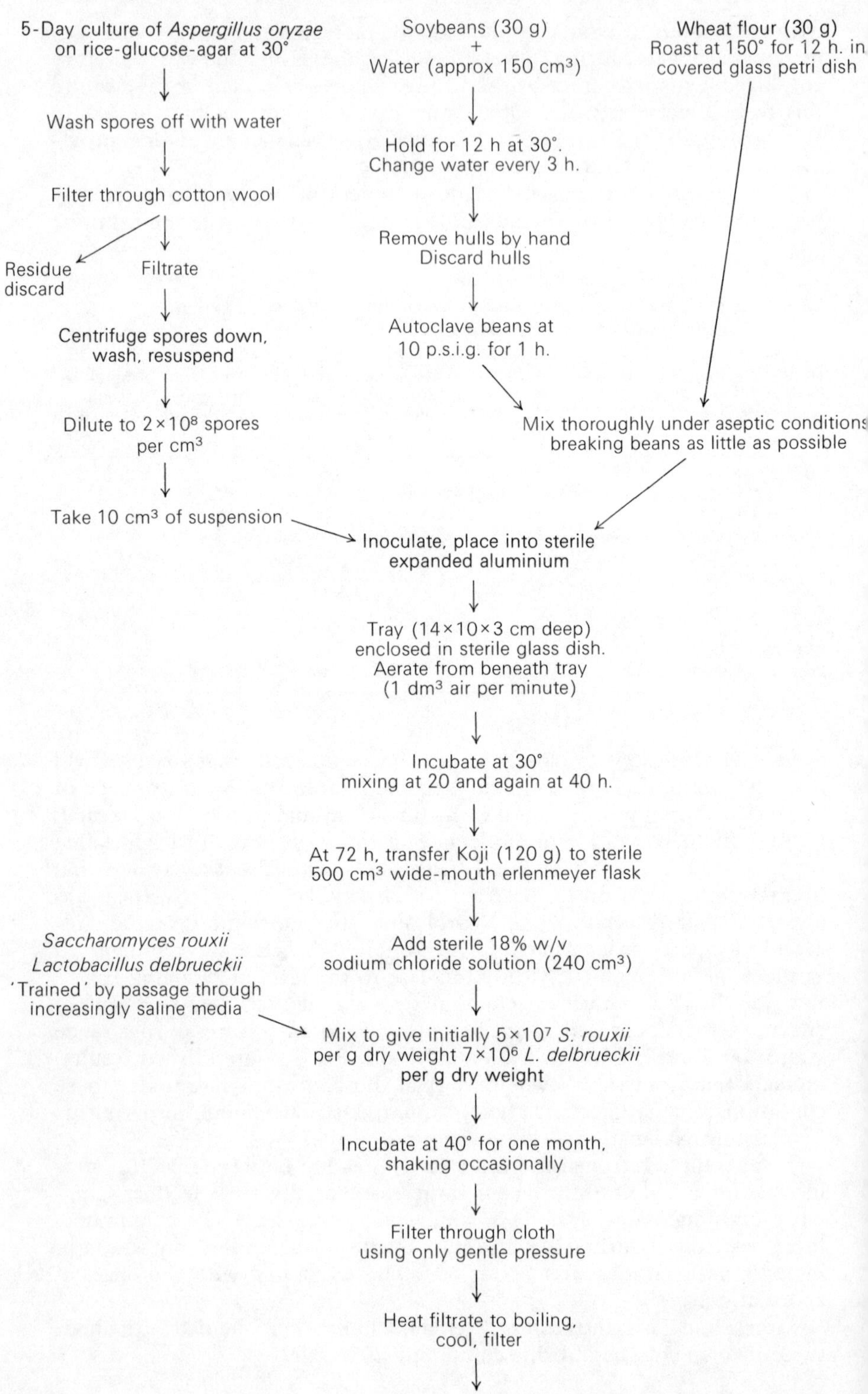

Moromi

The mixture of beans, wheat, mycelium and brine is called Moromi. Anaerobic conditions rapidly develop and prevent further fungal growth. The precise nature of the changes which then occur has been the subject of much controversy (Wood & Yong, 1973). Recent results clearly show that there is a two stage fermentation (Yong, 1971; Yong, F.M. and Wood, B.J.B., unpublished results) and substantiate the views of Yokotsuka (1960) and Onishi (1957*a*, *b*; 1958*a*, *b*; 1963) that a lactic acid fermentation takes place first. When the pH has dropped below about 5, there is a vigorous yeast fermentation involving *Saccharomyces rouxii*. In normal media, this yeast will grow well at neutral pH, but in 18% salt solution it will not grow until the pH is taken below 4.5 (Onishi, 1957*b*, 1958*b*). Changes in the composition of Moromi during soya sauce fermentation are shown in Table 13.2.

Table 13.2 Changes in composition of moromi during the soya sauce fermentation (Yong, 1971)

Time days	pH	Viable counts $\times 10^7$ per g dry wt		Total soluble nitrogen g % dry wt	Amino nitrogen g % dry wt	Ammonia nitrogen g % dry wt	Reducing sugars as glucose g % dry wt
		yeast	Lacto-bacillus				
0	6.5	4.6	0.7	0.81	0.20	0.02	0.65
1	6.5	3.8	0.9	1.14	0.40	0.02	1.6
2	6.5	4.0	2.5	1.12	0.35	0.02	2.2
3	6.5	2.4	2.6	1.54	0.46	0.02	3.2
4	6.0	2.5	2.5	1.31	0.45	0.02	4.1
5	—	2.3	1.1	—	—	—	—
6	5.5	2.1	1.2	1.39	0.51	0.02	4.8
8	—	1.0	0.2	—	—	—	—
10	5.0	0.2	0.1	1.33	0.51	0.02	5.0
14	5.0	7.5	0.05	1.20	0.51	0.02	5.0
18	4.5	7.3	0.05	1.21	0.57	0.02	5.2
22	4.5	7.0	0.01	1.27	0.53	0.03	5.1
26	4.5	5.5	0.01	1.38	0.71	0.03	5.7
31	4.5	4.9	0	1.46	0.41	0.03	6.0

In the traditional process employed in the Orient, the souring and yeast fermentation occurs at ambient temperature through the activity of chance contaminants. In consequence, this stage takes between 6 months and two years. Following a suggestion made by colleagues at the Singapore Institute of Standards and Industrial Research, we employed the rather elevated temperature of 40°C for this stage. By combining this with deliberate inoculation with pure cultures of yeast and *Lactobacillus* under otherwise aseptic conditions, we have routinely obtained soya sauce of excellent quality in one month from initiation of the Koji.

Chemical and semi-chemical soya sauces

Soya sauce made by the fermentation process described above, has a sweet, mellow aroma. While the soya sauce offered in good Chinese restaurants as well as some of the bottled soya sauce sold in shops in the United Kingdom, has this very characteristic smell, far too much of it is virtually devoid of aroma and is probably either chemical or semi-chemical soya sauce (Shinshiki Shoyu). From remarks made by Wolf &

Cowan (1971) it is obvious that the same situation holds true in the U.S.A. Chemical soya sauce is made by a simple acid hydrolysis of a mixture of wheat and soya meal, by refluxing with constant boiling hydrochloric acid for 12–16 h. The mixture is neutralized with sodium hydroxide to pH 4–5, then aged in hardwood casks and bottled (Minor, 1945). This product apparently finds no market whatsoever in Asia. The Shinshiki Shoyu does however find a market in the East, and has been estimated to represent 20% of the total Shoyu production in Japan (Yokotsuka, 1964). In this process, developed by the Noda Shoyu Company Ltd., defatted soyabeans are treated with a rather more dilute hydrochloric acid (7–8%, as against the 20% solution employed in the chemical process described above). After acid treatment, the mixture is neutralized as before with sodium hydroxide, mixed with wheat bran Koji, and fermented with yeast. Although a great improvement on the chemical soya sauce, this product still does not have the delicacy of flavour associated with Shoyu made by the classical methods.

13.4 Soy Milk, Tofu and Sufu (Chinese Cheese)

A milk made by soaking soyabeans in water, grinding, filtering and pasteurizing or boiling the filtrate to destroy the trypsin inhibitor, has long been a staple article of diet in the East. It is made both on a factory scale and at home. It has an attractive creamy colour and is very palatable. Unfortunately, precise directions for making it are hard to come by, since a lot seemingly depends on the characteristics of the particular batch of beans being used, and previous experience therefore plays a big part in determining the precise conditions to be used in making the milk.

Tofu is an unfermented curd made from soya milk by calcium or magnesium sulphate precipitation (Wang & Hesseltine, 1970; Gray, 1970). It will be noticed that this is rather different from the process used in producing a curd from animal milk, where acidification, either artificial, or through the action of lactic acid bacteria, is employed, or coagulation is effected by the digestive enzyme rennin. According to Wang & Hesseltine (1970) the process is as follows:

[i] Washed soyabeans are soaked in water at 25°C for 5–6 h.

[ii] Excess water is drained off and the beans are ground and pressed through cheese cloth. The insoluble materials are discarded and being rich in protein, make an excellent animal food after suitable heat treatment.

[iii] The milk is boiled and then cooled to 50°C.

[iv] Calcium or magnesium sulphate is added.

[v] The precipitated protein is strained off through cheese cloth, and pressed to remove as much liquid as possible. The liquid will still contain sucrose, peptides, amino acids *etc.* and should prove to be an excellent substrate for microbial growth, although we are not aware that it is so employed at present. The curd is tofu, contains calcium or magnesium from the precipitating salt and although bland is perfectly edible and nourishing and is reputed to be an important protein source in the diet of the military forces of the Chinese People's Republic, who are said to have learned a great

appreciation of the value of this simply prepared yet very concentrated source of protein, during their struggle for supremacy in China. Wang & Hesseltine (1970) report that it can be flavoured with Shoyu or Miso, or cooked in admixture with other foods such as meat, fish and vegetables.

If it is to be converted into Sufu, the Tofu is sliced into pieces about one inch cube. The cubes are then soaked for one hour in a solution containing 6% sodium chloride and 2.5% citric acid. This marinade is apparently to decrease bacterial growth during the subsequent fungal growth, and not all authors include it in their directions. The cubes are dried in an oven at 100°C for 10 min, then individually suspended in a cabinet maintained at 20°C and inoculated with a suitable mould. *Actinomucor elegans* seems to be the favoured mould according to Wang & Hesseltine (1970), although they note that other micro-organisms have been employed with success, notably *Mucor hiemalis, M. silvaticus* and *M. subtilissimus*.

Because of the low temperature employed, incubation takes some 3–7 days, depending upon the mould being used, and is continued until the cubes are completely covered with a firm layer of white or buff mould mycelium. The low temperature called for in these directions seems particularly strange when it is considered that this food is typical of China, Formosa, and similar countries.

The finished cubes are reported to have a bland flavour, although the mould is bound to have released extracellular proteinases and lipases in order to grow on the Tofu. The cubes are finally aged in a '5–10% brine to which 10% ethanol may be added in the form of rice wine or distilled liquor' (Nelson & Richardson, 1967). Other authors are rather ambiguous as to whether the brine or the wine added to it is 10% with respect to alcohol. Unless salt is dissolved in the wine, it is rather difficult to see how a solution possessing the specified characteristics can be produced by addition of wine to brine. Ageing takes place for one to two months, and although none of the papers examined make any reference to it, it would be most surprising if no yeast or bacterial growth took place during this time.

13.5 Miso

Miso production ranks third in the annual tonnage of soyabeans consumed in Japan. This, like Shoyu, is perhaps more important for adding and intensifying flavour than for its nutritional contribution. But, again like Shoyu, it would be surprising if its nutritional contribution were without significance; Shibasaki & Hesseltine (1962) quote a Japanese daily *per capita* consumption of 29.3 g Miso and 26.9 g Shoyu. As with soya sauce, Miso is prepared both on a factory scale and in the home, Shibasaki & Hesseltine (1962) show that domestic production in the early 1960's was roughly two thirds the size of factory production.

As will become clear, Miso production is very similar to soya sauce production in both technique and microbiology, but whereas Shoyu is pretty well restricted to beans + wheat as starting materials, the Miso producer can start with beans + rice, beans + barley, or beans alone. Further variety is introduced by varying the processing conditions;

Shibasaki & Hesseltine (1962) recognize a total of six major varieties of Miso in Japanese practice. It is not clear why a beans + wheat Miso cannot also be prepared, but perhaps the flavour of this mixture would be too strong, or be otherwise unacceptable to the consumer.

The selection and preparation of the soyabeans is of considerable importance, and to ensure uniformity in the final product, which is a paste, it is essential that the beans should behave in a reasonably uniform manner in water uptake, response to cooking *etc.*; in practice, this means ensuring that a batch should be of a single variety of bean. Traditionally, whole beans are soaked, steamed and blended as described later, but Shibasaki & Hesseltine (1962) report the preparation of an excellent product in a process in which they crushed the dry beans to grits in a roller mill, then washed the grits, soaked in water for 2.5 h, drained, steamed at 5 p.s.i.g. for 1 h, and employed this preparation of the beans. Clearly such a process both effects economies of time over processes requiring that the whole beans be soaked overnight, and also reduces the need for uniformity between beans.

Where rice is to be used in the Miso, a Koji is prepared by growing *Aspergillus oryzae* on soaked, steamed whole grains of the cereal. When rice is the cereal of choice, Shibasaki & Hesseltine (1962) emphasize that polished white rice must be used since brown rice has too hard and waxy a surface to give good mould growth. On the other hand, polished brown rice is apparently preferred for the production of the seed-koji. For the production of Mame Miso, which is made entirely from soya beans, the Koji is apparently prepared by growing the mould on part of the beans. Gray (1970) cites reports which indicate that wheat or wheat flour can be employed in Miso production; the method of producing the Koji is not indicated but since rice was also employed, it could well be that the Koji was produced on this grain.

While it is clear from the figures cited by Shibasaki & Hesseltine (1962) that barley Miso is fairly important, accounting for about a quarter of the total annual Miso production, we have been unable to find any published information on the production of the Koji for this variety of Miso.

The reader will recall that the Koji for soya sauce production was made by moulding the entire batch of vegetables (both wheat and soya beans) to be employed. For Miso production however, the bulk of the beans are not normally moulded, but are blended directly with the rice Koji at the beginning of the stage corresponding to the Moromi stage in Shoyu production.

The rice Koji is apparently ready for use some 50 hours after inoculation with the mould. Shibasaki & Hesseltine (1962) describe how the Koji is incubated in a room maintained at controlled temperature and humidity so as to provide optimum conditions of moistness and a temperature between 35–43°C within the trays of moulding rice. As with Shoyu Koji, the Miso Koji must be thoroughly mixed at least twice during the period of mould growth, in order to promote maximum growth of mould throughout the rice.

The mature Koji is then mixed with salt and steamed soya beans, the proportions of each depending upon the type of Miso to be produced (for details see Shibasaki & Hesseltine (1962)). Just as in the Moromi stage of

Shoyu production, the high (4–12% by weight) salt content and anaerobic conditions rapidly inhibit mould growth and permit the growth of desirable bacteria and yeasts. Development of the appropriate microflora is aided by adding a portion of Miso from a previous fermentation. The main bacterial species are lactic acid producers, while *Saccharomyces rouxii* is the yeast of choice. In a series of papers and a patent, Shibasaki & Hesseltine (1962) have reported the development of a fast, controlled Miso fermentation. They found that a satisfactory Miso could be produced from a mash fermented with *S. rouxii* alone. Apparently no lactic acid bacteria were present. This is rather surprising, since in Shoyu Moromi the yeast cannot grow until the pH has decreased to below 5. In our own studies (Yong, 1971) we found that the pH of a sterile soya sauce Moromi would gradually drop to below 4.5 although no micro-organisms could be detected in it, and it might be that something similar is happening here.

The anaerobic fermentation is generally complete more quickly than is the case with soya sauce, indeed it only takes one to two weeks for some of the lighter, sweeter, less salty varieties. On completion of this anaerobic fermentation, the product is finally blended to give a smooth paste and packaged for sale or use. Both Shibasaki & Hesseltine (1962), and Nelson & Richardson (1967) comment on the attractiveness of the flavour of carefully prepared Miso. It apparently finds use in soupmaking, in seasoning meat, fish and vegetables during cooking, and as a spread. American workers compare its consistency with that of peanut butter.

13.6 Other Oriental Food Fermentations

Little work has been carried out on other fermentations in the West, although they have been known to Westerners for quite a long time. The careful reviews by Hesseltine (1965), Hesseltine & Wang (1967) and Gray (1970) should be consulted by the reader who needs detailed information. For the purpose of the present review, we shall confine ourselves to briefly noting the nature of the various fermentations and products, and commenting upon points of particular scientific or practical interest.

Tempeh

Whereas Miso and Shoyu are primarily of importance as flavouring materials, Tempeh, like another Indonesian food, Ontjom, apparently makes a significant quantitative as well as qualitative contribution to the diet in areas where it is produced and consumed. The two differ both in substrate (soyabeans for Tempeh, peanut press cake for Ontjom) and in the mould used (*Rhizopus* spp. for Tempeh, *Neurospora sitophila* for Ontjom).

Hesseltine & Wang (1967) describe the production of Tempeh in a simple system which they developed for laboratory use. Martinelli & Hesseltine (1964) have described package and tray fermentations, and have also developed methods whereby prepared beans can be inoculated, packed into perforated plastic bags and stored in the deep-freeze until required; in the latter case mature Tempeh was obtained 36–38 h after removing a bag from the deep-freezer and placing it in a suitable incubator.

In the basic process, Hesseltine & Wang (1967) direct that dehulled soyabean grits (not flour, presumably to allow maximum penetration of the fungal hyphae) be soaked in tap water for 2 h at 25°C, cooked for 30 min, drained and cooled, inoculated with spores of *Rhizopus oligosporus*, lightly packed into shallow trays (they found Petri dishes to be very satisfactory) and incubated at 31°C for 20–24 h. At the end of the incubation, the soyabean grits are totally bound together by fungal mycelium into an apparently solid cake; so solid in fact that they could slice it almost like bread. They recommended either roasting the cake or frying it in vegetable oil as the cooking processes of choice.

It appears that a somewhat similar product can be made if the normal mould is replaced by the *Neurospora* usually employed in the Ontjom fermentation (Steinkraus, Lee & Buck, 1965); the resulting foodstuff is said to have an almond flavour.

Hesseltine (1965) also describes the production of yet another food of this type, the raw material being the solid residue left over after the extraction of soyabean milk. In certain areas of China, these residues are pressed into cakes and inoculated with a mould originally described as a *Mucor*, but reclassified by Hesseltine as that same *Actinomucor elegans* which is so important in the Sufu fermentation.

As with Tempeh the fermented cake is eaten after frying in vegetable oil, or it may be cooked with vegetables. This foodstuff is called Meitauza.

An interesting aspect of the Tempeh fermentation is the inability of *Rhizopus oligosporus* to utilize sucrose or the other oligosaccharides found in soyabeans (Sorenson & Hesseltine, 1966), although it can grow on maltose, glucose and fructose. This behaviour is in marked contrast to the vigour with which *Aspergillus oryzae* attacks sucrose, releasing invertase immediately upon germination. It appears that *R. oligosporus* must grow primarily at the expense of the lipids present in the bean, and to this end it secretes very active extracellular lipases and proteinases, as does *Neurospora sitophila*.

Wang, Ruttle & Hesseltine (1969) have reported that *Rhizopus oligosporus* produces an antibiotic active against many Gram-positive bacteria, including some typical intestinal Clostridia. They interpret this as contributing to the beneficial nutritional effects of Tempeh, and make a similar claim for the *R. chinensis* used in the Lao-chao fermentation of rice (Wang & Hesseltine, 1970).

Ontjom

A particularly attractive feature of this Indonesian fermentation is that it utilizes as substrate a material, the residue left after the oil has been pressed from peanuts, which would otherwise be a waste product. According to Hesseltine & Wang (1967), the press cake is broken up and mixed with water; after 24 h, residual oil has floated to the surface and can be skimmed off. The peanut solids are then washed several times, steamed, pressed into shallow trays and inoculated with material from a previous fermentation. The mould mycelium invades the entire cake and presumably binds it together as does the mould mycelium in Tempeh. The importance of good aeration is stressed. Unlike any of the fermentations discussed previously, sporulation of the mould is a normal and even desirable part of the fermentation, imparting to the finished product a pink colour. Gray (1970) is of the opinion that this is an undesirable feature, should there be an attempt to produce Ontjom for the Occidental market 'especially if [the consumer] knew that the distinctive colour was due to the presence of thousands of spores of a very common coloured mould'. This fear seems to be a little excessive, since the colour of blue cheese is an essential part of the attraction of that product, and likewise the brown colour of mushroom gills is due to the presence of spores, a fact which has surely been demonstrated by every child at some time, through the preparation of a spore-print.

The finished product can, it is stated, be cooked in several ways, for example frying, roasting and serving with ginger sauce, or roasting and covering with boiling water and adding salt and sugar. Hesseltine & Wang (1967) compare the flavour of the fried material with that of minced-meat.

This is perhaps the most extraordinary of the many remarkable food fermentations reported from the East. In the first place, the substrate employed is the peanut. Now it was in fungal infections of peanuts that aflatoxin formation was first recognized, yet here we have not the intact nuts but a press-cake being deliberately fermented. In the second place, the conditions employed in Indonesia do not seem to involve rigid microbiological control, since the inoculated material is often left in a 'shady place' to ferment. In other food fermentations, such as Miso, control is exercised by the use of salt, and the development of acidic conditions; no such restraints are imposed here, but it must be acknowledged that the product is eaten fairly soon after completion of the fungal fermentation. In Shoyu and other *Aspergillus* fermentations, even though carbohydrate-rich materials are present, excessive spore formation is very undesirable since the onset of sporulation is accompanied by very active de-amination of amino acids, giving rise to a strong smell of ammonia, yet in this case it is a normal part of the fermentation. Hesseltine & Wang (1967) point out that they know of no other case anywhere in the world where *Neurospora* is utilized in the preparation of food. Finally, the reported meaty flavour of the product is not so expected as is that of Shoyu with the latter's high input of glutamic acid from both beans and wheat flour. All in all, this is a fascinating fermentation posing many riddles for the biochemist and the microbiologist.

Hamanatto

This is another of the fermentations which employ soyabeans initially fermented with *Aspergillus oryzae*. A Malayan dish called Tao-Cho and one from the Philippines called Tao-Si seem to be somewhat similar to judge by the descriptions given by Hesseltine (1965) and by Gray (1970). The whole beans are washed, soaked in water overnight, steamed for 5–10 h at atmospheric pressure and cooled to around 30°C. The subsequent treatment may vary somewhat, since Hesseltine (1965) quotes Smith (1958) as requiring that the beans be inoculated over their entire surface with a Koji prepared from roasted wheat or barley, whereas Hesseltine & Wang (1967) mix the freshly cooled beans with parched wheat flour (2 parts beans to one part flour), then inoculate with a short or medium stalked *A. oryzae*, a process which more closely resembles that for soya sauce. According to Smith (1958), the beans are covered with green mycelium after 20 h incubation, are then dried in the sun for a day to a moisture content of 12%, afterwards mixed with ginger and immersed in brine. Hesseltine & Wang (1967) do not specify the time of incubation with the mould, but it seems reasonable to assume that it will be of the same order as that used in Shoyu preparation. The developed Koji is then added to 18% brine (Koji, 2.5 Kg; sodium chloride, 650 g; water, 3.6 l; ginger, *quant. suff.*). In both cases, the 'Moromi' is then aged under pressure for up to a year. There is no apparent information on microbial activities or biochemical changes during this ageing process, but it seems safe to predict that yeasts and lactic acid bacteria will play a part. The aged beans are removed from the brine and dried in the sun. At the completion of this process, the beans are reported to be very dark in colour, 'rather soft, about like raisins', and with a flavour like Miso or Shoyu but rather sweeter.

Hesseltine & Wang (1967) list various uses for Hamanatto, including snacks, and for adding flavour to foods such as beef, fish, and Cantonese-style lobster. We are intrigued by the fact that the beans retain their integrity so successfully during the prolonged ageing in brine, since in our experimental Shoyu Moromi, we find that the beans disintegrate almost completely during a month in brine at 40°C. This, probably reflects the combined actions of cellulase, amylases and proteinases from the Koji-mould on the structure of the bean. It seems reasonable to suggest that this difference is due to the seed-coat being left intact in the Hamanatto fermentation.

This would seem to be an Oriental foodstuff which could be especially suitable for development for European markets, as a premium product.

Ang-Kak and Lao-Chao

The materials discussed so far have been prepared from high-protein substrates. As an illustration of the versatility of Oriental food fermentations, we now consider these two, which employ rice as principal substrate.

Ang-Kak is colouring matter. It is made by permitting the mould *Monascus purpureus* to grow on polished rice which has been soaked overnight, drained, autoclaved, cooled and inoculated with a suspension of mould spores. Hesseltine (1965) stresses the importance attached to the water level being low enough so that the grains remain separate even after autoclaving. This promotes good aeration and allows the grains to be well mixed by shaking from time to time during the fermentation. We have also found this to be a very easy fermentation to carry through on a laboratory scale. The finished rice is deep red in colour and the pigment is present all the way through the grains. This latter observation, and the very friable, crumbly nature of the finished grain make it quite clear that the fungal hyphae must completely permeate the rice grains, and that a range of extracellular enzymes (presumably including amylases and possibly cellulase) must be produced. However, little seems to be known about the activities of the mould or about the nature of the pigments. The only study of the pigments that we can trace is the one by Nishikawa (1932) referred to by Hesseltine (1965), in which two components are reported, one red, monascorubrin ($C_{22}H_{24}O_5$), and the other yellow, monascoflavin ($C_{17}H_{22}O_4$). Since Ang-Kak can be used for colouring rice wine, these pigments, if not actually water soluble, must be soluble in 10% alcohol.

The formulae quoted would not be inconsistent with a very tentative hypothesis that the pigments could be xanthophylls, and since they are so readily produced in solid culture on rice, and also on Sabouraud Agar, they might repay investigation as food-colouring agents in the West.

Lao-Chao, in contrast to Ang-Kak, seems to be an actual foodstuff, *i.e.* a material consumed primarily for its own nutritional value, rather than as an accompaniment to other things. The only information on it in Western literature is in a paper by Wang & Hesseltine (1970). It is made from glutinous (waxy) rice, probably the rice used for rice puddings in British kitchens. Wang & Hesseltine (1970) have identified various organisms in commercial starters for the fermentation. Good results were obtained when they employed a mixture of two of these organisms, *Rhizopus chinensis* and a species of the yeast *Endomycopsis*. Steamed, cooled rice was inoculated, lightly packed into a bowl, covered and allowed to ferment for two to three days at room temperature. The finished product is described as being soft, juicy, sweet and mildly alcoholic (1–2% ethanol, and 20–30% reducing sugar measured as glucose), and as possessing a fruity aroma. It is eaten as it is, or cooked with other things in desserts.

The precise nature of the interaction of yeast and mould in the fermentation is not clear. They found that rice inoculated with the mould alone had a good aroma, but was dry and had a mouldy appearance. They are of the opinion that most of the amylase and all of the lipase and proteinase released into the mass come from the mould, and it is therefore rather surprising that the yeast should make such a difference to the texture of the finished product. They also note that the mould produces an antibiotic active against many Gram-positive bacteria, including some typical intestinal Clostridia, and speculate that this may contribute to the nutritional effect of Lao-Chao. It would be interesting to know if the lactic acid bacteria of the gut are sensitive or resistant to this agent.

13.7 Conclusions

These, then, are some of the many food fermentations practised in the Orient. They raise many questions in the mind of the biochemist and microbiologist, and afford many research challenges and opportunities. That the list is by no means exhaustive may be judged from the fact that Hesseltine in his pivotal 1965 review lists 73 fermentations involving yeasts or fungi (although some are perhaps local variations on the theme of Sufu for example) and over 80 further synonyms. Stanton & Wallbridge (1969) give an interesting list of fermentations grouped in an ordered fashion by substrate.

We are of the opinion that several of these oriental food fermentations are capable of commercial exploitation for the 'luxury' end of the food market in Europe. Continental European marketing of the products would probably be less difficult than marketing in the United Kingdom, not least because of the greater culinary use of a wide range of fungi in the former.

13.8 References

GRAY, W.D. (1970). The use of fungi in food and in food processing. *Chemical Rubber Co. Critical Reviews in Food Technology* **1**(2), 225–329.

GROFF, E. H. (1919). Soy sauce manufacture in Kwangtung, China. *Philippines Journal of Science* **15**, 307–16.

HESSELTINE, C.W. (1965). A millennium of fungi, food and fermentation. *Mycologia* **57**, 1–148.

HESSELTINE, C.W. (1966). Fermented products—Miso, Sufu, Tempeh. *Proceedings of International Conference on Soybean Protein Foods, Peoria, Illinois,* October 17–19, pp. 170–9.

HESSELTINE, C.W. & WANG, H.L. (1967). Traditional fermented foods. *Biotechnology and Bioengineering* **9**, 275–88.

HESSELTINE, C.W. & WANG, H.L. (1968). Oriental fermented foods made from soybeans. *The Ninth Dry Bean Research Conference, Colorado State University, Fort Collins, Colorado,* August 13–15.

KAWAMURA, S. (1967). Review of PL 480 work on soybean carbohydrates. *United States Department of Agriculture, ARS-71-35*, pp. 249–54.

MARTINELLI, A. & HESSELTINE, C.W. (1964). Tempeh fermentation; package and tray fermentations. *Food Technology* **18**, 167–171.

MINOR, L.J. (1945). Soya sauce processes and how they can be improved. *Food Industries* **17**, 758–60.

NELSON, J.H. & RICHARDSON, G.H. (1967). Moulds in flavour production. In *Microbial Technology*, ed. Peppler, H.J., pp. 82–106. New York: Reinhold Publishing Corporation.

NISHIKAWA, H. (1932). Biochemistry of filamentous fungi. I. Colouring matters of *Monascus purpureus* Went. *Journal of the Agricultural Chemical Society of Japan* **8**, 1007–15.

ONISHI, H. (1957*a*). Studies on osmophilic yeasts. I. Salt-tolerance and sugar-tolerance of osmophilic yeasts. *Bulletin of the Agricultural Chemical Society of Japan* **21**, 137–42.

ONISHI, H. (1957*b*). Studies on osmophilic yeasts. II. Factors affecting growth of soy yeasts and others in the environment of a high concentration of sodium chloride (1). *Agricultural Chemical Society of Japan* **21**, 143–50.

ONISHI, H. (1958*a*). Studies on osmophilic yeasts. IV. Changes in permeability of cell membranes of the osmophilic yeasts and the maintenance of their viability in the saline medium. *Agricultural Chemical Society of Japan* **23**, 332–9.

ONISHI, H. (1958*b*). Studies on osmophilic yeasts. V. Factors affecting the growth of soy yeasts and others in the environment of a high concentration of sodium chloride (2). *Agricultural Chemical Society of Japan* **23**, 351–8.

ONISHI, H. (1963). Osmophilic yeasts. *Advances in Food Research* **12**, 53–94.

SHIBASAKI, K. & HESSELTINE, C.W. (1962). Miso fermentation. *Economic Botany* **16** (3), 180–95.

SMITH, A.K. (1949). Oriental methods of using soyabeans as food, with special attention to fermented products. *United States Bureau of Agricultural and Industrial Chemistry A.I.C.* **234**, June, 1949.

SMITH, A.K. (1958). Use of United States soybeans in Japan. ARS-71-12, *United States Department of Agriculture, Washington, D.C.*

SORENSON, W.G. & HESSELTINE, C.W. (1966). Carbon and nitrogen utilisation by *Rhizopus oligosporus*. *Mycologia* **58**, 681.

STANTON, W.R. & WALLBRIDGE, A. (1969). Fermented food processes. *Process Biochemistry* **4** (4), 45–51.

STEINKRAUS, K.H., LEE, C.Y. & BUCK, P.A. (1965). Soybean fermentation by the Ontjom mould, *Neurospora*. *Food Technology* **19**, 1301.

WANG, H.L. & HESSELTINE, C.W. (1970). Sufu and Lao-Chao. *Journal of Agricultural and Food Chemistry* **18** (4), 572–5.

WANG, H.L., RUTTLE, D.I. & HESSELTINE, C.W. (1969). Antibacterial compound from a soybean product fermented by *Rhizopus oligosporus*. *Journal of the Society for Experimental Biology and Medicine* **3**, 579.

WOLF, W.J. & COWAN, J.C. (1971). Soybeans as a food source. *C.R.C. Critical Reviews in Food Technology* **2** (1), 81–158.

YOKOTSUKA, T. (1960). Aroma and flavour of Japanese soy sauce. *Advances in Food Research* **10**, 75–134.

YOKOTSUKA, T. (1964). Shoyu. *The International Symposium on Oilseed Protein Foods, Institute of Food Technologists. U.S.A.* May 11–15, pp. 31–48.

YONG, F.M. (1971). M.Sc. Thesis, *Studies on Soy Sauce Fermentation,* University of Strathclyde, Glasgow, Scotland.

YONG, F.M. & WOOD, B.J.B. (1973). Microbiology and biochemistry of the soya sauce fermentation. *Advances in Applied Microbiology* **17**, 157–194.

CHAPTER 14

Industrial Exploitation of Ergot Fungi

P. G. MANTLE

14.1 Ergot Fungi as Plant Pathogens

The ergot fungi are a group of plant parasitic Ascomycetes, members of which may be found in most parts of the world where suitable graminaceous hosts exist as part of the indigenous flora or are grown as an agricultural crop. Natural infection is restricted to the flower parts (anthers, lodicules and ovary) which are enclosed within the glumes. The sequence of events in the natural ergot life cycle, as illustrated by *Claviceps purpurea*, is illustrated diagrammatically in Fig. 14.1.

The ovary is the site at which parasitism becomes established. There is no penetration of the host plant beyond the distal portion of the rachilla. Penetration of the ovary by hyphae from germinating ascospores or conidia normally occurs during the few days before or after fertilization of the ovary. This tissue is permeated within one week by a thin white mycelium which bears the asexual fructification ('sphacelia stage') on its surface. Complete colonization of the ovary coincides with the appearance of host exudate which accumulates within the floral cavity and frequently overflows as 'honeydew'. The honeydew is rich in sugars, amino acids and inorganic ions and not only provides a nutrient environment for the continued growth of the parasite but also acts as a vehicle for dissemination of the sphacelial conidia, which are responsible for establishing the epidemic phase of the plant disease. About two weeks after the initial infection of the ovary the growth of sporulating tissue subsides and is replaced by a more compact type of growth which gives rise to a hard, sclerotial, non-sporulating tissue. This growth form proliferates to give the characteristic ergot sclerotium, the shape of which is partly determined by the shape of the host floret but is nevertheless diagnostic for the principal species of ergot fungi.

Sclerotia usually contain several alkaloids. The ergot sclerotium is usually larger than the host's seed which it replaces and may be harvested with the seed crop. Alternatively it falls to the ground where it may remain dormant for several months during winter, or other climatic conditions which are adverse for the flowering of a suitable host, before giving rise to stromata bearing the sexual (*Claviceps*) stage. The products of sexual

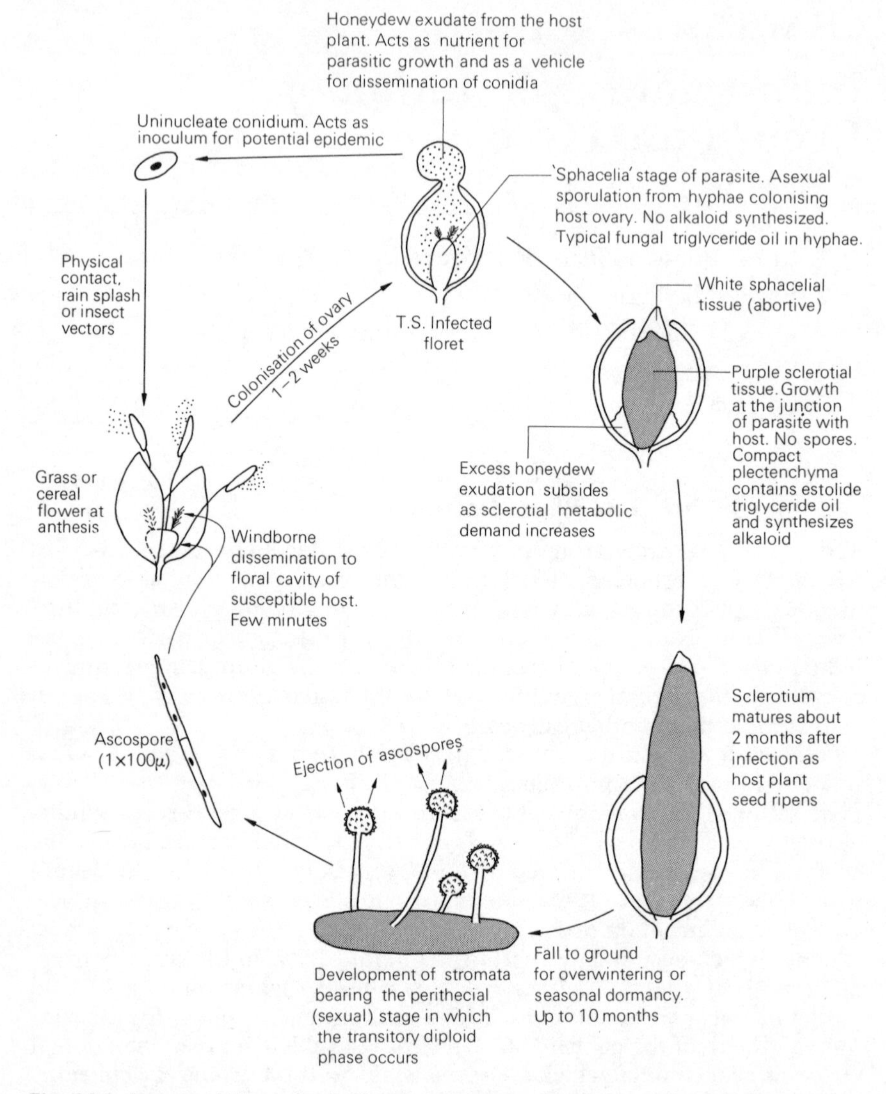

Fig. 14.1 Annotated life cycle of *Claviceps purpurea*. Generally representative of other ergot fungi.

reproduction are long needle-like ascospores which become windborne and may cause infection if the spores become deposited in the flowers of a susceptible host. Thus, fallen ergots may initiate the parasitic part of the life-cycle. This parasitic phase constitutes an important agricultural disease not only because it impairs the fertility of the host but also because the sclerotia are toxic to animals. Farm animals (grazing ergotized pasture or fed with contaminated grain) or humans (eating bread made from ergot-contaminated grain) will usually ingest alkaloids, most of which are pharmacologically potent and can act as poisons. The distribution and alkaloid content of the more important ergot fungi are summarized in Table 14.1.

14.2 Ergot Alkaloids

Several authoritative reviews of the ergot alkaloids are in the literature (Stoll, 1952; Hofmann, 1964; Stoll & Hofmann, 1965) and are particularly valuable for chemical aspects. The ergoline nucleus (Fig. 14.2) of the ergot alkaloids is biosynthetically derived from tryptophan and mevalonic acid, and the various naturally occurring alkaloids are either the relatively simple clavine alkaloids and short chain lysergic acid/isolysergic acid derivatives, or complex cyclic tri-peptide derivatives. Some of the clavine alkaloids have been shown to be biosynthetic precursors of lysergic acid and its derivatives. This subject has recently been reviewed by Thomas & Bassett (1972).

Pharmacological properties

The poisonous quality of many of the ergot alkaloids is only the combined expression of the principal pharmacological properties over a wide spectrum of possible activity (Table 14.2). These include central and sympathetic nervous stimulation, direct action on motor neural junctions of smooth muscle (uterus, arteriole walls, sphincter pupillae), and adrenalin and serotonin (5HT) antagonism. Such pharmacological properties are expressed as hypertension, clonic spasm, vomiting, salivation, inco-ordination, cyanosis of extremities leading to dry gangrene, infertility, agalactia, and tonic contraction of the uterus. Classical ergotism is thought to have resulted from both central and peripheral aspects of ergot alkaloid pharmacology. The gangrenous form has been confirmed clinically in both man and animals, but the convulsive form, recorded by Barger (1931) and recently described vividly in the popular account of an extraordinary syndrome at Pont-Saint-Esprit, France in 1951 (Fuller, 1969), has not received a convincing explanation in terms of the known central effects of *Claviceps purpurea* alkaloids.

Some naturally occurring alkaloids are potent hallucinogens as is also the semi-synthetic diethylamide of lysergic acid (LSD).

Geographical distribution

The extent of natural occurrence of those ergot fungi which contain, in the parasitic sclerotia, ergot alkaloids of potential industrial interest is summarized in Table 14.1. If a liberal view of the concept of what constitutes a species is taken, and this is more valid if the alkaloidal content is included as a diagnostic feature, it is seen that *Claviceps purpurea*, occurring on a wide range of temperate grasses, possesses the most immediately useful types of alkaloid. These alkaloids can occur in greater abundance than in most other species. Thus the greater ubiquity and immediate value of *C. purpurea* makes it the most desirable ergot fungus for exploitation.

Plant parasitism by *Claviceps paspali* is limited under natural conditions to *Paspalum* spp. of which about 200 are known in tropical and sub-tropical climates. Only about 30 of these species, most commonly *P. distichum* and *P. dilatatum*, have been recorded as susceptible hosts and their distribution is mainly sub-tropical. Although *C. paspali* is able to

Table 14.1 Some of the principal ergot fungi, their host range, alkaloid content and geographical distribution

Ergot fungus	Host range	Alkaloid content		Geographical distribution
		%	Principal components	
Claviceps purpurea (and several closely related species)	Many temperate pasture grasses. Cereals (rye, wheat, barley and hybrids	0.1–0.5	Ergotoxine group: Ergocornine Ergocryptine Ergocristine Ergotamine group: Ergotamine Ergosine Ergometrine	World-wide temperate regions
C. fusiformis	Bulrush Millet (*Pennisetum typhoides*)	0.3	Agroclavine Elymoclavine Chanoclavine	Central Africa
C. gigantea	Maize (*Zea mais*)	0.03	Festuclavine Pyroclavine Dihydroelymoclavine Chanoclavine	Mexico
Sphacelia sorghi	Sorghum (*Sorghum vulgare*)	0.3	Dihydroergosine Festuclavine Pyroclavine Dihydroelymoclavine Chanoclavine	Nigeria Botswana (Possibly another form in India)
C. paspali	*Paspalum* spp.	0·003	D-lysergic acid methylcarbinolamide D-lysergic acid amide	Italy Turkey Southern U.S.A. South Africa South America Australasia
C. sulcata	*Brachiaria brizantha*	0.000	—	Central Africa

Fig. 14.2 Principal alkaloids and precursors of ergot fungi.

ergoline

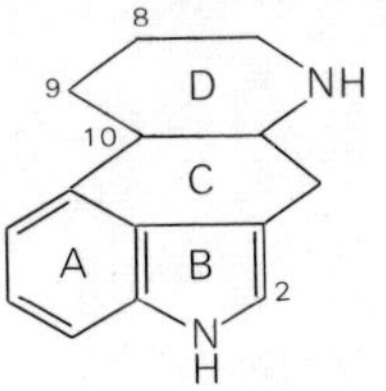
8
9
10
D
NH
C
A
B
2
N
H

D-lysergic acid (Δ9-10 lysergic acid)

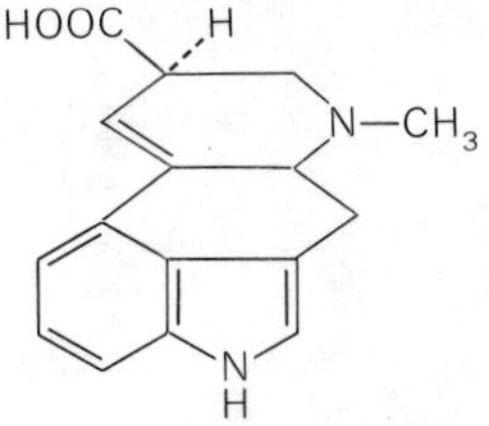
HOOC
H
N—CH3
N
H

tryptophan

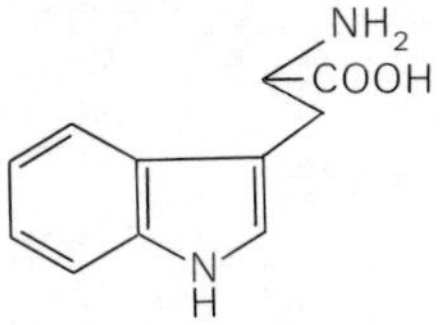
NH2
COOH
N
H

Isolysergic acid

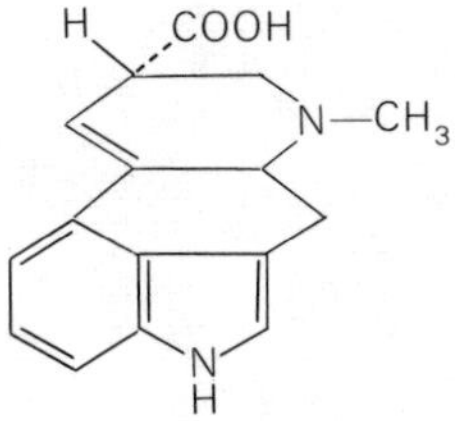
H
COOH
N—CH3
N
H

mevalonic acid

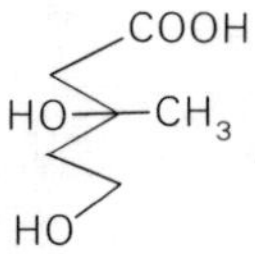
COOH
HO
CH3
HO

6-methyl-ergol-8-ene-8-carboxylic acid (Δ8-9 lysergic acid)

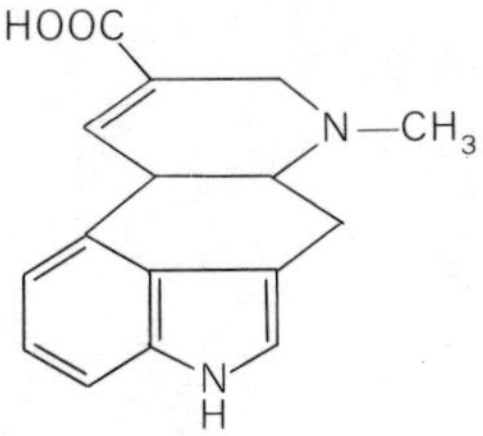
HOOC
N—CH3
N
H

Simple lysergic acid derivatives

ergometrine

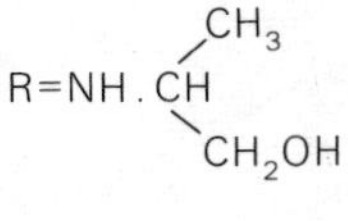
CH3
R=NH.CH
CH2OH

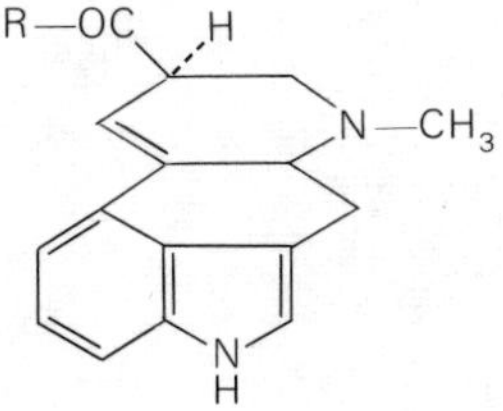
R—OC
H
N—CH3
N
H

lysergic acid methylcarbinolamide

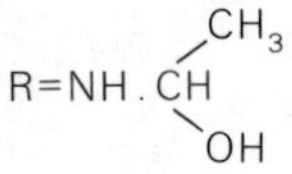
CH3
R=NH.CH
OH

lysergic acid diethylamide

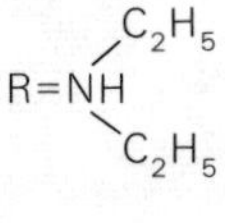
C2H5
R=NH
C2H5

Cyclic tripeptide derivatives

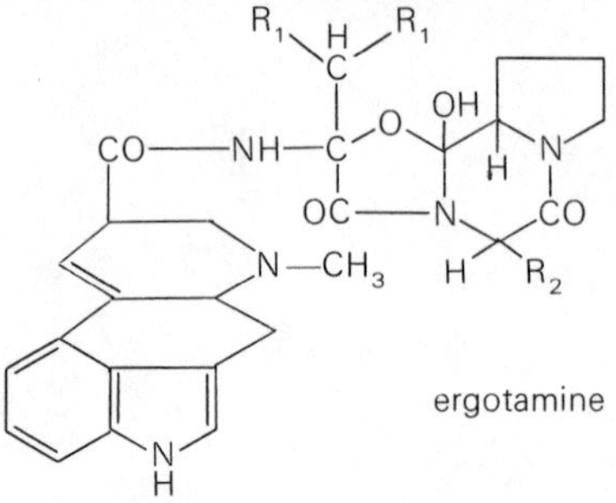

ergotamine $R_1 = H$

$R_2 = —CH_2—C_6H_5$

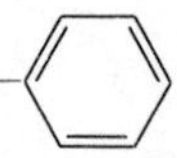

ergocryptine $R_1 = CH_3$

$R_2 = —CH_2—CH—(CH_3)_2$

dihydroergosine

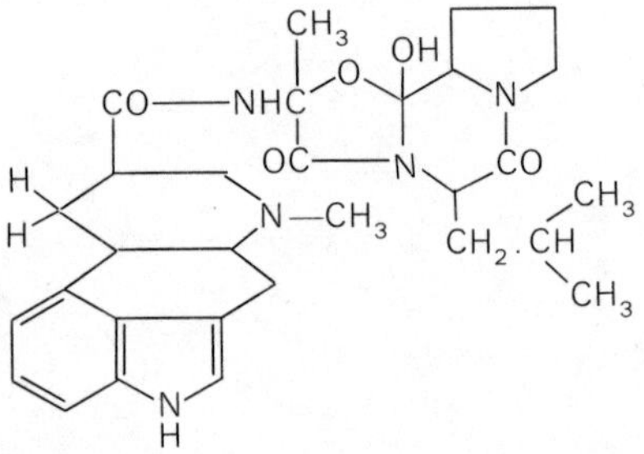

chanoclavine

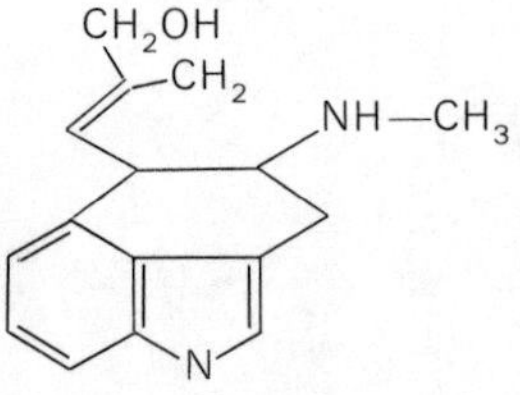

elymoclavine

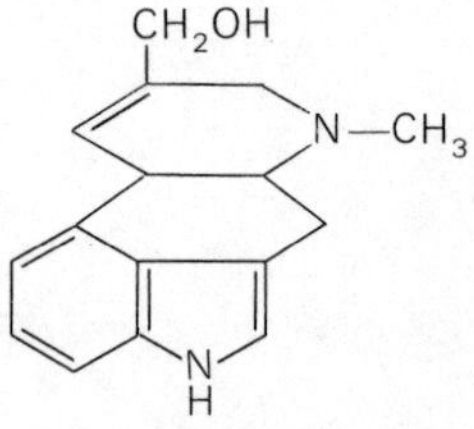

agroclavine

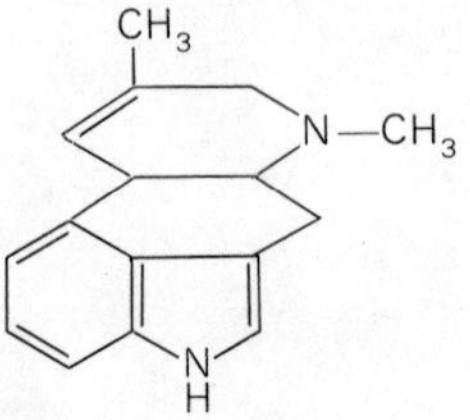

dihydroelymoclavine

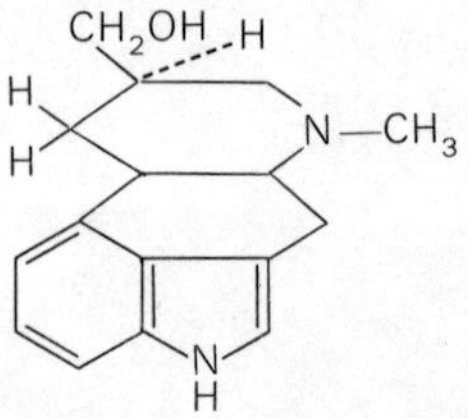

festoclavine (dihydroagroclavine) pyroclavine (dihydroagroclavine)

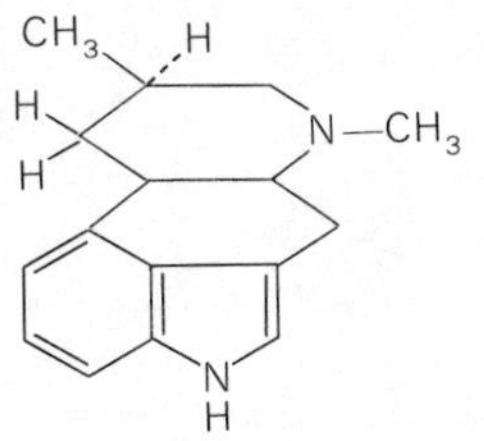

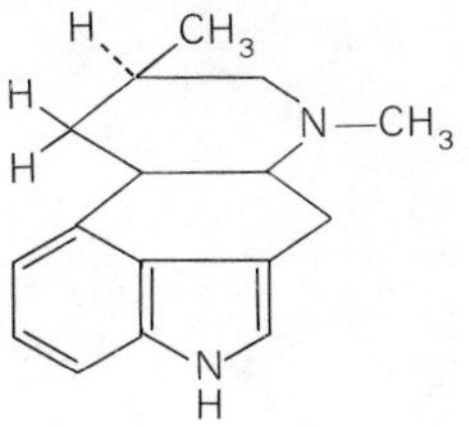

biosynthesize the lysergyl nucleus, making the species of potential value as a source of lysergic acid, the total alkaloid content of the natural sclerotia is usually lower than that of any other alkaloid-producing species.

Claviceps fusiformis, C. gigantea and *Sphacelia sorghi* have the most restricted host range (respectively millet, maize and sorghum). *C. fusiformis* produces only clavine alkaloids which, though pharmacologically potent, probably do not have any direct potential as pharmaceuticals. However, the biosynthetic role of the principal clavine alkaloids, namely agroclavine and elymoclavine, in the formation of lysergic acid derivatives gives *C. fusiformis* potential either as a source of precursors for addition to *C. purpurea* fermentations or as a component of a mixed culture fermentation. Sclerotia of *S. sorghi* and *C. gigantea*, though indigenous to different continents, are similar in containing, in addition to chanoclavine, reduction (dihydro-) products of some of those clavine alkaloids which are at a higher level of oxidation than chanoclavine. *S. sorghi* is unique in producing dihydroergosine (Mantle & Waight, 1968), a cyclic tri-peptide derivative of dihydrolysergic acid, in addition to and possibly biogenetically derived from dihydrogenated clavines (Mantle, 1968). Dihydrogenation of ergot alkaloids usually decreases their pharmacological activity and thus dihydroergosine probably has no place in therapeutics. However it would provide a convenient source of dihydrolysergic acid, by hydrolysis. A potential significance of dihydrolysergic acid will be discussed later.

Agricultural exploitation

Ever since the sclerotia of rye ergot entered the Materia Medica, the natural susceptibility of rye to the ergot disease has been exploited. Initially, this involved simply the selection of these curious deformed grains (mutterkorn, blé cornu, secale cornutum) to provide a basis for aqueous decoctions administered to hasten childbirth (Stearns, 1808). Doubtless, the same material was also used by unscrupulous persons in the hope of procuring abortions by employing the oxytocic properties of this medicament. Gradually, it became realized that the ergots were fungal tissue which developed as part of a plant disease syndrome. Consequent on increasing demand for medicinal ergot, crops of the most convenient and susceptible host, rye, were grown in areas suitable to the natural development of epidemics of the disease. Thus the product of disease became more valuable than the host. With the characterization of some of the principal *Claviceps purpurea* ergot alkaloids, first in 1918 (Stoll, 1918) and later extending into the 1930's, it became clear that selection of strains of

Table 14.2 Principal pharmacological properties, poisonous effects and therapeutic uses of naturally occurring and semi-synthetic ergot alkaloids

Alkaloid	Pharmacological spectrum	Poisonous effects	Therapeutic uses
Ergotamine group Ergotoxine group	Vasoconstriction Uterine contraction Central nervous stimulation Adrenergic blocking Serotonin antagonism	Peripheral vascular disturbance Foetal distress Agalactia Sympathetic excitation	Migraine relief (Ergotamine) Inhibition of lactation (Ergocryptine)
Ergometrine	Uterine contraction Serotonin antagonism	Foetal distress	As an oxytocic at end of 2nd stage of childbirth. Control of post-partum haemorrhage, and an aid to uterine involution.
D-lysergic acid methylcarbinolamide	Central nervous stimulation Uterine contraction (<Ergometrine)	Central excitation (More toxic than Ergometrine)	None
D-lysergic acid amide	Central nervous stimulation	Hallucinations	None
D-lysergic acid diethylamide	Central nervous stimulation Serotonin antagonism	Hallucinations	Some psychoses
1-methyl-D-lysergic acid butanolamide	Serotonin antagonism	Nausea	Migraine prophylaxis. Aspects of carcinoid syndrome.
Agroclavine Elymoclavine	Central nervous stimulation	Central excitation Agalactia Infertility	None
Dihydroergosine Dihydro-clavines	Weak	None	None
D-6-methylcyanomethylergoline	Hypothalamic stimulation	None	None yet

fungus having the potential for high alkaloid content, comprising preferred alkaloids, could increase the medicinal productivity of the rye crop. This led to more sophisticated methods of ergot farming whereby spores of the selected strain were applied to the flowering crop to initiate the disease. As the honeydew stage developed, the crop was traversed by mechanical devices designed to spread the honeydew to uninfected flowers.

Agricultural exploitation of ergot as a valuable plant disease suffers from several economic disadvantages. Capital costs of establishing rye crops include restricting the land to one crop per year, and the purchase of labour, seed, fertilizer and weed control. Laboratory preparation and field application of inoculum is expensive and the success in developing an epidemic is conditioned by favourable climatic conditions. Cool damp weather prolongs the flowering period and allows dissemination of conidia exuded in the honeydew. The climate also influences the development of the individual sclerotia. Harvesting is usually by the laborious process of hand-picking but, where the whole crop is harvested mechanically, the sclerotia must be separated from the grain, and subsequently marketed.

The modern tendency towards high labour costs may easily result in a labour-based commodity exceeding the costs of a fermentation process, except in climatically suitable areas of the world where the land has low fertility and where there is a cheap labour force. Of all cereals, rye is most suited to poor quality soil. As ergot farming consisted of creating epidemics of what otherwise was regarded as a dangerous disease, there was the added disadvantage of providing a potential source of infection for adjacent cereal crops, especially wheat, which flower later than rye. Pasture grasses are also at risk and, under adverse management, can develop an ergot content which is a hazard to grazing animals.

Fermentation potential

It is not surprising that attention was turned to *Claviceps purpurea* as soon as it was realized that filamentous fungi could be sources of valuable products elaborated during a fermentation process. In spite of considerable investigation in both academic and industrial laboratories, this organism has, in general, proved to be the most difficult of the ergot fungi. Whereas most strains of *C. purpurea* produce, as parasites, sclerotia containing alkaloids, only very few isolates have yet yielded more than traces of alkaloid in submerged culture. Those giving yields of about 1 mg total alkaloid cm^{-3} culture filtrate are unstable and during routine sub-culture readily lose the capacity for alkaloid production.

Preservation of stock cultures has thus become an important aspect in exploiting the limited success so far achieved with *Claviceps purpurea*. Mycelial fragments, produced in submerged culture or spores suspended in culture broth to which sterile glycerol (10%) has been added, are conveniently preserved in sealed glass ampoules under liquid nitrogen. Both viability and alkaloid yielding potential are well retained.

Evolution of successful processes

The most important landmark in the development of an economic fermentation process for the production of ergot alkaloids was the work of Chain, Tonolo and their co-workers (Arcamone *et al.*, 1961) in Rome.

They tried the most logical approach of using isolates of *Claviceps purpurea* from natural sclerotia which contained the desirable peptide alkaloids but these completely failed to yield alkaloids in submerged culture. Tonolo then chose to work with *C. paspali* which, though conveniently indigenous to Italy, occurs as sclerotia having an exceedingly low content of a rather uninteresting alkaloid (Table 14.1). Further, he employed the unusual approach of attempting to establish parasitic growth of *C. paspali* on germinating embryos of rye—a host which will not support parasitism of the ovary by *C. paspali*. Having obtained infection of the rye embryos and the development of a parasitic fungal tissue containing sclerotial-like cells, re-isolation of the fungus from this tissue yielded a strain which was not only a more vigorous parasite of rye embryos but also gave small yields (*c.* 20 $\mu g\ cm^{-3}$) of alkaloid in submerged culture. New strains were selected following evaluation of the potential of large numbers of colonies which developed from separate hyphal fragments, and these exhibited wide variation in their ability to produce alkaloid (5–500 $\mu g\ cm^{-3}$). The best strains developed a violet pigment during the production stage of fermentation. This was unusual as the parasitic sclerotia of this species are not naturally pigmented. Further strain selection and medium development led to an economically sound industrial process from which Farmitalia in Italy were able to market D-lysergic acid. Total alkaloid yields exceeding 5 $mg\ cm^{-3}$ consisted mainly of one substance, D-lysergic acid methylcarbinolamide, which was extracted with organic solvent after adjusting the pH value of the culture filtrate to generate the free bases in alkaline conditions. After hydrolytic cleavage of the amide, D-lysergic acid was used in Switzerland in the chemical synthesis of ergotamine and other peptide alkaloids (Stadler *et al.*, 1969), ergometrine (Stoll & Hofmann, 1943) and 1-methyl D-lysergic acid butanolamide (methysergide, Deserit ® Sandoz Ltd.). The latter, though not a natural product, was found experimentally (Fanchamps *et al.*, 1960) to be a specific and most potent antagonist to 5-hydroxy tryptamine (serotonin) and is used as a prophylactic against certain types of migraine and has value in terminal carcinoid therapy (Dubach & Gsell, 1962).

The *Claviceps paspali* fermentations were performed for 8–10 days in production stage in a simple semi-defined medium, the principal source of carbon being mannitol supplied at a higher concentration than the amount metabolized during the fermentation. Ammonium ions in conjunction with a tricarboxylic acid radical (*i.e.* succinate) were the best nitrogen source and the medium was completed with phosphate and magnesium ions together with the trace elements provided by tap water. This first successful industrial process for the semi-synthetic production of therapeutic ergot alkaloids, continued until the microbiological process was replaced by one developed by a Swiss pharmaceutical company (Sandoz Ltd.) whose research and development in the field of ergot alkaloids is outstanding (Hofmann, 1964). The process apparently used a new isolate of *C. paspali*, obtained from New Guinea (Kobel, Schreier & Rutschmann, 1964). In producing high yields of a double-bond isomer of lysergic acid (Δ8–9 lysergic acid or, more correctly, 6-methyl-ergol-8-ene-8-carboxylic acid, Fig. 14.2) this strain became a more simple source of D-lysergic acid by chemical rearrangement of the double-bond from the Δ8–9 position to the Δ9–10 position. This saved a hydrolytic step and also avoided patent restrictions of the previous process in which lysergic acid methylcarbinolamide was the fermentation product.

Claviceps fusiformis is probably the most convenient of the ergot fungi in that most isolates from natural sclerotia do produce alkaloid in surface cultures, but concomitant production of a viscous $\beta 1 \rightarrow 3$ glucan has usually impaired growth and alkaloid production in submerged cultures. Recent studies (Szczyrbak, 1972) on alkaloid production by a strain of the fungus isolated from ergots parasitic on millet in Senegal revealed a variant strain which, while accumulating glucan during the first few days in submerged culture, later hydrolysed the polysaccharide through the action of a $\beta 1 \rightarrow 3$ glucanase and a β-glucosidase which were detected coincident with the appearance of sclerotial-like cells.

For a large scale (400 dm^3) process, a sucrose/ammonium sulphate/mineral salts medium, with pH value controlled at 5.0 by automated additions of NaOH, supported agroclavine alkaloid yields of about 4 mg cm^{-3}. Half replacement culture, whereby half the culture was transferred to an equal volume of double strength medium every 3 days, raised agroclavine yields to 6 mg cm^{-3} at the end of each 3-day period. Action of the glucanase maintained a low viscosity culture filtrate, which not only improved aeration by allowing an adequate dissolved oxygen concentration to be achieved at a lower impeller speed but also made the culture easier to filter at harvest. Furthermore, by avoiding the complication of emulsions, this enzyme facilitated the n-butanol extraction of alkaloid from the broth. High alkaloid titre was associated with a sclerotial type of growth giving short frequently-septate mycelial fragments composed of uninucleate cells which, when treated with a lipid stain (Sudan IV) showed an extensive distribution of lipid throughout the cytoplasm (Banks, Mantle & Szyzyrbak, 1974). The sclerotial triglyceride oil was not only more abundant than that within the cells of thin sporulating sphacelial hyphae but its fatty acid composition was also quantitatively distinctive, being rich in palmitate whereas young sphacelial hyphae were rich in linoleate (Mantle & Szczyrbak, 1972).

Principles and problems of Claviceps purpurea

The technology of alkaloid production by *Claviceps paspali* and *C. fusiformis* has thus largely been elucidated, but the most desirable of all the *Claviceps* spp., *C. purpurea*, still presents enormous difficulties. The first successful ($>$1 mg alkaloid cm^{-3}) laboratory scale process for alkaloid production by *C. purpurea* was reported by Tonolo (1966). The strain of fungus originally isolated from wild triticale in Spain had gone through several years of investigation in the laboratory (Mantle & Tonolo, 1968) and was atypical in failing to sporulate in culture. After it had been reselected many times to produce its highest yields of alkaloid the strain was found to have lost its pathogenicity. In spite of continued research with this strain, no significant increases in yield have occurred. Alkaloid production was associated with sclerotial-like cells, rich in a triglyceride oil which was similar in fatty acid composition to that found to be characteristic of *C. purpurea* sclerotial tissue *in vivo* and *in vitro* (Mantle, Morris & Hall, 1969). Ricinoleic acid is the principal component and forms the basis of complex estolide triglycerides (Morris & Hall, 1966) which only occur in *C. purpurea.* Strain variability resulted in the continuous segregation of alkaloid-non-producing variants which, in the same fermentation conditions, grew only to give the filamentous sphacelial hyphae whose lower triglyceride oil content contained no ricinoleate (Mantle, P.G., unpublished results).

Tonolo (1966) reported alkaloid production in a mannitol (20%)/ peptone (3%) medium but similar yields could also be obtained on a sucrose (30%)/ammonium citrate/ mineral salts medium. A medium very similar to the latter was described by the Farmitalia research group (Amici *et al.,* 1965, 1966) for the patented process for alkaloid production, principally ergotamine, by a strain whose origin was not disclosed but which has many similarities to that used by Tonolo. Amici *et al.* (1967*a*) and Arcamone *et al.* (1970) reported in detail the progress of the fermentation, both in shake flasks and on an 800 dm^3 stirred fermenter scale. They have since obtained several of the peptide alkaloids, in yields of a similar order of magnitude, from strains of *Claviceps purpurea* of Italian and Swiss origin grown in shake flasks (Amici *et al.,* 1969). Alkaloid produced in *C. purpurea* fermentations tends to be located in higher proportion within the mycelium (Amici *et al.,* 1966, 1969) whereas most of that produced in *C. paspali* and *C. fusiformis* fermentations is released into the broth. This may simply be the result of the lower water solubility of ergotamine and the ergotoxine group (ergocristine, ergocryptine, ergocornine). *C. purpurea* may also produce in fermentation a much higher proportion, relative to the peptide alkaloid yield, of chanoclavine, Δ8–9 and Δ9–10 lysergic acids than accumulate within the parasitic sclerotial tissues during growth on the plant (Mantle, 1969*a, b*). This may indicate that it has not yet been possible to simulate sufficiently closely in axenic culture, the satisfactory conditions, particularly with regard to nutrition, which support the normal parasitic growth and its associated alkaloid production.

Ergot fungi do not have special nutrient requirements for growth *in vitro,* and even for alkaloid production it is not necessary to use exotic growth factors. Optimum carbon sources are usually sucrose, glucose or mannitol, while nitrogen may be ammonium or amino in origin. Correct inorganic phosphate concentration is most important. However, the rapid exhaustion of phosphate ions from the medium of batch fermentations does not coincide with cessation of growth but may eventually be a growth-limiting factor essential in the production of an excess of a secondary metabolite (Amici *et al.,* 1969). Magnesium sulphate concentration does not appear to be as critical as that of phosphate, and metal ions other than iron and zinc are not especially required. Thus in the formulation of inexpensive media, and ammonium is a sufficiently cheap nitrogen source, it may be possible to utilize a certain amount of crude carbon source provided that allowance is made for nutrient contaminants.

Perhaps the most distinctive feature of *Claviceps purpurea* in relation to the present subject under discussion, is that as a common plant pathogen it is almost invariably able to synthesize alkaloid, which reaches a concentration of about 0.2–0.5% within the sclerotial tissues, whereas in axenic culture it is rare to obtain sclerotial tissue and even rarer to obtain

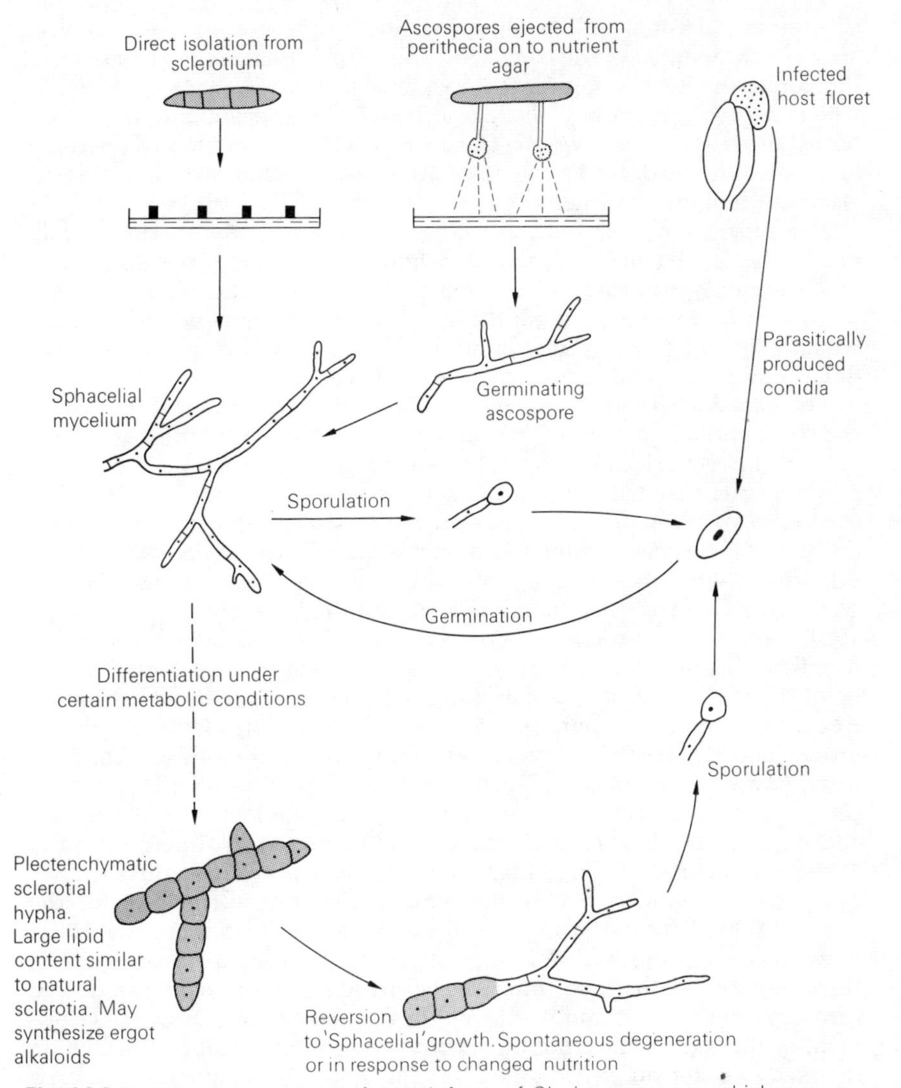

Fig. 14.3 Variation and sequence of growth forms of *Claviceps purpurea* which can occur *in vitro*.

significant yields of alkaloid (Fig. 14.3). It seems that the plectenchymatic sclerotial growth form is a pre-requisite for alkaloid biosynthesis. Sclerotial plectenchyma consists either (in stationary culture) of a compact tissue, often pigmented pink/purple, composed of hyphae whose constituent cells are nearly isodiametric and filled with lipid, or (in submerged culture) of separate short, branched hyphae consisting of cells which are much wider than the usual sphacelial cells and contain approximately three times as much triglyceride oil. Although the absolute amount of oil present in the tissues varies from strain to strain, the present author has not yet found any exception to invalidate the hypothesis (Mantle & Tonolo, 1968) that in *C. purpurea* alkaloid biosynthesis only occurs in sclerotial cells. The identity of sclerotial cells is always shown by the presence of ricinoleic acid as a component of the triglyceride oil. However, it is not valid to infer the converse, for, in nature, it is possible to find *C. purpurea* sclerotia which do not contain any alkaloid although the sclerotia are, in general terms, otherwise quite typical. Consequently, the formation of sclerotial tissue *in vitro* need not always imply that there will also be a coincident elaboration of ergot alkaloid.

The conclusion, therefore, that sphacelial cells of *Claviceps purpurea* do not synthesize alkaloid, but that only sclerotial cells can, leads to the simple, though gravely neglected, deduction that a study of the whole parasitic phase may be a valuable, if not the best, way of obtaining the information which can lead to the formulation of an axenic culture process for the differentiation to, and growth of, sclerotial cells.

The parasitic situation provides for the regular transition, within about one week of the first signs of infection, from a white sporulating tissue to a purple compact sclerotial tissue. Little is known of the composition of the host's nutrient contribution to the developing parasite but it is clear that sucrose is the carbon source (Bassett *et al.*, 1972) and that it is utilized through the complex action of a β-fructofuranosidase (Dickerson, 1972). Subsequent growth is by the continuing addition of the standard type of sclerotial tissue and there is no reversion to the sphacelial form (Nisbet, 1974). Thus the resultant sclerotium is a uniform structure consisting distally of the first-formed tissue. The last-formed tissue is located at the point of attachment of the sclerotium to its host. Alkaloid is distributed throughout the sclerotium and this suggests that, as sclerotial cells are formed and acquire their characteristic morphology, they also synthesize a certain amount of alkaloid. Further synthesis may be restricted by the tissue having been pushed away from the point of attachment to the host and thus becoming starved of nutrients as a result of the interpolation of new sclerotial tissue having a high metabolic demand. Within this complex host/parasite interaction may be found the cultural principles for the successful exploitation of *C. purpurea* as a regular source of alkaloids in large scale fermentations. Such principles may be found accidentally in the common industrial empirical approach of strain-selection and medium development, but studies on the metabolism of parasitic development of the fungus have not only already provided valuable information (Corbett, Dickerson & Mantle, 1974) but have also confirmed the view that this approach has potential.

Genetical aspects of ergot fungi have received relatively little attention. This is partly due to the fact that the sexual process has not been obtained *in vitro* but only occurs in the stromata which develop from natural parasitic sclerotia. Some recent experiments (Mantle, P.G., unpublished results) have been performed with *Claviceps purpurea* strains which individually produce sclerotia with a distinctive morphology and alkaloid content. An inoculum consisting of a mixture of two pathogenic strains can yield sclerotia which are composed of a mixture of the two strains of mycelium. However, it is not yet clear whether the component mycelia can both contribute to the primordial stroma which may eventually, after a period of cold treatment, develop from the sclerotium. Even if this did happen it would be necessary for hyphae of the component strains separately to contribute either an antheridium or an ascogonium in order that dikaryotic ascogenous hyphae, carrying both genomes, could develop. The formation of asci from this heterozygous dikaryon would form a basis for strain improvement through breeding. Although this is potentially attractive there is no evidence that *C. purpurea* is yet suited to genetical analysis or mating techniques involving the conventional sexual process.

However, it has been concluded (Spalla *et al.*, 1969) that heterokaryosis is an important feature of strains which produce high yields of alkaloid *in vitro*, and that the occurrence of alkaloids in parasitic sclerotia is related to the heterokaryotic status of the mycelium (Amici *et al.*, 1967*b*). Evidence in support of these views has been presented (Amici *et al.*, 1969) but cannot be regarded as conclusive. The present author does not consider heterokaryosis to be an absolute requirement for sclerotial formation and associated alkaloid production either in axenic culture or in the parasitic state in view of successful experiments with single spore isolates *in vitro* and *in vivo*. While natural sclerotia are frequently a heterogeneous mixture of mycelia derived from several infecting hyphae, there is no evidence that inter-strain hyphal anastomosis occurs. However, the diversity of morphology amongst colonies isolated from different parts of the same sclerotium, together with the apparent 'segregation' of some strains into sub-strains having different morphology, highlights an important area for research and this would include a reappraisal of the genetical status of *Claviceps purpurea* mycelia.

Single uninucleate spore isolates (homokaryons) of *Claviceps fusiformis* have been able to give very high yields (6 mg cm^{-3}) of clavine alkaloids in fermenters (Banks, Mantle & Szczyrbak, 1974), showing that heterokaryosis is not a significant factor for this species.

Mutagenic agents have been found to increase the yield of ergot alkaloids but the spectrum of alkaloids produced by an ergot strain is usually relatively stable (Kobel & Sanglier, 1970) irrespective of the host plant species, and is not readily changed by mutation. However, there would be some value in a mutagen which changed the principal alkaloid within the cyclic tripeptide group.

The role of certain ergot alkaloids in inhibiting pituitary prolactin secretion has recently received further attention (Varga *et al.*, 1972) which indicates a possible application in human medicine. 2 Bromo-ergocryptine was compared with testosterone in a double-blind clinical study on the suppression of post-partum lactation in 60 non-breast-feeding mothers. The ergot alkaloid was as effective as the steroid in relieving acute mammary enlargement and may have the advantage of less risk of thromboembolism. Such application of ergocryptine would add further incentive for direct fermentation production of this alkaloid by *Claviceps purpurea*, the 2-bromoderivative being a subsequent chemical modification.

Possible potential value of other ergot fungi

Claviceps gigantea which, as its name implies, produces the largest of the ergot sclerotia has been included in Table 14.1 to illustrate diversity of the occurrence of dihydro-ergot alkaloids. *C. sulcata* is included as a contrasting species which appears to be alkaloid-free.

Sphacelia sorghi, which almost certainly will be found to be a species of

Claviceps when the sexual stage is described, is a name which may include more than one ergot fungus depending on whether it is applied to the African or Asiatic ergot-like pathogens of Sorghum. The form which is indigenous to Nigeria is quite distinct from any of the ergot fungi so far described and contains quite large amounts of dihydroergosine (Mantle & Waight, 1968) in addition to the dihydroclavine types (Mantle, 1968) which occur otherwise only in *C. gigantea* (Agurell, Ramstad & Ullstrup, 1963). The identities of all the minor ergoline metabolites of *S. sorghi* have not yet been elucidated but alkaloid production (mainly dihydroergosine) exceeding 500 μg cm^{-3} has been obtained from a selected strain after 20 days growth in surface culture on a sucrose/asparagine/mineral salts medium (Mantle, 1973). *S. sorghi* also produces a viscous glucan which has so far prevented successful submerged fermentation. The potential value of dihydroergosine relates to the potent pharmacological properties of D-6-methyl-8-cyanomethylergoline as an anti-implantation agent in rats and mice in the absence of the central toxicity of many other ergot alkaloids (Rezebek, Semonsky & Kucharczyk, 1969; Seda *et al.*, 1971; Mantle & Finn, 1971). It is possible that such a substance might find therapeutic value in birth control or in the inhibition of lactation in higher animals or in the human. The synthesis of the cyanomethylergoline derivative is via dihydrolysergic acid (Semonsky & Kucharczyk, 1968) and a potential source of this substance would be through the hydrolysis of dihydroergosine.

During the past decade, several new ergot fungi have been identified and new alkaloids have been discovered. Large numbers of ergot isolates from sclerotia obtained from many parts of the world have been screened for alkaloid production. The ubiquity of the ergot fungi must still allow for new types to be discovered and for new strains having great potential as fermentation subjects to be obtained. There is still a place for expeditions to collect ergot fungi particularly from unpopulated and unexplored areas of the world.

Hazards of fermentation production

The hazards in ergot alkaloid production fall into one of two categories, human and microbial. There are the hazards, implicit in a microbiological process for the production of pharmacologically active substances, which demand that the normal precautions must be taken to prevent accidental contamination of personnel with the product. Further, the increased availability of lysergic acid derivatives and/or D-lysergic acid, resulting from processes of improved efficiency, can facilitate unauthorized synthesis of the powerful hallucinogen LSD. Although the mode of action of LSD is a fascinating subject for psychotomimetic research, and the substance may even have therapeutic value in the treatment of some psychoses, its abuse has caused and may continue to cause untold suffering. Thus strict legislative controls and factory security may need to accompany development of more efficient fermentation production of ergot alkaloids.

In another sense there are microbiological hazards in performing the process with regard to strain instability (especially the chance of losing production in *Claviceps purpurea*) which may only be overcome by initially laying down a large number of replicated master cultures, sufficient for several years' use, in ampoules under liquid nitrogen. Routine strain selection and a continuous search for new isolates may be necessary for a manufacturer to remain in this business as competitors seek new fermentation products to circumvent patent restrictions.

The advantages of a fermentation process over field production can all be expressed in economic terms. Alkaloid yields in fermenters as a percentage of the mycelial dry weight, may exceed 20% whereas natural sclerotia contain less than 1%, or even less than 0.01%. In submerged culture fermentation the process can be under precise control, production is not limited to seasonal variations and the whole manufacturing process up to the marketing of the product can be performed on the same site. Although the most directly useful alkaloids, ergometrine, ergotamine and the ergotoxine group are still the most difficult to produce by direct microbial fermentation, it must surely be only a matter of time, not forgetting considerable research and development effort, before a successful process is realized. The reluctance of certain pharmaceutical companies to discuss their progress in this field of research and development not only indicates the significance of the topic but may also imply that real progress is being made. It may be fair to conclude that large scale fermentations for the production of moderate yields of ergotamine are already potentially available but that the yields have not yet reached values which satisfy the economics in competition with the part-microbiological part-synthetic process currently employed by Sandoz Ltd. In a controlled fermentation there are several possible manipulations which could not apply to field production of parasitic sclerotia. Clavine alkaloids have been shown to be incorporated into lysergic acid derivatives and thus it may be possible either to add clavine alkaloid to, say, an ergotamine fermentation or, if suitable common cultural conditions can be found, to grow two complementary *Claviceps* spp. together.

Certain clavine alkaloids have been shown to be elaborated by members of the Fungi Imperfecti (Spilsbury & Wilkinson, 1961; Agurell, 1964), but, as yet, no lysergic acid derivatives have been found as metabolites of fungi other than ergot fungi. Thus it seems unlikely that other fungi have any potential concerning the production of ergot alkaloids.

This chapter has concentrated on the ergot alkaloids as the principal justification for exploitation of ergot fungi but it should be remembered that ergot sclerotia contain a wide range of known substances, some of which were first discovered in ergot. It is probable that other new and important substances can still be discovered in ergot and as such may provide further grounds for exploitation.

14.3 Drugs as Metabolites of other Fungi

Following the foregoing discussion of potentials and hazards in the exploitation of ergot fungi it is appropriate to mention some of the other potent pharmacologically active metabolites of filamentous fungi. Many of these drugs are so toxic as to be unlikely to exhibit any useful biological properties at otherwise sub-toxic dosages. Others which are less toxic have already been found to have additional properties and the fungi which produce these metabolites may yet be shown to have a potential as sources of drugs which are useful either directly or indirectly after chemical modification.

Among the metabolites of filamentous micro-fungi are found some of the most potent carcinogens. The aflatoxins elaborated during saprophytic growth of *Aspergillus flavus* on plant products such as moist groundnuts or cereal seed are probably the best known and have been shown to cause liver tumours in animals and may be involved in the aetiology of certain forms of cancer in some human populations in Africa. The aflatoxins have been recently extensively reviewed (Detroy, Lillehoj & Ciegler, 1972). Ochratoxin and citrinin, metabolites of several *Penicillia* and *Aspergilli*, cause kidney disease in pigs which are fed mouldy grain (Krogh, 1972). Zearalenone, a metabolite of *Fusarium graminearum*, causes oestrogenism in pigs fed corn moulded by this fungus (Mirocha, Christensen & Nelson, 1972). Sesquiterpenoid trichothecanes, produced by a variety of Imperfect Fungi, have been involved in mycotoxicoses but have also demonstrated interesting antifungal and cytostatic properties. Facial eczema in sheep is caused by sporidesmin, a complex indolic substance having a sulphur bridge across a diketopiperazine ring, which occurs in the spores of *Pithomyces chartarum* growing on senescent grass leaves in New Zealand pastures and can also be produced in artificial culture.

Toxic metabolites, more closely related to the ergot alkaloids, are found in the carpophores of some Basidiomycetes. Psilocin and psilocybin (Hofmann *et al.*, 1959) are indole derivatives which show hallucinogenic properties similar to lysergic acid diethylamide. Mexican Indians used the sacred mushrooms which contain these drugs in certain religious ceremonies and both substances may be found in the fruit bodies and laboratory cultures of *Psilocybe mexicana* and *P. cubensis* (Brack *et al.*, 1961; Catalfomo & Tyler, 1964; Leung & Paul, 1969). Other notable agaric metabolites include muscarine (*Amanita* spp., *Inocybe* spp., *Clitocybe* spp.) which may cause death from heart or respiratory failure. *Amanita muscaria* fruit bodies also contain muscinol, an insecticidal substance also having psychoactive properties (Waser, 1967). Tricholomic acid, which is present in *Tricholoma muscarium*, is a powerful fly-killer (Takemoto, 1961) and also possesses an unusual quality as a flavouring agent which is active at much lower concentrations than sodium glutamate. The deadly phallotoxins and amatoxins in fruit bodies of *Amanita* spp. cause irreversible damage to liver cells. These two groups of toxic peptides appear to affect different target organelles, the more toxic amatoxins affecting the nucleus and the phallotoxins affecting the endoplasmic reticulum (Wieland & Wieland, 1972). The fungi which produce the *Amanita* toxins are, in nature, mycorrhizal associates and in common with many other mycorrhizal Basidiomycetes grow rather slowly in laboratory culture. Thus Basidiomycetes are frequently unsuitable as sources of metabolic products whereas most of the Fungi Imperfecti have growth rates *in vitro* which may be exploited industrially as biosynthetic systems.

14.4 References

AGURELL, S. (1964). Costaclavine from *Penicillium chermesinum*. *Experientia* 20, 25–6.

AGURELL, S.L., RAMSTAD, E. & ULLSTRUP, A.J. (1963). The alkaloids of maize ergot. *Planta Medica* 11, 392–8.

AMICI, A.M., MINGHETTI, A., SCOTTI, T. & SPALLA, C. (1965). *British Patent 998254.*

AMICI, A.M., MINGHETTI, A., SCOTTI, T., SPALLA, C. & TOGNOLI, L. (1966). Production of ergotamine by a strain of *Claviceps purpurea* (Fr.) Tul. *Experientia* 22, 415–16.

AMICI, A.M., MINGHETTI, A., SCOTTI, T., SPALLA, C. & TOGNOLI, L. (1967*a*). Ergotamine production in submerged culture and physiology of *Claviceps purpurea. Applied Microbiology* **15**, 597–602.

AMICI, A.M., SCOTTI, T., SPALLA, C. & TOGNOLI, L. (1967*b*). Heterokaryosis and alkaloid production in *Claviceps purpurea. Applied Microbiology* **15**, 611–15.

AMICI, A.M., MINGHETTI, A., SCOTTI, T., SPALLA, C. & TOGNOLI, L. (1969). Production of peptide ergot alkaloids in submerged culture by three isolates of *Claviceps purpurea. Applied Microbiology* **18**, 464–8.

ARCAMONE, F., CHAIN, E.B., FERRETTI, A., MINGHETTI, A., PENNELLA, P., TONOLO, A. & VERO, L. (1961). Production of a new lysergic acid derivative in submerged culture by a strain of *Claviceps paspali* Stevens and Hall. *Proceedings of the Royal Society Series B* **155**, 26–54.

ARCAMONE, F., CASSINELLI, G., FERNI, G., PENCO, S., PENNELLA, P. & POL, C. (1970). Ergotamine production and metabolism of *Claviceps purpurea* strain 275 FI in stirred fermenters. *Canadian Journal of Microbiology* **16**, 923–31.

BARGER, G. (1931). *Ergot and ergotism.* London: Gurney and Jackson.

BASSETT, R.A., CHAIN, E.B., CORBETT, K., DICKERSON, A.G.F. & MANTLE, P.G. (1972). Comparative metabolism of *Claviceps purpurea in vivo* and *in vitro. Biochemical Journal* **127**, 3P.

BRACK, A., HOFMANN, A., KALBERER, F., KOBEL, H. & RUTSCHMANN, J. (1961). Tryptophan als biogenitsche vorstufe des psilocybins. *Archiv der Pharmazie* **294**, 230–4.

CATALFOMO, P. & TYLER, V.E. (1964). The production of psilocybin in submerged culture by *Psilocybe cubensis. Lloydia* 27, 53–63.

DETROY, R.W., LILLEHOJ, E.B. & CIEGLER, A. (1972). Aflatoxin and related compounds. In *Microbial Toxins* vol. VI. New York: Academic Press.

DICKERSON, A.G. (1972) A β-D-fructofuranosidase from *Claviceps purpurea. Biochemical Journal* **129**, 263–72.

DUBACH, U.C. & GSELL, O.R. (1962). Carcinoid syndrome: alleviation of diarrhoea and flushing with 'Deseril' and Ro 5-1025. *British Medical Journal* 1390–1.

FANCHAMPS, A., DOEPFNER, W., WEIDMANN, H. & CERLETTI, A. (1960). Pharmakologische charakterisierung von Deseril, einem Serotonin-Antagonisten. *Schweizerische Medizinische Wochenschrift* **90**, 1040–6.

FULLER, J.G. (1969). *The day of St. Anthony's fire.* London: Hutchinson.

HOFMANN, A. (1964). *Die Mutterkorn-alkaloide.* Enke, Stuttgart.

HOFMANN, A., HEIM, R., BRACK, A., KOBEL, H., FREY, A., OTT, H., PETRZILKA, T. & TROXLER, F. (1959). Psilocybin und psilocin, zwei psychotrope Wirkstoffe aus mexikanischen rauschpilzen. *Helvetica Chimica Acta* **42**, 1557–72.

KOBEL, H. & SANGLIER, J.J. (1970). Qualitative change of alkaloid spectrum by mutation in *Claviceps purpurea.* Abstract. *1st International Symposium on the Genetics of Industrial Microorganisms.* Prague.

KOBEL, H., SCHREIER, E. & RUTSCHMANN, J. (1964). 6-methyl-$\Delta^{8,9}$-ergolen-8-carbon saure, ein neues ergolinderivat aus kulturen eines stammes von *Claviceps paspali* Stevens et Hall. *Helvetica Chimica Acta* **47**, 1052–64.

KROGH, P. (1972). Nephropathy caused by mycotoxins from *Penicillium* and *Aspergillus. Journal of General Microbiology* **73**, XXXIV.

LEUNG, A.Y. & PAUL, A.G. (1969). The relationship of carbon and nitrogen nutrition of *Psilocybe baeocystis* to the production of psilocybin and its analogs. *Lloydia* **32**, 66–71.

MANTLE, P.G. (1968). Studies on *Sphacelia sorghi* McRae, an ergot of *Sorghum vulgare* Pers. *Annals of Applied Biology* **62**, 443–9.

MANTLE, P.G. (1969*a*). Development of alkaloid production *in vitro* by a strain of *Claviceps purpurea* from *Spartina townsendii. Transactions of the British Mycological Society* **52**, 381–92.

MANTLE, P.G. (1969*b*). Studies on *Claviceps purpurea* (Fr.) Tul. parasitic on *Phragmites communis* Trin. *Annals of Applied Biology* **63**, 425–34.

MANTLE, P.G. (1973). Production of ergot alkaloids *in vitro* by *Sphacelia sorghi. Journal of General Microbiology* **75**, 275–81.

MANTLE, P.G. & FINN, C.A. (1971). Investigations on the mode of action of D-6-methyl-8-cyanomethylergoline in suppressing pregnancy in the mouse. *Journal of Reproduction and Fertility* **24**, 441–4.

MANTLE, P.G., MORRIS, L.J. & HALL, S.W. (1969). Fatty acid composition of sphacelial and sclerotial growth forms of *Claviceps purpurea* in relation to the production of ergoline alkaloids in culture. *Transactions of the British Mycological Society* **53**, 441–7.

MANTLE, P.G. & SZCZYRBAK, C.A. (1972). Factors affecting clavine alkaloid production in submerged cultures of *Claviceps fusiformis*. *Journal of General Microbiology* 73, XXII.

MANTLE, P.G. & TONOLO (1968). Relationship between the morphology of *Claviceps purpurea* and the production of alkaloids. *Transactions of the British Mycological Society* **51**, 499–505.

MANTLE, P.G. & WAIGHT, E.S. (1968). Dihydroergosine: a new naturally occurring alkaloid from the sclerotia of *Sphacelia sorghi* McRae. *Nature, London* **218**, 581–2.

MIROCHA, C.J., CHRISTENSEN, C.M. & NELSON, G.H. (1972). F-2 (Zearalenone) estrogenic mycotoxin from *Fusarium*. In *Microbial Toxins* vol. VII. New York: Academic Press.

MORRIS, L.J. & HALL, S.W. (1966). The structure of the glycerides of ergot oils. *Lipids* **1**, 188–96.

REZABEK, K., SEMONSKY, M. & KUCHARCZYK, N. (1969). Suppression of conception with D-6-methyl-8-cyanomethylergoline (I) in rats. *Nature, London* **221**, 666.

SEDA, M., REZABEK, K., MARHAN, O. & SEMONSKY, M. (1971). Stimulation of gonadotrophin secretion with the pregnancy inhibitor D-6-methyl-8-cyanomethylergoline (I). *Journal of Reproduction and Fertility* **24**, 263–5.

SEMONSKY, M. & KUCHARCZYK, N. (1968). Ergot alkaloids XXX. Synthesis of D-6-methyl-8-ergolin-1-ylacetic acid and some of its derivatives. *Collection of Czechoslovak Chemical Communications* **33**, 577–82.

SPALLA, C., AMICI, A.M., SCOTTI, T. & TOGNOLI, L. (1969). Heterokaryosis of alkaloid producing strains of *Claviceps purpurea* in saprophytic and parasitic conditions. In *Fermentation Advances*. New York: Academic Press.

SPILSBURY, J.F. & WILKINSON, S. (1961). The isolation of festuclavine and two new clavine alkaloids from *Aspergillus fumigatus* Fres. *Journal of the Chemical Society* 2085–91.

STADLER, P.A., GUTTMANN, S., HAUTH, H., HUGUENIN, R.L., SANDRIN, E.D., WERSIN, G. & HOFMANN, A. (1969). Die synthese der alkaloide der ergotoxin-gruppe. *Helvetica Chimica Acta* **52**, 1549–64.

STEARNS, J. (1808). Account of the Pulvis Parturiens, a Remedy for quickening Childbirth. *Medical Repository, New York. 2nd Hexade* **5**, 308–9.

STOLL, A. (1918). *Swiss Patent 79879.*

STOLL, A. (1952). Recent investigations on ergot alkaloids. *Fortschritte der Chemie Organischer Naturstoffe* **9**, 114–74.

STOLL, A. & HOFMANN, A. (1943). Partial synthese von alkaloiden vom typus des ergobasins. *Helvetica Chimica Acta* **26**, 944–65.

STOLL, A. & HOFMANN, A. (1965). The ergot alkaloids. In *The Alkaloids* vol. 8. New York: Academic Press.

SZCZYRBAK, C.A. (1972). Production of clavine alkaloids *in vitro* by *Claviceps fusiformis*. *Ph.D. Thesis. University of London.*

TAKEMOTO, T. (1961). On the insecticidal component of *Tricholoma muscarium*. *Japanese Journal of Pharmacy and Chemistry* **33**, 252–4.

THOMAS, R. & BASSETT, R.A. (1972). The biosynthesis of ergot alkaloids. In *Progress in Phytochemistry* vol. 3. London: Interscience.

TONOLO, A. (1966). Production of peptide alkaloids in submerged culture by a strain of *Claviceps purpurea* (Fr.) Tul. *Nature, London* **209**, 1134.

VARGA, L., LUTTERBECK, P.M., PRYOR, J.S., WENNER, R. & ERB, H. (1972). Suppression of puerperal lactation with an ergot alkaloid: a double-blind study. *British Medical Journal* 743–4.

WEILAND, T. & WEILAND, O. (1972). The toxic peptides of *Amanita* species. In *Microbial Toxins* vol. VIII. New York: Academic Press.

Additional References

BANKS, S.T., MANTLE, P.S. & SZCZYRBAK, C.A. (1974). Large-scale production of clavine alkaloids by *Claviceps fusiformis*. *Journal of General Microbiology* **82**, 345–61.

CORBETT, K., DICKERSON, A.S. & MANTLE, P.S. (1974). Metabolic studies on *Claviceps purpurea* during parasitic development on rye. *Journal of General Microbiology* **84**, 39–58.

NISBET, L.J. (1974). Differentiation of *Claviceps purpurea* during parasitic and axenic culture. *Ph.D. Thesis, University of London.*

CHAPTER 15

Biodeterioration and Biodegradation by Fungi

H. O. W. EGGINS and D. ALLSOPP

15.1 Introduction

The biosphere operates essentially on a cyclical basis, both in terms of the individual organisms and also in regard to the chemical elements of which they are composed. The overall elemental cycles are driven by the extra-terrestrial addition of solar energy. For some elements, such as carbon and nitrogen, biological cycling is of major importance for the transformations of significant proportions of such elements, whilst for other biologically active elements, for instance zinc, biological cycling may be of small significance in regard to the total quantities of the element present on the earth's surface.

The fungi play a most active role in the recycling of elements. They are ubiquitous as transformers and utilizers of the dead remains of other organisms. They are important as parasites, but their impact on the rest of the biosphere is most striking when they are playing their saprophytic role. Frequently they are referred to within the term *micro-organisms* where they lie uneasily with their bed-fellows the bacteria. Generally it is the bacteria which are thought to be the most important of the two main groups of micro-organisms, possessing within their group as a whole, a wider range of transformatory activities than the fungi, and apparently, due to the difficulties of estimating amounts of mycelial material in natural substrates as opposed to counting the unicells of typical bacteria, appearing to be present in smaller amount than the bacteria in many natural substrates.

These concepts of the fungi ignore their real importance, where estimates of mycelial protoplasm per gramme of a typical woodland soil show that fungi are at least of equal activity with the bacteria. One reason for the abundance of the fungi lies in the ability of a number of them to utilize polymeric carbohydrate compounds, notably cellulose, which are such characteristic products of the primary synthesizers, namely the green plants. The fungi are equipped with extracellular hydrolytic enzymes to break down such substances, and through their hyphal systems are also able to penetrate and rapidly colonize substrates, probably using their translocating powers to redistribute essential nutrients.

It is to the primary importance of the fungi as utilizers of cellulose and other polymeric carbohydrates that these organisms owe their central role in the recycling of elements, for cellulose is the most abundant of the substances which arise from the primary producers, and is the main single form of carbon which is recycled.

Man has distorted the elemental cycles to provide materials and foodstuffs for his own use. In doing this he has tried to retard the natural recycling activities of the range of organisms which would otherwise continue the cycling of these materials. Rodents, insects, bacteria and fungi are amongst the biodeteriogens which compete with man for his foodstuffs and materials (Butler & Eggins, 1964). Fungi affect a wide range of such substances, and their importance is emphasized by their being the major deteriogens of such cellulosic substances as wood and the vegetable fibres, which are still materials of great economic significance. Again, fungi are serious deteriogens of other polymeric carbohydrates such as the starches, which have played, and continue to play, a central role as foodstuffs for man and his livestock.

Apart from cellulosic materials and starches the hyphal fungi attack a wide range of materials of diverse chemical composition. Such attacks cause considerable losses both to developed economies as well as to those of the developing countries. Indeed there is much evidence to show that biodeterioration losses are much more serious in the latter countries (Butler & Eggins, 1966), although control measures are frequently not sufficient to contain such problems. The difficulties of assessing economic losses (Eggins, 1967), on which the decisions for control techniques are based, will be discussed later, as will the increasing importance of biodeterioration as a world scale problem.

The fungi play another important part in relation to the materials that man has abstracted from the natural elemental cycles. This role is the converse economically of biodeterioration, although the biological processes are the same. Where the attack is of use to man's economy the process is referred to as biodegradation. In biodeterioration it is the organism which helps to transform a material into a terminal waste material, thus acting in a pragmatically negative way; clearly this is a process which should be retarded as much as possible. However, the fate of many organic materials, once they have been used and discarded as a terminal waste, is ultimately to be returned to the elemental cycles through the agency of organisms, the hyphal fungi playing a major role in this activity. Now that the human population with its demand for materials and concomitant terminal waste production is increasing in such striking proportions, the build-up of wastes both in a concentrated form in the form of refuse, as well as its haphazard distribution in the form of litter, is of ever-growing concern. The biological processes which help to remove these wastes are described as biodegradation, acting in a pragmatically positive way in man's economy, and being processes which can be accelerated with advantage. As an example, the attack by *Chaetomium globosum* on fresh newsprint in store would be described as biodeterioration, its break-down of the cellulose composing the paper creating waste; the same biological processes of *C. globosum* acting on used newsprint littering a woodland or in a biological waste fermenter would be described as biodegradation. These distinctions are of practical significance and use.

15.2 Methods of Attack

Organisms can attack materials causing mechanical and chemical damage and soiling (Hueck, 1965).

Mechanical damage

Mechanical damage occurs where the organism physically damages a material. The gnawing of wood by rodents and the penetration of thin plastic sheets by termites are examples of this form of damage, taken from the biodeterioration field generally. Although it is well known from plant pathology that fungi can mechanically penetrate cuticles and cell walls by development of hyphal stylets and that such penetration of thin cellulosic films has been demonstrated *in vitro*, it is unlikely that such mechanical penetration would by itself be of any significance. At best it could be an initiating mechanism for chemical damage, which is by far the most important method of attack.

Chemical damage

This form of damage is caused by all four major groupings of the fungi. Here the organism damages the material by *chemical assimilation*, using the material, or parts of the material, as a nutrient substrate. The damage is caused primarily by the actual disappearance of the material, it being synthesized into fungal mycelium, with a portion of the material in the majority of instances being transformed into CO_2.

Many fungi are able to cause much damage due to their ability to produce extracellular enzymes. Cellulose is able to be decomposed to cellobiose via long—1–4 anhydroglucose chains by the extracellular enzymes of many Ascomycetes, Fungi Imperfecti and Basidiomycetes, especially the Homobasidiomycetes. The fungal stages of cellulose decomposition occur within the cell where cellobiase breaks down cellobiose to utilizable glucose units. Lignin is thought to be decomposed mainly by Basidiomycete fungi, notably members of the Polyporales and Agaricales. Although the details are not clear, breakdown may be by the release of aromatic units from the lignin molecule. Further breakdown is probably carried out by a wider range of fungi. Hemicelluloses are a varied group of polysaccharides, and upon hydrolysis by extracellular enzymes yield hexoses, pentoses and uronic acids. The enzymes tend to be specific to particular hemicelluloses.

It can be seen therefore that the main structural units of plants and materials from them are all susceptible to the extracellular enzymes of fungi.

Such chemical assimilatory damage mainly manifests itself in economic terms either by straightforward weight loss, which is of particular importance in stored food products, or by strength loss as is shown by fungal attack on wood and fibres.

Chemical damage can also be caused by *chemical dissimilation*, where a fungus causes the main damage to a material by dissimilating objectionable chemical substances into a material. Examples are the production of mycotoxins by fungi growing on food materials, for example aflatoxin produced by *Aspergillus flavus*, or the stains which can occur on textiles and plastics resulting from pigments diffusing from hyphae, *e.g.* from *Fusarium* spp. on cotton and *Penicillium janthinellum* on nylon. Obviously there must be some chemical assimilatory activity of the fungus to support its dissimilating activity, but the damage caused by assimilation of the material or associated detritus may in some cases be negligible compared to the presence of a dangerous mycotoxin or obvious stain. Total levels of hyphal growth associated with a material may be very small in comparison with the degree of economic damage caused.

Soiling

The final category of damage is *soiling*, where the presence of the fungal mycelium is the main cause for objection. Here the particular abilities of the fungi to cover large areas with their hyphae are of note, together with the widespread characteristic of the fungi to form well-coloured spores either of bright colours or of sombre yet distinct browns and black, *e.g. Pullularia* (*Aureobasidium*) *pullulans* and *Cladosporium* spp. Such soiling frequently occurs in situations where there is no competition from the other major group of micro-organisms, the bacteria. Thus the soiling of electrical equipment by widespread fungal hyphae is caused both by the characteristic of the fungi to produce an extensive mycelium supported by low concentrations of nutrients, as well as by the ability of such mycelium to bridge air gaps between supporting substrates. This latter characteristic is of particular concern, as the alterations in conductivity and breakdown of insulants is frequently due to this ability.

The presence of macroscopically obvious mycelium is exemplified by the dark hyphae of fungi which can infest standing timber as well as growing on the surface of freshly worked timber. Such fungi, *e.g. Ceratocystis* spp., *Phialophora* spp. cause a characteristic 'blue stain' resulting from the presence of their hyphae, and this staining can result in severe losses in the timber industry. The damage caused by chemical assimilation of the timber by these 'blue stain' fungi is usually negligible.

Soiling due to the presence of evident coloured sporulation is often considered to be of greater importance than is really the case. However, it frequently is of concern, and therefore must be considered with due attention. Again it may occur where there is little or no associated bacterial damage, due to the ability of the fungi to grow at lower RH than the bacteria. The soiling of interior paint work can be caused by dark spored fungi, *e.g. Cladosporium, Phoma* and *Pullularia* spp., whilst similar damage can be caused to external paint work where the humidity is not sufficiently high or prolonged enough to permit the growth of algae (Skinner, 1971).

15.3 Ecology

Understanding the nutrition and morphology of the fungi is essential to understanding their ecology and those aspects of their ecology which are described as biodeterioration and biodegradation.

Humidity

Unlike the bacteria and the algae, the fungi as a group are entirely heterotrophic. Usually the fungi require a film of water around their hyphae for growth to occur and thus they compete, frequently unsuccessfully, with the other groups of micro-organisms. However, as has been mentioned earlier, some fungi are able to grow mycelially at RH well below that of the other groups (Ayerst, 1969). Even if the RH is below that which permits hyphal extension, a fungus will frequently react to such a lowering of the RH by sporulating, a reaction virtually denied to the other competing groups.

The water relations of the components of a fungal mycelium are at present very imperfectly understood. Nevertheless, it is a fact of common observation that the ability of a fungus to affect the local water relations of a substrate have had profound economic effects, giving rise to remedial measures, costing many millions of pounds per year. This particular effect is that of the Basidiomycete *Serpula* (*Merulius*) *lacrymans* being able to colonize dry timber in buildings, giving rise to the condition known as 'dry rot'. This is achieved by the fungus initiating growth in timber with a high water content and then using the energy and nutrients gained to produce rhizomorphs which can explore dry timber many metres away from the initial colony. Although such rhizomorphs can transport water, further water is produced at the site of action by the utilization of the timber with the evolution of CO_2. Such water is frequently observed in the form of droplets of water on the mycelium, hence the specific epithet of *lacrymans*. This mode of colonization is not, however, restricted to this specific fungus, and the characteristic morphology and physiology of the hyphal fungi should not be ignored in considering their ecology in relation to water availability, particularly as this may be decisive in determining the ability of a fungus to exploit a substrate successfully in relation to the other microbial groups.

Unlike other micro-organisms, only a minority of fungi can successfully sporulate in submerged aqueous conditions; amongst the hyphal fungi many of the lower Phycomycetes regularly reproduce under submerged conditions, whilst recent work has shown that an important group of the Fungi Imperfecti are secondarily adapted to freshwater submerged conditions, and some economically important wood deteriorating Ascomycetes regularly fruit in submerged marine conditions (Jones & Irvine, 1971).

Most hyphal fungi do require to grow on a surface to sporulate, and that includes the group referred to immediately above. Bearing in mind their humidity requirements for growth, this frequently brings them into competition with the algae if the surface is illuminated. In non-illuminated situations the main competitors in high humidity may well be the bacteria or yeasts, yet both groups lack a characteristic advantage possessed only by the hyphal fungi. This advantage ecologically is that the fungal hyphae can penetrate into a substrate, exploiting solid substrates by their ability to translocate nutrients, and possibly oxygen, to fresh situations. Such exploitation probably enables an extensive mycelium to balance nutrient-wise a substrate which in itself is rather inhospitable. Thus it is suspected that Basidiomycetes can successfully penetrate into solid timber because of their ability to translocate nitrogen, either from an external source, or from old, inactive portions of the mycelium. Less extreme forms of exploitation do enable fungi to sporulate at surfaces and to successfully compete with the surface flora.

Such growth of fungi into substrates often enables growth and breakdown to continue when adverse humidities may prevail at the surface. Examples of economic significance is the internal exploitation of painted, incorrectly prepared or constructed, timber joinery (Building Research Station, 1966).

Temperature

Like other micro-organisms the hyphal fungi include organisms which are able to grow at biological temperature extremes. Although the number of species which grow at these extremes is small, their economic effects can be considerable.

Psychrophilic fungi are well known and are responsible for certain pathological conditions. The minimum temperature for growth appears to be around 5°C and several species are able to put on significant growth below 0°C. As can be imagined some of these fungi cause considerable problems in refrigerated food stores, for example species of *Cladosporium* and *Sporotrichum* (Cochrane, 1958).

However, much more attention should be paid to those fungi which can be isolated at temperatures below the usual incubation temperature of 25°C. It is suspected that results obtained at this temperature, chosen for technical reasons rather than as some average of ambient temperatures, may mask precise examination of fungal activity in many situations; in temperate regions temperatures for long periods are much below 25°C although above 0°C. Careful consideration of the temperature regime of many materials in use will show very considerable fluctuations, not only from season to season, day to day, but often between night and day. The effects on the fungi are not well understood, but there can be striking differences between insulated and shaded structures in the temperate regions, a piece of wood joinery on the sunny side of a building fluctuating 25°C within twenty-four hours, with temperatures of over 40°C being achieved (Eggins, 1968). Habitats are thus created where thermophilous and thermophilic fungi are able to grow, and it is important to note from a biodeterioration viewpoint that a majority of the known thermophilic fungi are cellulolytic (Eggins & Malik, 1969). Their role in the deterioration of cellulosic substrates is still being determined, although it is suspected that for many exposed timber structures there will be a range of fungi corresponding to the extensive range of temperatures which may be experienced.

Thermophilic fungi, together with the Actinomycetes, can cause serious deterioration of hay stacks, if these are incorrectly constructed with over-damp hay. The presence of the fungi can cause unpleasant odours but more importantly the metabolic heat can eventually give rise to a fire risk. When biodegradation of converted waste cellulosic materials is considered the insulating properties of accretions of these materials help to provide ideal conditions for the growth of thermophilic fungi.

Such fungi can be seen to be able to play an important role in the disposal of town and agricultural waste, but unfortunately the economic climate does not appear encouraging at the present time for this unique group of fungi.

pH

It is probable that fungal cells are well buffered as little change is reported in the endogenous respiration with variation of the external medium in a pH range of 5–8. Exogenous respiration and growth however are affected by changes in the external pH although the detailed mechanisms operating at a sub-cellular level have yet to be clarified. The marked effect of changes in pH on the colonization and decay of a substrate is of great significance in biodeterioration and biodegradation studies (Allsopp, 1973); aspects of this problem being further complicated where biocides are present, as these compounds often have optimum activity in a narrow pH range which may not cover all eventualities of field conditions.

Eh

When considerations are made of a material in mass it can be seen that the quantities of available oxygen may vary greatly. Certain processes involving essentially aerobic fungi such as solid state fermentation and composting must incorporate provision of a constant supply of oxygen. This may be performed by pumping in air, constantly moving the substrate, or by using a substrate with an open texture. The aerobic nature of many deteriogens has been the key to their control in that the preservation of some materials (notably stored foodstuffs) can be effected by changing the surrounding atmosphere, by increasing for example, the concentration of carbon dioxide (Hall, 1968). Preservation may thus be carried out without risk of tainting the materials.

An atmosphere of 10% CO_2 is usually sufficient to retard or inhibit the growth of many bacteria and fungi, the effect being enhanced as temperature is lowered. An atmosphere of 20% CO_2 coupled with temperatures just above freezing is an ideal environment for the transport of fruit which would be damaged by inhibitory concentrations of CO_2 or very low temperatures if used independently.

Osmotic pressure

Raising the osmotic potential of a material by addition of salt or sugar or both to a level higher than that occurring in the cytoplasm is a well tried method of microbial inhibition and is widely used in foods such as meats, fruits and jams. Sugar concentrations in the regions of 50–70% are usually adequate; salt being added to a 20–25% level. Such treatment may not be totally effective, especially in damp storage, as some osmophilic and halophytic organisms are known among the bacteria, yeasts and fungi, and for example the osmophilic yeasts *Saccharomyces rouxii* and *S. mellis* and the mould fungi of the *Aspergillus glaucus* series.

Interaction

Allusion has already been made to the general abilities of the hyphal fungi in competition with other microbial groups. It is necessary, however, to consider some of the problems which arise with competition between individual fungi in attempting to understand the implications of biodeterioration.

Successional activities in fungi are stimulated either by placing an uncontaminated material in contact with a contaminated substrate, by submerging the substrate in natural water, or by exposing the material to the air spora. In all these instances appropriate growth parameters must occur. Frequently in water and soil where it may be possible for the organisms to obtain all their nutrients in the natural solution, it is suspected that the substrate may provide no more than a free surface on which attachment may take place. Nevertheless, on many substrates a succession is soon set up with the substrate being utilized for growth. In many situations only a few fungi may be responsible for gross deterioration, but simple isolation experiments may show a considerable range of micro-organisms. It is necessary for effective control measures, particularly when in the future more specific fungicides may be used in increasing quantities, to determine which fungi are primary deteriogens and which merely grow on the breakdown products of these deteriogens. Such facts are only likely to be elucidated by sound application of ecological principles —considering the effects of sporulation, age of mycelium and selectivity of isolation media and incubation conditions (Eggins, 1968).

15.4 The Biodeterioration of some Important Materials and the Part Played by Fungi

Cellulosic materials

These materials which are temporarily abstracted from natural carbon cycles for man's use, as discussed earlier, form one of the most important substrates which is prone to biodeterioration. The major types of cellulosic materials are wood, used as timber or as wood products such as paper and board, and other plant fibres such as cotton, flax, jute, hemp and sisal used as rope, cordage, textiles and packing and filling materials. Some indication of the amounts of cellulosic materials produced each year is given in Table 15.1 (Allsopp, 1973).

The method and extent to which a cellulosic material is attacked by fungi depends on several factors in addition to those which apply to organisms generally such as humidity, temperature, availability of micronutrients and pH. The physical form of the cellulose substrate is of importance; a large solid mass being less easily penetrated by the fungus

Table 15.1 Annual world production of cellulosic materials

Material	Production $\times 10^6$ tonnes (1968)
Cotton fibre	11.3
Jute fibre	2.7
Flax fibre	0.6
Other fibres	1.6
Paper pulp	23.3
Industrial roundwood	1.2 ($\times 10^9$ m^3)

itself, water, nutrients and oxygen than a divided or dissected substrate. Even with dense cellulosic materials such as timber, the fungal hypha shows itself as a highly specialized organ of penetration and ramification, able to explore all possible lines of least resistance. Other specialization is seen in fungi which are able to re-use nitrogen by transporting cytoplasmic constituents or their breakdown products from old to young cells. These fungi are thus able to explore substrates having very low carbon: nitrogen ratios such as many heartwoods.

The make-up of the cellulosic material may vary depending upon its source and materials other than cellulose such as lignin being present. The cellulose component may become more varied in its nature due to changes resulting from processing into manufactured articles. Subsequent fungal decay may be hastened by the creation of areas of amorphous cellulose on a mainly crystalline macro-molecule, and on a larger size scale, by the roughening of fibrils and fibres which may facilitate fungal penetration at these sites of damage. Other processes may occur which slow the onset of fungal decay, such as the removal of contaminant nutrients, drying, weathering and the application of chemical preservatives.

Whatever factors operate to lengthen or shorten the life of cellulosic materials, be they natural or brought about by man, the ultimate fate of such material is to be broken down by burning or biological attack.

Until recent years, the fungal breakdown of cellulosic materials has been regarded as a pragmatically negative process, but recently the process has been utilized to man's advantage where waste materials are concerned. Materials with a high cellulose content such as town waste (dustbin or trash-can waste) can be composted on a large scale when broken up to provide a low-volume product suitable for use as a soil conditioner or sanitary landfill. These processes may not always be the most attractive or economic methods of waste disposal, but with modern equipment the cellulose decomposing activities of fungi may be harnessed to good effect.

The decay of cellulosic materials by fungi is probably the most familiar and widely-recognized form of biodeterioration and although there is still great scope for work to be carried out into the enzymology of the process, the least known major area is probably that of the processes involved in the colonization and penetration of substrates, *i.e.* the detailed ecology of cellulose decay.

Non-cellulosic materials

An account of the biodeterioration of non-cellulosic materials must of necessity appear to be somewhat fragmented, owing to the wide range of nutrients available, the variety of physical form and the uses to which these materials are put. There is, however, an underlying unity in the field and laboratory evaluation of the susceptibility to biodeterioration of such materials if an ecological approach is taken, and as many factors as possible concerning the interaction of the material and living organisms are considered. With materials which are unlike human nutrients such as cutting oils or plastics, not having the inherent vice of say stored foodstuffs which are not only accepted as fit for humans but also for micro-organisms, there is always a tendency to consider any deteriogens as simple chemical impurities rather than complex living organisms which grow, reproduce, die and carry out life functions such as assimilation, respiration and excretion as parts of an often varied life cycle. Considered below are some non-cellulosic materials and the problems at present associated with them.

PLASTICS The widespread use and high resistance to biological attack of these materials generally, coupled with their lightness has led to problems in litter situations, where they remain in view on land and water surfaces. This problem may well be solved with the present advent of photo-degradable and biodegradable plastics, which are geared to the actual service life of the product. There are problems however concerning the biodeterioration of plastics. The clean, smooth nature of many plastic surfaces is impaired by even slight amounts of mould growth, amounts which would not be noticed or thought to be objectionable on other materials. Such surface growth often draws on surface debris for nutrients, the plastic itself being unaffected. More serious problems occur where a fungus is able to utilize components of the plastic material. The component most likely to be attacked is the plasticizer, such attack leading to the material becoming brittle, cracked and discoloured (Eggins *et al.*, 1971). Careful consideration of in-use conditions, the exact type of plastic to be employed and the use of fungicides are the present approaches to this problem. Although the total amount of biodeterioration of materials such as plastics is much smaller than that associated with cellulosic materials, cases of decay are often harder to recognize and often occur in severe situations which are difficult to observe and remedy, for example in buried plastic covered cables and plastic insulants in electrical equipment used in humid tropical conditions.

GLASS Although glass is not utilized by fungi as a nutrient source, considerable damage may be caused by fungi. Owing to their deterioration being readily obvious, the main types of glass items studied to date have been the lenses of optical instruments such as binoculars, cameras, gun-sights, microscopes and surveying instruments. Damage is caused by the growth of fungi, living on accumulations of dirt on the lens surface or nearby. This growth either obscures vision by its presence or more seriously by etching the lens surface by its corrosive secretions. This problem is usually encountered in the tropics, and some difficulty has been experienced in its prevention. If total sealing of an instrument is practical at the time of manufacture, internal growth of fungi may be prevented, but the most vulnerable surfaces are those exposed to the atmosphere and these are still open to attack, especially in humid and dirty field conditions. The incorporation of toxic compounds within or on the surface of optical instruments can be difficult, as the optical properties may be impaired. Regular inspection, cleaning and good storage conditions are essential where such high-cost materials are involved. Fungi which have been implicated in such situations are species of *Aspergillus, Penicillium* and *Scopulariopsis* (Nagamuttu, 1967).

ELECTRICAL EQUIPMENT Electrical equipment may be subject to similar types of fungal colonization as mentioned concerning glass, particularly in tropical use (Wasserbauer, 1967). Associated problems are losses in insulating properties of components and also direct short circuits via fungal hyphae. The use of non-nutrient components, fungicides, and perhaps most important, the total sealing of assemblies after manufacture are the steps usually taken when such equipment is at risk. Where equipment must be ventilated due to heat generation (which in itself may lead to encouragement of fungi in humid conditions), the cooling air should be filtered as much as ventilation requirements allow.

FUELS Hydrocarbon fuels such as kerosene and lubricants such as cutting oil emulsions are susceptible to fungal attack. Growth may occur throughout an emulsion, or in the case of fuels, at a fuel/water interface. Whether in bulk store, or in use, for example in aircraft wing tanks, a water layer often forms. The water has its origin in the original supply or is formed from condensed atmospheric moisture which may form when an aircraft fuelled in the humid tropics climbs into cold air. The actual consumption of fuel by fungi is not significant, but other aspects of this problem are. Matted mycelium may break off and clog fuel lines or filters, and the corrosion of several kinds which may be initiated by fungal growth may weaken the tanks or surrounding structures. Aircraft wing tanks are subject to more rigorous and frequent inspection now these phenomena are recognized. The most common deteriogen appears to be the 'kerosene fungus', *Cladosporium resinae*.

Growth of fungi and bacteria in cutting oil emulsions may lead to several problems. The properties of the oil may be changed, leading to shortened tool life and poorer finish, operator skin infection risks become greater, and an odour problem may be present. Good housekeeping and the use of biocides compatible with fuels and oil emulsions are generally employed to combat this form of biodeterioration.

PAINT AND PAINT FILMS Paints may be susceptible to fungal deterioration either in can or as an applied film. Growth in can may lead to gassing, discolouration and a change in the properties of the paint. Films may be rendered brittle and cracked by fungal growth (*c.f.* plastics biodeterioration) and may be marked by dark-sporing fungi such as *Cladosporium*, *Phoma* and *Pullularia* spp. The latter manifestation of growth often passes unrecognized as fungal attack, being regarded as a surface accumulation of dirt or as bleeding through of substances from the material beneath the paint film. In-can and film deterioration is to a great extent combated by the use of fungicides incorporated at the time of manufacture of the paint.

LEATHER AND GLUE A proteinaceous material such as a hide is prone to considerable attack during its manufacture into leather, especially during the very early stages, where water, temperature, nutrient and infection conditions are conducive to microbial growth, especially that of proteolytic bacteria. The cooling of fresh hides, salting and correct handling, coupled with the use of biocides can effect great savings in this area. In-use decay of leather is often due to fungi which may damage and mark leather, especially where storage conditions are poor. Good storage and the use of suitable fungicides are the usual preventive measures taken to prevent decay of finished leathers.

Animal glues before application are high in both protein and water. If such materials cannot be used and dried quickly, fungicides may be used to good effect. In use, decay is possible if glued areas become wet. Waterproof surface treatment and the use of a fungicide in the glue may therefore be desirable where damp conditions may be encountered.

DRUGS AND COSMETICS Hazards here are the development of odours and discolouration, changes in texture and chemical properties, and the possibility of user infection by contaminated products (Butler, 1968). The manufacture of drugs under sterile and clean conditions usually limits decay to conditions of poor storage at a later time. Cosmetics often contain nutrient ingredients which are prone to decay and care must be exercised in the addition of any fungicides, bearing in mind the ultimate use of such products. In addition, clean efficient processing, good storage and packing are the usual solutions to such problems. There is however some difficulty in that manufacturers supplying a product surrounded by an aura of advertising which heralds its innocuous and 'pure' nature may be understandably reluctant to admit the existence of such problems.

STORED PRODUCTS Any product may be stored, but the term is usually applied to foodstuffs. It is within this field that the greatest amount of study has been given to the environmental control of biodeterioration. By regulation of temperature, humidity, ventilation, light and atmospheric gases, materials usually prone to rapid decay may be kept in a wholesome state without recourse to chemical preservation. The ideal conditions vary from material to material. The main problem, once the correct conditions for any material are known, is to ensure that the conditions prevail in all parts of a store as a material in bulk may maintain its own environment and take time to come into equilibrium with the storage environment.

Prevention of biodeterioration

Materials may be protected against biodeterioration in several major ways. A normal environment factor may be adjusted in order that the organism may be inhibited. Examples of this method are cooling and refrigeration, storage in high CO_2 level, changing osmotic pressure by addition of salt or sugars, lowering pH by pickling, reducing water and oxygen by drying and vacuum packing. Such methods are mainly applied in the food industry, where the addition of preservatives is of necessity restricted. With non-food materials, a range of other methods of prevention of biodeterioration are available in addition to some of those mentioned above. These methods tend to be independent of external control of environmental factors as in-use conditions vary greatly and service life is obviously much longer than that of foodstuffs. Protective treatments for materials may be classified as either active or passive.

Passive treatments are such as to have no toxic action on organisms, but provide a physical barrier to them or render the material unavailable for their nutrition. Such treatments are designed to protect the material against structural attack; they do not prevent surface growth by organisms not actually utilizing the material. A physical barrier may be coated around a material, an example being encapsulation with a synthetic polymeric film. Some materials, *e.g.* cotton cellulose, may be chemically modified, by acetylation, although such methods may be costly and lead to undesirable physical changes in the material.

Active treatments to prevent biodeterioration are designed to have a direct effect on the growth and development of organisms. The treatment is designed to have effect throughout the material, and thus not only structural damage but also surface growth is minimized. Compounds which have a toxic action on fungi fall into this category. The range of fungicides available is diverse and large and is subject to continual development and change. The characteristics of a hypothetical ideal fungicide are given below:

The ideal fungicide must:

(a) Be toxic to all possible deteriogens or block their action.
(b) Not be toxic to humans or animals and be safe to apply.
(c) Not introduce undesirable colours in the material protected.
(d) Be compatible with any other treatments given to the material.
(e) Not weaken the material or hasten weakening by other agencies (*e.g.* heat or light).
(f) Be resistant to leaching.
(g) Be eventually degradable, this being geared to the life of the material determined by other factors such as wear.
(h) Have a competitive cost.

It can be seen therefore that toxicity alone is only one of many factors to be considered, and although the characteristics listed above belong to an ideal fungicide it is encouraging to note that in the development of new fungicides more factors are being studied which will facilitate a closer approximation to be made to an ideal situation.

15.5 Methodology of Investigations

Koch in 1882 enunciated his postulates which he suggested should be used to identify the causative organism of a disease. As Hueck (1965) has so usefully compared pathology with hylopathology (or biodeterioration) so it follows that Koch's postulates are applicable to biodeterioration problems. However, modifications must be made because of the greater vegetative uniformity of the hyphal fungi compared to other groups.

Clearly, if the principal deleterious effect is the mere presence of fungal growth on the surface of a material, microscopic examination should be sufficient to determine the species involved, if sporulation is occurring. Lloyd (1965) described the uses of transparent adhesive tape for easily removing such samples from surfaces for microscopic examination, the adhesive maintaining in position fragile structure.

The scanning electron microscope is becoming more widely used for surface examination, but the levels of magnification are such that for some workers the observations seem to be more misleading than helpful!

The agar sausage made up with various media can be useful for exploring in more detail the micro-organisms growing on surfaces, but this relies on incubation and examination of the fungi which subsequently grow.

This latter form of examination leads on to a consideration of the requirements of the general colonization of materials where fungal growth may not be particularly obvious. Techniques are required which are able to determine at a particular point in time the actively growing fungi on a particular substrate, the frequency of fungi, their reaction with one another, and most importantly their individual reaction with the substrate.

The above questions can prove very difficult to answer because frequently fungi damage materials by growing within them, where visual examination is virtually impossible and reliance must be placed on sampling and cultural techniques for determining both qualitatively and quantitatively the fungal activity. This is complicated by the already-mentioned lack of vegetative specificity; indeed, visual characterization of the vegetative stages of the hyphal fungi will usually do no more than reduce the possible 130 000 species of fungi into one of three groupings, although Nobles (1964) has described in detail the hyphal characteristics of a number of wood destroying Basidiomycetes.

Sampling techniques which may be used for hyphal fungi are fraught with dangers, for frequently they have been adapted from bacteriology, where the technique is designed to sample unicells. The hyphal fungi, by definition, may produce a complicated mycelial system which is not capable of being accurately sampled as are a distribution of unicells. However, the fungi may become single or few-celled structures when they sporulate and the sampling of these unicellular structures can produce very misleading results. Results may be misleading because a sporulating fungus may thus give a much higher count than an associated non-sporulating fungus; spores may also maintain viability long after the producing vegetative hyphae are moribund and therefore give a misleading prolonged dominance.

The dilution plate technique, in view of the above factors, is of limited use in determining the role of the hyphal fungi. The use of the Warcup technique and estimating fungi in terms of percentage frequency is usually more revealing.

Micro-environments—substrate composition

It has been fairly said that isolation techniques frequently disclose only what has, what is, and what may be capable of growing on a particular substrate. In biodeterioration problems it is usually essential to be able to distinguish between an organism or organisms responsible for deteriorating the substrate and those fungi which are merely secondary organisms. Frequently, the problem is that a substrate is rarely homogenous, but is composed of innumerable micro-environments. Thus a piece of wood will exhibit in its cellular structure different habitats not only within tissues such as rays, sapwood and heartwood, but also between cells and even the parts of the cell walls. There will also be differences in pH and, particularly for any solid substrate more than a few millimetres thick, changes in Eh. In the typical external environment many substrates experience great changes in temperature, as has already been mentioned. All these factors must be taken into account in isolating the hyphal fungi, and a universal medium and one temperature of isolation is rarely the answer.

Isolation must reflect the above requirements. Selective and demonstrative media should be used where possible (Booth, 1971) together with a range of temperatures which reflect the practical situation. Once the fungi are isolated they should be inoculated on to test substrates and Koch's postulates followed through.

15.6 Biodegradation

Economic manufacturing and production usually give rise to wastes. In primitive societies where there have been low densities of population and production such wastes usually present no disposal problems. As human populations grow haphazard discarding of the related increase in waste materials is usually curtailed by the community for both aesthetic and economic reasons. There are many historical instances where legislation has been introduced to curb the haphazard disposal of waste for presumably primarily aesthetic, subjective reasons, long before any public health or economic nuisance aspects were either recognized or considered. Such legislation usually related to urban areas.

Increasing manufacture and production has two main effects on waste aspects, both primarily economic. Firstly, the development of larger specialist manufacturing units concentrates process wastes causing disposal problems. Secondly, competition for raw materials emphasizes the need for a reduction in the production of unnecessary process wastes and, if possible, the use of a process waste as the raw material of another manufacture.

Thus the historical increase in human populations results in both a concentration of terminal wastes which result from the use and discarding by the population of manufactured materials, and also in the concentration of process wastes.

Such an increased production of both process and terminal wastes by a community results in the development of disposal systems, where the wastes resulting from manufactured materials are concentrated as refuse into a clearly defined and organized system. The handling of wastes in the form of refuse within an organized system should result in a clear reduction in the aesthetic, health and economic problems which are associated with wastes.

It is well known, however, that developed communities do not place all their wastes into refuse disposal systems even when these are provided efficiently. Instead a proportion of the waste of the community is haphazardly discarded in low concentration into the environment as litter. This litter is usually generated by individuals from terminal wastes, whilst process wastes do not usually contribute.

Understanding the origins of litter, as opposed to refuse, aids in the examination of the characteristics of anthropocentric litter and of its role as biological litter.

Refuse

Refuse as a class of substance can be divided into three main groups; domestic, industrial and agricultural. In each group the mycelial fungi may play an important part in degrading these terminal wastes, thus allowing them to re-enter the natural cycles of some of their constituents.

Table 15.2 shows the average contents of domestic refuse in the United Kingdom (H.M.S.O., 1971). As will be seen, organic materials, principally wood derivatives, make up the major component. Thus a large proportion

Table 15.2 Average content of United Kingdom refuse (1968).

Component	Percentage
Dust, ash and cinder	22
Paper and cardboard	37
Glass	9
Metal	9
Plastics	1
Rags and textiles	2
Vegetable and organic material	18
Miscellaneous	2

of the material is capable of microbiological recycling. Disposal methods have relied on microbial recycling in regard to land fill (with or without prior pulverization) and composting, but not, of course, in incineration. Landfill relies on the natural breakdown of materials without any acceleration. Hyphal fungi are known to play an important part in this breakdown particularly the frequently occurring thermophilic fungi, many of which are active cellulose decomposers, for example *Chaetomium thermophile* and *Humicola lanuginosa*. Landfill is decreasing rapidly in importance as sites for in-filling become scarcer and at a greater distance from areas of high production of domestic refuse. Alternatives both rely on an acceleration of recycling. Composting relies on the natural microflora of the refuse and systems have been designed from the simple windrow to highly developed continuous composting plants. Excellent descriptions of the basis of these systems will be found in Gray (1966). Although the starting point is usually very rapid growth of bacteria growing on easily assimilable materials it is the cellulose decomposing fungi, particularly the thermophiles, which play an important part although there is controversy regarding the role of the cellulolytic Actinomycetes. All these organisms decompose the cellulosic fractions eventually producing humus as the main end product. Although such systems appear attractive by producing a useful soil conditioner as an end product, unfortunately economic considerations at the present time have precluded the large scale adoption of these systems.

Processes are being developed for the upgrading of various industrial wastes by hyphal fungi. Of particular interest are the food and timber industry wastes. A considerable body of work has been carried out using these materials as carbon source with the addition of inorganic nitrogen or urea as the starting point for the production of fungal protein. Hyphal mycelial protein has certain advantages over single cell protein produced by the bacteria and yeasts, as is mentioned elsewhere (see Chapter 12).

Agricultural wastes, both plant and animal, have traditionally been composted, where, again because of their high cellulose content the hyphal fungi have been of considerable importance. Numerous papers have been published on the succession which occurs in these composting processes, *e.g.* Chang & Hudson (1967). Today, however, the economics of agricultural intensification and specialization frequently do not permit of the correct use of these composts, and instead of an asset, as in the past, these manures become a liability. Fulbrook *et al.* (1973) have shown the importance of viewing these wastes possibly as a source of animal feeds. Considerable quantities of animal slurries are wasted, and much straw burnt, which could usefully be combined and transformed with the aid of cellulolytic fungi.

Litter

As a broad generalization litter results from the activities of individuals and this immediately gives two characteristics of litter, namely that separate items are generally of small size in relation to human artefacts, and that their distribution is usually linked to the distribution of human activities. Bulky items of terminal waste are usually placed in a refuse system or remain in a concentrated form close to man's activities; an obvious exception to this is the haphazard discarding of vehicles and agricultural machinery.

Materials which form terminal wastes which are haphazardly distributed by individuals are usually packaging materials which may be of wood, paper, rubbers, plastics, glass, metals or ceramics.

The growth of litter production into a problem of international proportions in recent years is linked to four developments. Firstly, the massive growth of the human population capable of producing litter. Secondly, the design of transport systems enabling this population to travel briefly into far larger areas than are used for everyday activities. Thirdly, the increasing proneness to use packaging materials as a result of the need to protect ever more expensively manufactured articles. Fourthly, the invention of cheap, persistent materials which can be used for packaging.

Anthropocentrically, litter is primarily an aesthetic, subjective problem and many communities have specific legislation prohibiting such litter, although this is difficult to enforce. In urban areas litter may be removed by mechanical means. Outside these areas this is impossible, and the procedure relied upon to remove anthropocentric litter in the environment, generally is exactly those activities of organisms which operate naturally in biological litter breakdown. In such situations anthropocentric and biological litter are coincident. Aesthetic problems in the environment generally arise when the natural mechanisms of litter breakdown are insufficient to remove from sight litter produced by man in large quantities, or when man-made litter is resistant to natural biological breakdown.

It is important to note that this intrusion of litter of manufactured origin into the biological litter system is the result of human activities which today are usually specified as illegal and the problems arising are the direct result of this illegal action.

Before discussing the biological breakdown of the common components of litter, comment should be made regarding the involuntary distribution of waste chemicals, as opposed to materials, in the environment. It is generally accepted that 'chemical' used in this sense implies that the chemical is distributed virtually at the molecular level in solution or suspension, whilst 'material' implies discrete, macroscopic aggregates.

The hyphal fungi play an important role in the biodegradation of waste paper as would be expected. The only other litter where their effects may be of importance are the plastics. There is a considerable amount of literature on the effects of the mycelial fungi on plastics, but the consensus of opinion is that the polymers are mainly resistant, with the exception of some of the polyurethanes (Kaplan *et al.,* 1968), whilst some of the plasticizers are susceptible (Eggins *et al.,* 1971). The rate of breakdown, however, is so slow that the contribution to the recycling of such plastic litter will be very slow.

The problems of control of biodeterioration and biodegradation is closely associated with the economic effects of the processes. As a measure of the importance of the hyphal fungi this cannot be ignored. Much work is now being carried out on this evaluation and it is being shown that in both aspects the problems are growing. The hyphal fungi express their biology uniquely in both ways.

15.7 References

ALLSOPP, D. (1973). Some aspects of the colonisation and decay of fungicidally-protected cotton textiles by soil fungi. *Ph.D. Thesis. University of Aston in Birmingham.*

AYERST, G. (1969). The effects of moisture and temperature on growth and spore. germination in some fungi. *Journal of Stored Products Research* 5, (2), 127–41.

BOOTH, C. (1971). Introduction to general methods. *Methods in microbiology* Vol. 4, 1–47. London: Academic Press.

BUILDING RESEARCH STATION (1966). Prevention of decay in window joinery. *Digest 73* (2nd series) 6 pp.

BUTLER, N.J. (1968). The microbiological deterioration of cosmetics and pharmaceutical products. *Biodeterioration of materials* pp. 269–80. London: Elsevier.

BUTLER, N.J. & EGGINS, H.O.W. (1964). Biodeterioration: with special reference to the microbiological deterioration of foods. *Proceedings of the International Food Industries Congress, 1964.*

BUTLER, N.J. & EGGINS, H.O.W. (1966). Microbiological deterioration in the tropical environment. *Microbiological deterioration in the tropics* (*S.C.I. Monograph 23*) pp. 3–13.

CHANG, Y. & HUDSON, H.J. (1967). The fungi of wheat straw compost. 1. Ecological studies. *Transactions British Mycological Society* 50 (4), 649–66.

COCHRANE, V.W. (1958). *Physiology of fungi.* New York: Wiley.

EGGINS, H.O.W. (1967). The economics of biodeterioration. *Environmental Engineering* No. 29: 15–6.

EGGINS, H.O.W. (1968). Ecological aspects of biodeterioration. In *Biodeterioration of materials* pp. 22–7. London: Elsevier.

EGGINS, H.O.W. & MALIK, K.A. (1969). The occurrence of thermophilic cellulolytic fungi in pasture land soil. *Antoine van Leeuwenhoek* 35 (2), 178.

EGGINS, H.O.W., MILLS, J., HOLT, A. & SCOTT, G. (1971). Biodeterioration and biodegradation of synthetic polymer. In *Microbial aspects of pollution* pp. 267–79. Ed. Sykes, G. and Skinner, F.A. London: Academic Press.

FULBROOK, F.A., BARNES, T.G., BENNETT, A.J., EGGINS, H.O.W. & SEAL, K.J. (1973). Upgrading cellulytic waste for recycling. *Surveyor* 12 Jan., 24–7.

GRAY, K.R. (1966). Accelerated composting. *British Chemical Engineering* 11 (8), 851–3.

HALL, E.G. (1968). Atmosphere control in storage and transport of fresh fruit and vegetables. *Food Preservation Quarterly* 28 (1–2), 2–8

H.M.S.O. (1971). *Refuse disposal. Report of working party on refuse disposal 1971.* Department of the Environment, H.M.S.O. London.

HUECK, H.J. (1965). The biodeterioration of materials as part of hylobiology. *Material und Organismen* 1, 5–34.

JONES, E.B.G. & IRVINE, J. (1971). The role of marine fungi in the biodeterioration of materials. In *Biodeterioration of materials, Vol.* 2 pp. 422–31. London: Applied Science.

KAPLAN, A.M., DARBY, R.T., GREENBERGER, M. & ROGERS, M.R. (1968). Microbial deterioration of polyurethane systems. *Developments in Industrial Microbiology* Vol. 9, pp. 201–15. Society for Industrial Microbiology. Washington, D.C.

LLOYD, A.O. (1965). An adhesive tape technique for the microscopical examination of surfaces supporting, or suspected of supporting, mould growths. *International Biodeterioration Bulletin* 1, 10–3.

NAGAMUTTU, S. (1967). Moulds on optical glass and control measures. *International Biodeterioration Bulletin* 3 (1), 25–7.

NOBLES, M.K. (1964). Identification of cultures of wood-inhabiting hymenomycetes. *Canadian Journal of Botany* 43, 1097–139.

SKINNER, C.E. (1971). Laboratory test methods for biocidal paints. In *Biodeterioration of materials, Vol. 2* pp. 346–54. Applied Science, London.

WASSERBAUER, R. (1967). Czechoslovak research into microbiological corrosion of electrical equipment. *International Biodeterioration Bulletin* 3 (1), 1–2.

CHAPTER 16

Mycotoxins

W. H. BUTLER

16.1 Introduction

Following the discovery of penicillin, a metabolite of *Penicillium notatum*, much work has been done in searching for possible antibacterial agents. On the other hand, the mycotoxins, although recognized for many years received little attention until the early 1960's when the aflatoxins were isolated. Prior to that time most of the work was done in Russia and the United States.

Mycotoxicoses are defined by Forgacs & Carll (1962) as 'poisonings of the host which follow entrance into the body of toxic substances of fungal origin'. These include ingestion of poisonous mushrooms (Basidiomycetes), scab-infected grain (*Claviceps purpurea*), grain, seeds and dead and decaying leaves and stems as hay and straw, on which there is saprophytic growth of fungi. At the present time there is no known industrial hazard from mycotoxins except that arising from agricultural practice. Many species of fungi have been isolated which produce toxic or carcinogenic products (Brook & White, 1966; Enomoto & Saito, 1972). In the present review a brief account will be given of some of those mycotoxicoses which result in loss of livestock and possibly present a public health risk to man.

16.2 Toxins Produced on Growing Crops

The oldest known food-borne disease is ergotism which results from the ingestion of rye infected with *Claviceps purpurea*. The clinical syndromes associated with ergotism have been recognized for over 2000 years although the direct association of the disease with rye was made in about 1670. The active principles are alkaloids of lysergic acid which are produced in the sclerotium of the fungus which has replaced the grain of rye (see Chapter 14). Improved agricultural practice has eliminated ergotism from those countries where rye is grown although the alkaloids are still used therapeutically for their effects upon uterine muscle.

In the North Central States of the United States there have been sporadic outbreaks for many years of diseases of pigs and horses associated with the ingestion of barley infected with *Fusarium graminearum* (Barley

Scab) (see Curtin, 1968). The mycotoxins associated with the *Fusarium* species have been studied extensively by Japanese workers. This has followed cases of vomiting and diarrhoea in man after eating scabby grain. Many toxin producing strains of *Fusarium* were isolated in these studies including *Fusarium graminearum* and *Fusarium nivale*. Culture filtrates were shown to be toxic to mice and the active principles were shown to be trichotheanes (Ueno, 1971).

16.3 Toxins Produced by Fungi Growing on Leaves and Stems

Stachybotryotoxicosis

This mycotoxicosis occurs as a result of the fungus *Stachybotrys atra* growing on straw and mainly affects horses although other animals, including man, can be affected. The first cases in horses were reported from Russia in 1931 (Forgacs & Carll, 1962). In horses two forms of stachybotryotoxicosis are recognized. The typical form results from continued ingestion of small amounts of the toxin while the atypical form results from massive dosage. In the typical form there is suppression of the bone marrow leading to agranulocytosis and thrombocytopenia and death from overwhelming infection. The atypical form is usually fatal and appears to produce central nervous system disorders. These are described in detail by Forgacs & Carll (1962).

Forgacs *et al.* (1958) described the experimental production of the disease in horses as well as calves, pigs, sheep and mice, all of which are susceptible. Drobotko (1945) described a dermal toxicity in man with irritation of the respiratory mucous membranes. This was noticed in workers investigating the disease in horses and was reproduced in other members of the laboratory on exposure to the crystalline toxin. A leukopenia developed following the application to the skin and it was concluded that the toxin was readily absorbed.

Facial eczema in ruminants

Facial eczema is the name given to a disease of sheep which was originally recognized in New Zealand. The disease occurred when sheep grazed on pasture infected with the fungus *Pithomyces chartarum* (formerly *Sporidesmium bakerii*) (Percival & Thornton, 1958). Synge & White (1959) isolated a crystalline substance called *sporidesmin* whose structure was elucidated by Hodges *et al.* (1963). In sheep, both in the field and experimentally, the main lesions occur in the liver. The animals are jaundiced and the livers show a marked biliary proliferation and fibrosis with focal necrosis of parenchymal cells (Mortimer & Taylor, 1962). The photosensitization, which causes the facial eczema, is a result of the jaundice and failure to excrete phylloerythrin.

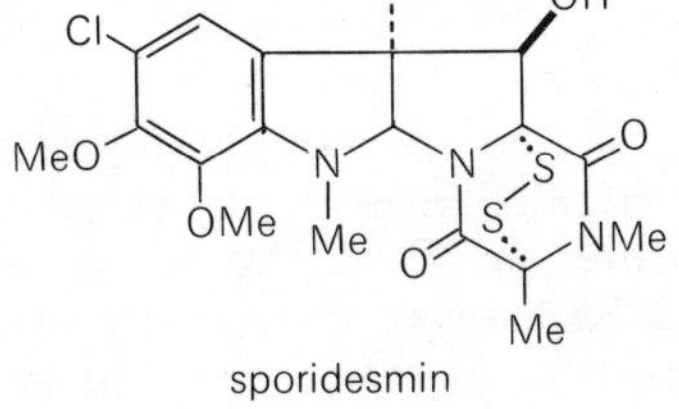

sporidesmin

Sporidesmin is toxic to many species—rabbits (Clare, 1959) and guinea pigs (Synge & White, 1959), both of which have severe lesions of the liver. Rats are also susceptible to the toxin but do not show the same liver lesion (Rimington *et al.*, 1962). The only lesion which could be demonstrated in the rat was that of an increased vascular permeability (Slater, Sträuli & Sawyer, 1964). Long-term feeding has failed to demonstrate any carcinogenic activity.

Lupinosis

Lupinosis of sheep is a disease of long standing characterized by chronic liver damage which has been recognized in many parts of the world. In 1880 Kühn isolated *Phomopsis leptostromiformis* (= *Cryptosporium leptostromiformis*) from lupins and suggested that fungal toxins were responsible for the disease although this was not proved (quoted by van Warmelo *et al.*, 1971).

Following an outbreak of lupinosis of sheep in South Africa in 1969, in which 530 ewes died with severe liver lesions, *Phomopsis leptostromiformis* was isolated from the lupins grazed by the sheep. Subsequently sheep, dosed with culture material, developed jaundice and had severe fatty infiltration of the liver indicating that lupinosis of sheep is a mycotoxicosis. The active principle has not been characterized.

16.4 Toxins Produced from Grains and Seeds

Alimentary toxic aleukia (A.T.A.) in man

In a review by Mayer (1953) it is stated that Sarkisov of the All Union Scientific Experimental Laboratory for the Study of Toxic Fungi (U.N.I.L.) U.S.S.R. isolated *Fusarium sporotrichioides* from grain which caused this disease. Sporadic outbreaks have been reported in Eastern Siberia since 1913 and are reviewed by Mayer (1953). It has been shown that the fungus will grow on grain which has over-wintered on the ground and has been covered with snow. The fungus will grow at −10°C and the optimum temperature for toxin production is 1.5–4°C. So far the nature of the toxin is not known.

Clinically the disease is very similar to that of stachybotryotoxicosis. The early stages are characterized by ulceration of the oral mucosa and a gastroenteritis. This is followed by a series of blood dyscrasias. The third stage seems to be characterized by secondary infections which are often fatal. The patient who survives this period slowly recovers but a period of 2 months is required for the bone marrow to return to normal.

In no experimental animal has the full human syndrome been produced. Horses, guinea pigs, dogs and monkeys are susceptible to the toxin but cattle, sheep and chickens appear to be resistant.

Yellowsis rice disease

Fermentation by fungi of rice, soya and barley has been used in Japan for centuries in the production of beverages and condiments (see Chapter 13). Most of these procedures are based upon the use of various strains of *Aspergillus*. Kinosita & Shikata (1965) have reviewed the subject and point out that it was not until 1891 that there was a suggestion of toxicity related to mouldy rice. Since then many fungi have been isolated which produce toxins.

Following World War II large amounts of rice were imported into Japan. From this rice, much of which was yellow, many strains of *Penicillium* were isolated, of which three in particular were toxic, *P. citreo-viride*, *P. citrinum* and *P. islandicum* (Miyake & Saito, 1965).

The toxins of *Penicillium islandicum* have received considerable attention. Uraguchi *et al.* (1961*a*) have described both the acute and long term toxicity to rats and mice. Uraguchi *et al.* (1961*b*) extracted the mouldy rice and isolated two toxic substances, luteoskyrin and a chlorine containing peptide, islanditoxin, whose structures have been determined (see Saito, Enomoto & Tatsuno, 1971). Acutely in both the rat and mouse luteoskyrin produces a centrilobular necrosis while islandiotoxin produces periportal necrosis. Prolonged feeding of luteoskyrin induces hepatic carcinoma while it is doubtful if islanditoxin is a carcinogen.

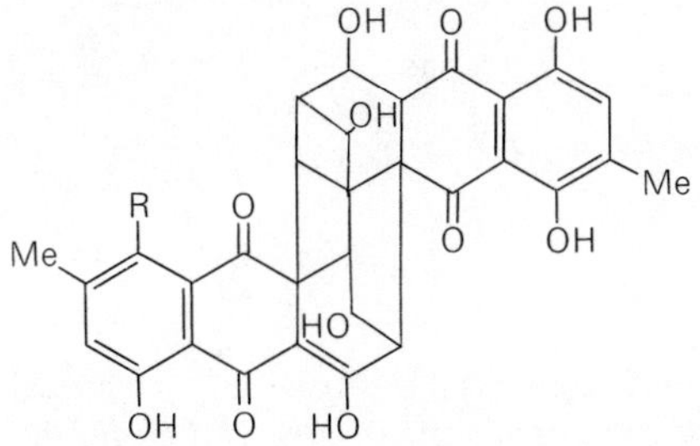

luteoskyrin

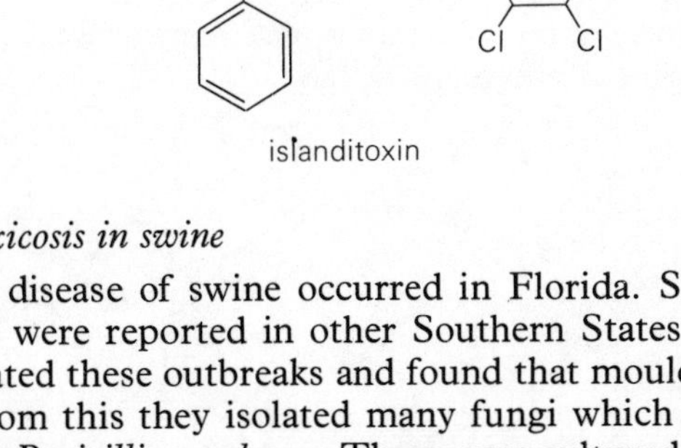

islanditoxin

Mouldy corn toxicosis in swine

During 1952 a disease of swine occurred in Florida. Shortly after this similar diseases were reported in other Southern States. Burnside *et al.* (1957) investigated these outbreaks and found that mouldy corn feed was responsible. From this they isolated many fungi which included *Aspergillus flavus* and *Penicillium rubrum*. These were cultured and the extracts shown to be toxic; the *P. rubrum* extracts being more toxic than those from *A. flavus*. The extracts of both fungi were demonstrated to be toxic to goats, horses, swine and mice. Both toxins produced massive liver necrosis and haemorrhages while the toxin of *P. rubrum* also caused severe renal tubular necrosis and haemorrhage.

The toxins produced by *Penicillium rubrum* received somewhat more attention than those of *Aspergillus flavus*. The toxins designated rubratoxin A and B were isolated by Townsend, Moss & Peck (1966) and characterized by Moss *et al.* (1968, 1969). The toxicity of the rubratoxins has been discussed fully by Moss (1971) who concluded that although the compounds undoubtedly contribute to the toxicity of mouldy feed to animals their toxicity was such that they represented a minor component in the mycotoxin problem.

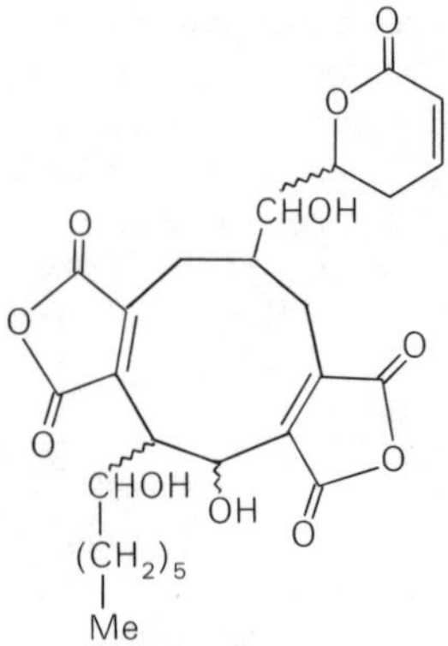

rubratoxin

Aflatoxins

The aflatoxins are a group of related difurano-coumarin secondary metabolites produced by certain strains of *Aspergillus flavus* and *A. parasiticus*. These compounds were isolated from groundnut meal, following outbreaks of liver disease among pheasants, turkeys, pigs and calves during 1960 in Great Britain. The great interest in these compounds has been stimulated by the observation that they are hepato-carcinogenic to several species and that they may be present in some human food.

Four aflatoxins were isolated and their chemical structures determined (Asao *et al.*, 1965). These are termed aflatoxin, B_1 and G_1, being the most abundant, and their dihydro derivatives B_2 and G_2, occurring in lesser amounts. The compounds have a blue fluorescence in ultraviolet light which forms the basis of most of the simple chemical assay procedures used at present to detect the aflatoxins in food.

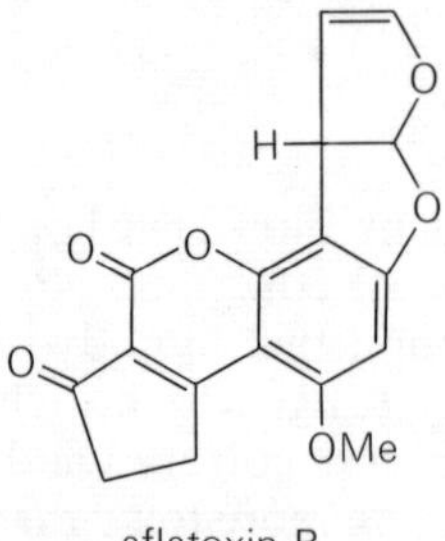

aflatoxin B_1 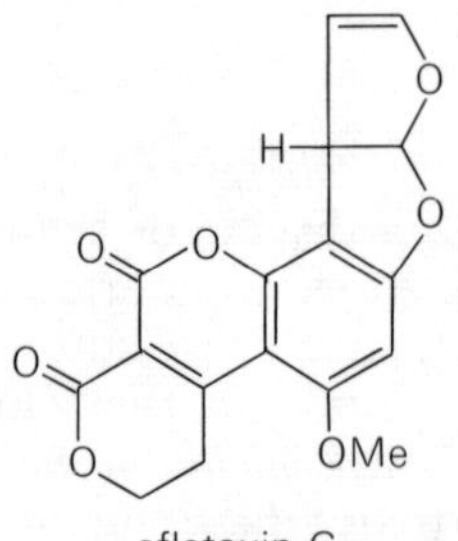 aflatoxin G_1

The original outbreak of disease attributable to the aflatoxins was associated with contaminated peanut meal. However, subsequent investigation has shown that toxin production is possible on a wide range of substrates such as cotton seed, maize and rice (Hesseltine *et al.*, 1966). The optimum temperature for growth of the aflatoxin-producing strains of *Aspergillus flavus* is 30°C with a relative humidity of 75% or greater. At moisture contents of 10–20% *Aspergillus flavus* will grow on peanuts which have been damaged by insects or rodents; this usually occurs after harvesting. In any sample of contaminated peanuts, only a small number of infected nuts are required to render a whole batch of meal prepared from them highly toxic to animals.

The aflatoxins are acutely toxic to most species although there is considerable variation in their susceptibility. The LD_{50} for aflatoxin B_1 for ducklings is 0.335 mg kg^{-1} and for female rats 17.9 mg kg^{-1}. The organ most consistently affected in all species is the liver but others notably the kidney and adrenals, show acute damage. The nature of the acute liver lesion has been reviewed by Butler (1970) and Newberne & Butler (1969).

The first report that peanut meal contaminated with aflatoxin was carcinogenic to the rat was that of Lancaster, Jenkins & Philp (1961). Subsequently, using meals assayed for aflatoxin it was demonstrated that 0.5 ppm in the diet induced 100% incidence of hepatic carcinoma in the rat (Butler & Barnes, 1963). Using pure B_1, Wogan & Newberne (1967) demonstrated 100% incidence of hepatic carcinoma following dietary levels of 0.015 ppm. The aflatoxins have also been shown to induce extra hepatic neoplasm notably of the glandular stomach and the kidney (Butler & Barnes, 1966; Butler, Greenblatt & Lijinsky, 1969).

At the present time the carcinogenic action of the aflatoxin has been demonstrated in rats, ducks, trout and ferrets.

The recognition of the carcinogenic activity of the aflatoxins has raised many questions concerning the public health hazard presented by fungal contamination of food. Direct evidence that aflatoxin causes human liver disease is difficult to obtain. However, there is increasing evidence that in those areas of the world where there is an increased incidence of hepatic carcinoma the population is exposed to aflatoxin (Alpert, Hutt & Davidson, 1968; Shank, 1971; Campbell & Salamat, 1971). As a result of the experimental data on the carcinogenicity of the aflatoxins the WHO have recommended a maximum level of 0.3 ppm aflatoxin in supplemented food for young children in those regions where protein malnutrition is common.

Sterigmatocystin

Following the recognition of the carcinogenic potential of the aflatoxins, Scott (1965) recovered many strains of toxic fungi from cereal and legume crops in South Africa. From these sterigmatocystin which had been characterized by Bullock, Roberts & Underwood (1962), was isolated from

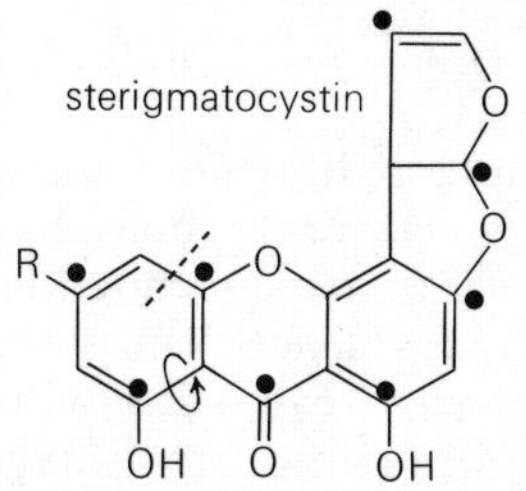

sterigmatocystin

Aspergillus versicolor and *A. nidulans* (Holzafsfel *et al.*, 1966). Sterigmatocystin is acutely toxic to rats and monkeys (Purchase & Van der Watt, 1969; Van der Watt & Purchase, 1970). The compound has also been shown to induce hepatic carcinoma in the rat (Purchase & Van der Watt, 1970). Surveys of food supplies are being undertaken in Mozambique to determine the extent of contamination by sterigmatocystin.

Citrinin and ochratoxin

Citrinin was first isolated from *Penicillium citrinum* in 1931 and its structure was determined by Brown *et al.* (1949) but little attention was paid to its toxic properties. In Denmark a toxic nephropathy in swine has been associated with the feeding of mouldy cereals for more than 40 years. More recently *P. viridicatum* and nephrotoxic compounds have been isolated from barley known to cause the procine nephropathy. The compound has been characterized as citrinin (Friis, Hasselager & Krogh, 1969). By feeding pure citrinin to rats and swine a complete reproduction of the nephropathy was achieved. *P. citrinum* was also isolated from yellow rice imported into Japan from Thailand in 1951. Rice, contaminated with the fungus, was shown to produce a nephropathy in mice and citrinin was isolated from the rice (see Saito *et al.*, 1971). No field cases of renal disease were reported by the Japanese workers.

In South Africa during the screening of cereals for toxigenic fungi (Scott, 1965) *Aspergillus ochraceus* was isolated and shown to produce a toxin called ochratoxin (Van der Merwe *et al.*, 1965). Ochratoxin as well as being hepatotoxic was shown to be primarily a nephrotoxin in the rat (Purchase & Theron, 1968) but is probably not carcinogenic. In the study undertaken in Denmark into the swine nephropathy, ochratoxin was also isolated from the contaminated barley. Indeed Krogh (1973) has reported levels of 27 ppm ochratoxin and 2 ppm citrinin in the affected barley. The result of the nephropathy is a delay in the pigs reaching the correct killing weight and hence economic loss to the producer. Tissue residues of the toxins in the bacon are also being investigated. These findings demonstrate clearly that the hazards from mycotoxins are not confined to those areas of the world where there is a poor standard of agricultural practice.

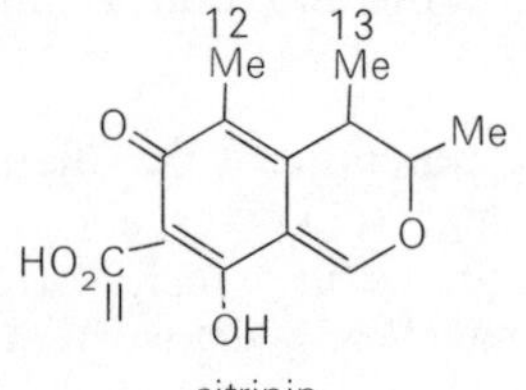

citrinin

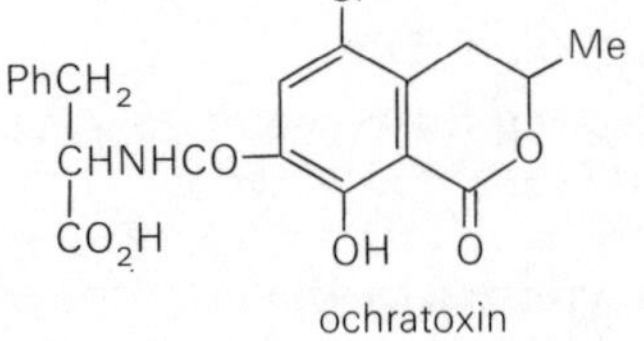

ochratoxin

16.5 Conclusions

From the preceding sections it is readily apparent that mycotoxins are widespread and present a hazard to both human and animal health. Improved agricultural practices have resulted in the effective elimination of ergotism. At present the main problem is from fungi attacking stored grains and resulting in not only severe economic and food loss but also a hazard to public health. Improved harvesting and storage of food to prevent fungal growth should enable both the quantity and quality of food to be improved. As a result of the recognition of aflatoxin many countries have instituted control measures setting limits to the aflatoxin content of food. Harvesting and storage practices are being modified in producing countries to improve the quality of the product. These measures, following the realization of the extent of the problem of mycotoxins, may result in the prevention of future hazards to health.

16.6 References

ALPERT, M.E., HUTT, M.S.R. & DAVIDSON, C.S. (1968). Hepatoma in Uganda. *Lancet* **i**, 1265–7.

ASAO, T., BUCHI, G., ABDEL-KADAR, M.M., CHANG, S.B., WICK, E.L. & WOGAN, G.N. (1965). The structures of aflatoxin B_1 and S_1. *Journal of the American Chemical Society* **87**, 882–6.

BROOK, P.J. & WHITE, E.P. (1966). Fungus toxins affecting mammals. *Annual Review of Phytopathology* **4**, 171–94.

BROWN, J.P., CARTWRIGHT, N.J., ROBERTSON, A. & WHALLEY, W.B. (1949). The chemistry of fungi. IV. The constitution of the phenol, C_{11}, $H_{16}O_3$, from citrinin. *Journal of the Chemical Society* 859–67.

BULLOCK, E., ROBERTS, J.C. & UNDERWOOD, J.G. (1962). Studies in mycological chemistry. XI. The structure of isosterigmatocystin and an amended structure for sterigmatocystin. *Journal of the Chemical Society* 4179–83.

BURNSIDE, J.E., SIPPEL, W.L., FORGACS, J., CARLL, W.T., ATWOOD, M.B. & DOLL, E.R. (1957). A disease of swine and cattle caused by eating mouldy corn. II. Experimental production with pure cultures of moulds. *American Journal of Veterinary Research* **18**, 817–24.

BUTLER, W.H. (1970). Liver injury induced by aflatoxin. In *Progress in Liver Disease,* Vol 3, pp. 408–18. Ed. Popper, H. and Schaffner, F. New York: Gruen and Stratton.

BUTLER, W.H. & BARNES, J.M. (1963). Toxic effects of groundnut meal containing aflatoxin to rats and guinea pigs. *British Journal of Cancer* **27**, 699–710.

BUTLER, W.H. & BARNES, J.M. (1966). Carcinoma of the glandular stomach in rats given diets containing aflatoxin. *Nature, London* **209**, 90.

BUTLER, W.H., GREENBLATT, M. & LIJINSKY, W. (1969). Carcinogenesis in rats by aflatoxin B_1, G_2 and B_2. *Cancer Research* **29**, 2206–11.

CAMPBELL, T.C. & SALAMAT, L. (1971). Aflatoxin ingestion and excretion by humans. In *Mycotoxins in Human Health* pp. 271–280. Ed. Purchase, I.F.H. London: Macmillan.

CLARE, N.T. (1959). Photosensitivity diseases in New Zealand. Susceptibility of New Zealand white rabbits to facial eczema liver damage. *New Zealand Journal of Agricultural Research* **2**, 1249–56.

CURTIN, T.M. (1968). Mycotoxicoses. *Illinois Veterinarian* **10**, 23–7.

DROBOTKO, V.G. (1945). Stachybotryotoxicosis: a new disease of horses and humans. *American Review of Soviet Medicine* **2**, 238–42.

ENOMOTO, M. & SAITO, M. (1972). Carcinogens produced by fungi. *Annual Review of Microbiology* **26**, 279–312.

FORGACS, J. & CARLL, W.T. (1962). Mycotoxicoses. *Advances in Veterinary Sciences* **7**, 273–382.

FORGACS, J., CARLL, W.T., HERRING, A.S. & HINSHAW, W.R. (1958). Toxicity of *Stachybotrys atra. Transactions of the New York Academy of Sciences* **20**, 787–808.

FRIIS, P., HASSELAGER, E. & KROGH, P. (1969). Isolation of citrinin and oxalic acid from *Penicillium viridicatum* Westling and their nephrotoxicity in rats and pigs. *Acta Pathologica et Microbiologica Scandinavica* **77**, 559–60.

HESSELTINE, C.W., SHOTWELL, O.C., ELLIS, J.J. & STUBBLEFIELD, R.D. (1966). Aflatoxin formation by *Aspergillus flavus. Bacteriological Reviews* **30**, 795–805.

HODGES, R., DONALDSON, J.W., TAYLOR, A. & WHITE, E.P. (1963). Sporidesmin and sporidesmin B. *Chemistry and Industry* 42–3.

HOLZAFSFEL, C.W., PURCHASE, I.F.H., STEYN, P.S. & GOUWS, L. (1966). The toxicity and chemical assay of sterigmatocystin, a carcinogenic mycotoxin, and its isolation from two new species of fungi. *South African Medical Journal* **40**, 1100–1.

KINOSITA, R. & SHIKATA, T. (1965). On toxic mouldy rice. In *Mycotoxins and Foodstuffs*, pp. 111–32. Ed. Wogan, G.N. Boston: M.I.T. Press.

KROGH, P. (1973). Natural occurrence of ochratoxin A and citrinin in cereals associated with field outbreaks of swine nephropathy. *Journal of Pure and Applied Chemistry* **35**.

LANCASTER, M.C., JENKINS, F.P. & PHILP, J.MCL. (1961). Toxicity associated with certain samples of groundnuts. *Nature, London* **192**, 1095–7.

MAYER, C.F. (1953). Endemic panmyelotoxicosis in the Russian grain belt. I. The clinical aspects of alimentary toxic aleukia (ATA). A comprehensive review. *Military Surgeon* **113**, 173–89.

VAN DER MERWE, K.J., STEYN, P.S., FOURIE, L., SCOTT, DE B. & THERON, J.J. (1965). Ochratoxin A: a toxic metabolite produced by *Aspergillus ochraceus. Nature, London* **205**, 1112.

MIYAKE, M. & SAITO, M. (1965). Liver injury and liver tumours induced by toxins of *Penicillium islandicum* growing on yellowed rice. In *Mycotoxins and Foodstuffs*, pp. 133–46. Ed. Wogan. G.N. Boston: M.I.T. Press.

MORTIMER, P.H. & TAYLOR, A. (1962). The experimental intoxication of sheep with sporidesmin, a metabolic product of *Pithomyces chartarum*. I. Clinical observations and findings at post-mortem examination. *Research in Veterinary Science* **3**, 147–60.

MOSS, M.O. (1971). The rubratoxins, toxic metabolites of *Penicillium rubrum* Stoll. In *Microbial Toxins*, vol. 6, pp. 381–407. Ed. Ciegler, A., Kadis, A. and Ajl, S.J. London and New York: Academic Press.

MOSS, M.O., ROBINSON, F.V., WOOD, A.B., PAISLEY, H.M. & FEENEY, J. (1968). Rubratoxin B a proposed structure for a bis-anhydride from *Penicillium rubrum* Stoll. *Nature, London* **220**, 767–70.

MOSS, M.O., WOOD, A.B. & ROBINSON, F.V. (1969). The structure of rubratoxin A: a toxic metabolite of *Penicillium rubrum. Tetrahedron Letters* 367–70.

NEWBERNE, P.M. & BUTLER, W.H. (1969). Acute and chronic effects of aflatoxin on the liver of domestic and laboratory animals. A Review. *Cancer Research* **29**, 236–50.

PERCIVAL, J.C. & THORNTON, R.H. (1958). Relationship between the presence of fungal spores and a test for hepatoxic grass. *Nature, London* **182**, 1095–6.

PURCHASE, I.F.H. & THERON, J.J. (1968). Acute toxicity of ochratoxin A to rats. *Food and Cosmetics Toxicology* **6**, 479–83.

PURCHASE, I.F.H. & VAN DER WATT, J.J. (1969). Acute toxicity of sterigmatocystin to rats. *Food and Cosmetics Toxicology* **7**, 135–9.

PURCHASE, I.F.H. & VAN DER WATT, J.J. (1970). Carcinogenicity of sterigmatocystin. *Food and Cosmetics Toxicology* **8**, 289–95.

RIMINGTON, C., SLATER, T.F., SPECTOR, W.G., STRÄULI, U.D. & WILLOUGHBY, D.A. (1962). Sporidesmin poisoning in the rat. *Nature, London* **194**, 1152–3.

SAITO, M., ENOMOTO, M. & TATSUNO, T. (1971). Yellowed rice toxins. Luteoskyrin and related compounds, chlorine-containing compounds and citrinin. In *Microbial Toxins* vol. 6, pp. 299–380. Ed. Ciegler, A., Kadis, S. and Ajl, S.J. London and New York: Academic Press.

SCOTT, DE B. (1965). Toxigenic fund isolated from cereal and legume products. *Mycopathologia et Mycologia applicata* **25**, 213–222.

SHANK, R. (1971). Dietary aflatoxin loads and the incidence of human hepatocellular carcinoma in Thailand. In *Mycotoxins in Human Health*, pp. 245–62. Ed. Purchase, I.F.H. London: Macmillan.

SLATER, T.F., STRÄULI, U.D. & SAWYER, B. (1964). Sporidesmin poisoning in the rat. I. Chemical changes. *Research in Veterinary Science* **5**, 450–72.

SYNGE, R.L.M. & WHITE, E.P. (1959). Sporidesmin: a substance from *Sporidesmin bakeri* causing lesions characteristic of facial eczema. *Chemistry and Industry* 1546–7.

TOWNSEND, R.J., MOSS, M.O. & PECK, H.M. (1966). Isolation and characterization of hepatotoxins from *Pencillium rubrum. Journal of Pharmacy and Pharmacology* **18**, 471–3.

UENO, Y. (1971). Toxicological and biological properties of fusarenon-X, a cytotoxic mycotoxin of *Fusarium nivale* Fn-2B. In *Mycotoxins in Human Health*, pp. 164–78. Ed. Purchase, I.F.H. London: Macmillan.

URAGUCHI, K., TATSUNO, T., TSUKIOKA, M., SAKAI, Y., KOBAYASHI, Y., SAITO, M., ENOMOTO, M. & MIYAKE, M. (1961*a*). Toxicological approach to the metabolites of *Penicillium islandicum* Sopp growing on yellowed rice. *Japanese Journal of Experimental Medicine* **31**, 1–18.

URAGUCHI, K., TATSUNO, T., SAKAI, F., TSUKIOKA, M., SAKAI, Y., YONEMITSU, O., ITO, H., MIYAKE, M., SAITO, M., ENOMOTO, M., SHIKATA, T. & ISHIKO, T. (1961*b*). Isolation of two toxic agents, luteoskyrin and chlorine-containing peptide, from the metabolites of *Penicillium islandicum* Sopp, with some properties thereof. *Japanese Journal of Experimental Medicine* **31**, 19–46.

VAN WARMELO, K.T., MARASAS, W.F.O., ADELAAR, T.F., KELLERMAN, T.S., VAN REUSBERG, I.B.J. & MINNE, J.A. (1971). Experimental evidence that lupinosis of sheep is a mycotoxicosis caused by the fungus *Phomopsis leptostromiformis*

(Küln) Bubak. In *Mycotoxins in Human Health*, pp. 185–193. Ed. Purchase, I.F.H. London: Macmillan.

VAN DER WATT, JJ. & PURCHASE, I.F.H. (1970). The acute toxicity of retrorsine, aflatoxin and sterigmatocystin in Vervet monkey. *British Journal of Experimental Pathology* 51, 183–90.

WOGAN, G.N. & NEWBERNE, P.M. (1967). Dose response characteristics of aflatoxin B_1 carcinogenesis in the rat. *Cancer Research* 27, 2370–6.

Species Index

Subject Index